Dear Geoffrey

Here is the book I mentioned in my letter. Many thanks for your help in organizing the original get-together.

Yours sincerely,

George Janossy

Immunodeficiency in HIV Infection and AIDS

EC/FERS/MRC Workshop on Immunodeficiency in HIV-1 Infections, Windsor, Surrey, May 3–4, 1991

Immunodeficiency in HIV Infection and AIDS

Editors

G. Janossy, London

B. Autran, Paris

F. Miedema, Amsterdam

89 figures and 34 tables, 1992

KARGER

Basel · München · Paris · London · New York · New Delhi · Bangkok · Singapore · Tokyo · Sydney

Library of Congress Cataloging-in-Publication Data
EC/FERS/MRC Workshop on Immunodeficiency in HIV-1 Infections
(1991: Surrey, England)
Immunodeficiency in HIV infection and AIDS/editors: G. Janossy, B. Autran, F. Miedema.
"EC/FERS/MRC Workshop on Immunodeficiency in HIV-1 Infections, Windsor, Surrey, May 3–4, 1991."
Includes bibliographical referencces and index.
1. AIDS (Disease) – Pathophysiology – Congresses.
2. HIV infections – Immunological aspects – Congresses.
I. Janossy, G. II. Autran, B. (Brigitte) III. Miedema, F. (Frank)
IV. Commission of the European Communities.
V. European Federation of AIDS Research.
VI. Medical Research Council (Great Britain) VII. Title.
[DNLM: 1. Antigens, CD4 – congresses. 2. Biological Markers – congresses.
3. CD4–CD8 Ratio – methods – congresses. 4. HIV Infections – diagnosis – congresses.
WD 308 E17i 1991]
RC607.A26E374 1991 616.97'92079–dc20
ISBN 3–8055–5547–4

Drug Dosage

The authors and the publisher have exerted every effort to ensure that drug selection and dosage set forth in this text are in accord with current recommendations and practice at the time of publication. However, in view of ongoing research, changes in government regulations, and the constant flow of information relating to drug therapy and drug reactions, the reader is urged to check the package insert for each drug for any change in indications and dosage and for added warnings and precautions. This is particularly important when the recommended agent is a new and/or infrequently employed drug.

Printed in Switzerland on acid-free paper by Thür AG Offsetdruck, Pratteln
ISBN 3–8055–5547–4

Contents

Functional Immunopathology of HIV-1 Infection

Foreword

As the project leader of the Concerted Action (CA) on 'Immunology and Immunopathology of HIV-Related Diseases' from the Basic Science Program of the European Commission Working Party on AIDS, it is my pleasure to present the proceedings of the workshop which took place at the Anugraha Hotel, Windsor, Surrey, on May 3–4, 1991. This is one of the last European workshops that have been organized through the patronage of this CA, the aim of which has been to bring together scientists of different backgrounds working in this area – especially immunologists, pathologists and virologists – to enhance their collaboration and to allow them to place their research in a global European perspective.

The idea of the CA goes back to the fall of 1984 but, due to administrative red tape, it actually got started only in the spring of 1988. From the beginning, its objectives were: (1) to analyze the pathobiology of HIV infection at the cellular and tissular level, in its functional compared with its morphological aspects; (2) to reach a better understanding of the immunological abnormalities, and (3) to investigate the immune response of HIV-infected individuals in a pathophysiological perspective as well as in view of prognosis assessment or of vaccine development. After less than 4 years, substantial progress has been made in these areas by the groups associated with the CA: collaborations have been established, results have been obtained and shared at meetings and workshops of various formats, from special meetings of a few research groups to global workshops gathering all the participants in the CA. This has resulted in numerous publications in peer-reviewed journals, in journals' special issues, or in books presenting the proceedings of some workshops. Such are 'Modern Pathology of AIDS and Other Retroviral Infections' (Racz P, Haase AT, Gluckman JC, eds), Karger, Basel 1990; 'Accessory Cells in HIV and Other Retroviral Infections' (Racz P, Dijkstra CD, Gluckman JC, eds), Karger, Basel 1991; or the currently in press 'Cytotoxic T Cells in HIV and Other Retroviral Infections', also to be published by Karger.

Currently, productive links have been established with other CA, and a real European network has been set up, the most significant results of which will continue to come out in the following months and years. In their own words, this has been a positive experience and of considerable importance to many a scientist who has participated and to their work: it has permitted to rapidly establish collaborations, to gain access to material which is difficult to obtain by other means, and to rapidly disseminate ideas, as yet unpublished data and new techniques. The CA, which will cease by mid-1992, has thus reached its major aim, which was to permit mutual cross-fertilization of European investigators with different backgrounds and to reinforce the network of European groups dedicated to stop AIDS.

The workshop on 'Immunodeficiency in HIV-1 Infections', organized by George Janossy (London), Brigitte Autran (Paris) and Frank Miedema (Amsterdam) is but the living evidence of such achievements. The data discussed here stem from the latest studies, reported by molecular and cellular immunologists and virologists, on the pathophysiology of HIV infection and the pathogenesis of AIDS. They are confronted to the needs of clinical immunologists to develop new and 'user-friendly' immunological markers based on the insights brought by these results for monitoring the progression of HIV infection.

The fact that the workshop was also cosponsored by two national organizations, the Dutch AIDS Foundation and the Medical Research Council AIDS Directed Programme, as well as by the fledgling European Federation of AIDS Research (EFAR) bespeaks auspiciously for the future of European collaborative AIDS research. It has been argued that such large-scope CA are not useful anymore, that CA should by now address more focused research projects performed by limited numbers of cooperating laboratories. Actually, if this were so, national AIDS organizations would be sufficient and CA would no longer be required. After 3 years as the project leader of a 'large' CA, I remain convinced that they certainly still have to play a key role for international scientific collaboration. The group size of the CA meetings and the productivity of discussions lends a business-like atmosphere to these expert workshop. It has been truly reassuring to observe the collaborative spirit among the European participants during the successive workshops organized by the CA in Paris, Hamburg and Pavia – and the most recent authoritive contribution by the invited American scientists at the Anugraha workshop.

Last, I am sure that the workshop's organizers would have liked to thank me for providing support for their endeavour, but I feel certainly more compelled to thank them for the good use they made of our tax money! Thank you George, Brigitte and Frank, and thank you to all the participants and authors for the great job the reader will get now the pleasure to discover!

Jean-Claude Gluckman
Hôpital de la Pitié-Salpêtrière
Paris, France

Preface

The clinical application of the expanding knowledge about the immunological consequences of HIV infection is one of the most pressing tasks. Immunologists have always been eager to participate in such a challenge. In 1981 one of the first papers about the clinical description of a new syndrome, AIDS, was published [1]. In this paper Leu-3a, a CD4 monoclonal antibody, was already utilized documenting the loss of a 'helper' T-cell subset as a principal feature of the disease. Soon a whole range of CD4 antibodies, including those submitted to International Leukocyte Typing Workshops, was exploited to prove that the CD4 molecule itself was part of the receptor for the newly discovered LAV/HTLV-III [2, 3]. Nine years later a revised classification system for HIV infection and AIDS surveillance has been drafted by the Center of Disease Control (CDC), Atlanta, Ga., USA, in which the CDC recommends the widening of the AIDS surveillance case definition by placing all HIV-infected persons with less than 200 CD4+ lymphocytes/mm^3 into the AIDS category [4] in addition to the clinical criteria established earlier [5]. The implementation of these new regulations is currently being discussed in the USA. Such a new concept would give appropriate emphasis to the laboratory findings by simplifying the classification of cases where the clinical findings are somewhat ambiguous. It could also be reconciled with the recent proposal for a WHO Staging System [6] where the 'laboratory axis' runs parallel with the clinical staging and both scores are recorded. Nevertheless, in Europe a stand equivalent to the CDC policy in adopting the laboratory parameter of low CD4 counts as an AIDS classifying criteria has not yet been accepted.

Similarly, the surprisingly severe immunosuppression seen in asymptomatic patients during relatively early stages of HIV-1 infection has been documented some while ago [7–9] but its significance from the point of view of patient management and preventive therapy has not yet been given appropriate emphasis. These assays have now been sufficiently simplified as whole blood methods [10], and there is consensus among the scientific community that they are suitable for routine use. AIDS is clearly a severe disease of immunoregulation which is initiated by HIV-1 infection and maintained by secondary immunopathological changes which aggravate and perhaps even determine disease development. Thus, patients need to be investigated not only by virological means but also by suitable methods which probe the relevant immune functions. The aim is therefore to measure these defects in 'user-friendly' tests which form an integral part of modern patient management and trial design.

At the Anugraha meeting, as part of the introduction to these issues, Dr. John Kagan of Division of AIDS, NIAID, NIH, USA, and Professor Ian Weller of Academic Department of Genito-Urinary Medicine, Middlesex Hospital, London, have discussed the concepts of the so-called 'surrogate markers' for AIDS development. There are two issues related to this topic.

First, these so-called 'surrogate markers' are essential for progress in this field. HIV-1 infection, a chronic disorder, leads to established AIDS after a long period of about 6–10 years. Consequently, very long observation is needed to assess the outcome of drug trials evaluated by clinical criteria alone. This problem has paradoxically been further aggravated by the success of therapy. Prophylaxis against *Pneumocystis carinii* pneumonia (PCP) has been widely introduced in individuals with HIV-1 infection [11, 12]. As a result, among the reported cases of AIDS within the groups of homosexual men the proportion with PCP as AIDS defining diagnosis has decreased from 62% in 1988 to 46% in 1990 [4]. This is referred to as a 'vanishing end-point'. Antiretroviral therapy such as AZT has also been widely initiated, further delaying the onset of illnesses that are included in the 1986 AIDS definition [5] leading to more 'distal end-points'. Thus, the utility of the clinical AIDS definition in identifying the numbers of patients with late-stage HIV disease is likely to decrease over the next few years. This trend of 'disappearing end-points' inhibits efficient trial assessment at a time when a host of new drugs need to be investigated rapidly. As these drug trials would require more proximal rather than distal end-points, an early evaluation with well-chosen relevant sur-

Table 1. Requirements for surrogate markers of AIDS progression

	I. Relation to (patho)physiology – 'natural history' marker
1	The marker is related to key events we know (or suspect) about pathogenesis
2	It is an important marker: its aberration leads to severe malfunction
3	It correlates with disease progression
4	It can be tested – even during severe illness
	II. Relation to therapy
1	Effective therapy changes the marker correspondingly
2	Failure to influence the marker implies failure of therapy
3	Active drug shows a dose-response curve with the marker
4	The marker is drug-independent (i.e. value X signifies similar prognosis in patients on or off therapy)
	III. Technical issues
1	The normal range for the marker is well defined
2	Low variations except when influenced by known factors
3	It is known when a change in the marker can be regarded significant
4	Low inter-laboratory variability due to ease of standardization
5	'User-friendly' and inexpensive
6	Quality control schemes are in operation

rogate markers would help. This could decrease trial costs by reducing the length of trials and the sample sizes needed for obtaining significant results.

The other area where the introduction of similar concepts are warranted is resource allocation for therapy. If costs are allocated only on the basis of the number of patients with full blown clinical AIDS and the planners remain uninformed about unhospitalized patients with low CD4 counts, the cost pressures for prophylactic and antiretroviral therapy cannot be readily identified. This arrangement is relatively disadvantageous for treatment centres which practice patient monitoring and preventive medicine. Again, in such situations the acceptance of well-documented surrogate markers, such as the CD4 count, may solve problems by precisely guiding patient-oriented resource allocations.

Second, the main concepts about the optimal surrogate markers are summarized in table 1. Some of these requirements are difficult to fully satisfy in clinical practice. It is particularly hard to test the most recent candidates for progression markers on untreated patients, because antiretroviral therapy is now recommended for all HIV seropositive persons with CD4 count $< 500/mm^3$. These issues increase the significance of analysing the various surrogate markers on established cohorts of untreated patients by utilizing stored serum and cellular samples when these are available frozen for a retrospective analysis. Probably the most important issue is that as the detailed knowledge about the pathogenesis of AIDS unfolds, these new findings provide a scientific justification for novel parameters of potential clinical importance. In addition to CD4 counts, the activation of the CD8 lymphocyte system, the poor lymphocyte performance leading to severe anergy and recently documented phenomena of cell death with apoptosis are all potentially applicable methods for disease monitoring in patients undergoing therapy.

At the workshop on the 'Immunodeficiency in HIV-1 Infections', held at the hotel 'Anugraha', Windsor, Surrey, both the practical issues of immunodiagnosis in HIV-1 infection and the recent advances in the pathogenesis of the disease have been discussed. Together with our invited American colleagues who represented leading teams in their selected areas, the participants of the workshop have followed a collaborative European trend initiated by Herrn Kurt Körber in Hamburg by awarding the coveted 'Green Rosette' price to Professor L. Montagnier and his co-workers, and purported by the Concerted Action on the 'Immunology and Immunopathology of HIV-Related Diseases' led by Professor J.-C. Gluckman on behalf of the European Commission Working Party on AIDS. The current conference has been supported by the European Federation of AIDS Research (EFAR) as well as by the Dutch AIDS Foundation and the Medical Research Council AIDS Directed Programme of the United Kingdom.

We are especially grateful to Prof. J.-C. Gluckman for helping us to organize this meeting and also to Mrs. Vanessa Lipton for efficient and cheerful administrative assistance. We are also greatly indebted to Dr. Angela Williams (Medical Research Council), Dr. Marianne von den Berg (Dutch Advisory Council on Health Research) and to Dr. Heinz Gretz (Körber Foundation) who not only supported the administrative action but also participated in person.

George Janossy
The Royal Free Hospital
School of Medicine, London, UK

References

1 Gottlieb MS, Schroff R, Schanker HM, Weisman JD, Fan PT, Wolf RA, Saxon A: *Pneumocystis carinii* penumonia and mucosal candidiasis in previously healthy homosexual men: evidence of a new acquired cellular immunodeficiency. N Engl J Med 1981;305:1425–1431.

2 Klatzmann D, Champagne E, Chamaret S, Gruest J, Guetard D, Hercend T, Gluckman JC, Montagnier L: T-lymphocyte T4 molecule behaves as the receptor for human retrovirus LAV. Nature 1984;312:767–768.

3 Dalgleish AG, Beverley PC, Clapham PR, Crawford DH, Greaves MF, Weiss RA: The CD4 (T4) antigen is an essential component of the receptor for the AIDS retrovirus. Nature 1984;312:763–767.

4 CDC: 1992 revised classification system for HIV infection and expanded AIDS surveillance case definition for adolescents and adults. MMWR 1991;40:787.

5 CDC: Classification system for human T-lymphotropic virus type III/lymphadenopathy associated virus infections. MMWR 1986;35:334–339.

6 WHO: Interim proposal for a WHO staging system for HIV infection and diseases. Weekly Epidemiol Rec 1990;65:221–224.

7 Shearer GM, Salahuddin SZ, Markham PD, et al: Prospective study of cytotoxic T lymphocyte responses to influenza and antibodies to human T lymphotropic virus III in homosexual men. J Clin Invest 1985;76:1699–1704.

8 Giorgi JV, Fahey JL, Smith DC, Hultin LE, Cheng HL, Mitsuyasu RT, Detels R: Early effects of HIV on CD4 lymphocytes in vivo. J Immunol 1987;138:3725–3730.

9 Miedema F, Petit AJ, Terpstra FG, Schattenkerk JK, de Wolf F, Al BJ, Roos M, Lange JM, Tanner SA, Goudsmit J, et al: Immunological abnormalities in human immunodeficiency virus (HIV)-infected asymptomatic homosexual men. HIV affects the immune system before $CD4^+$ T helper cell depletion occurs. J Clin Invest 1988;82:1908–1914.

10 Schellekens PT, Roos MT, de Wolf F, Lange JM, Miedema F: Low T-cell responsiveness to activation via CD3/TCR is a prognostic marker for acquired immunodeficiency syndrome (AIDS) in human immunodeficiency virus-1 (HIV-1)-infected men. J Clin Immunol 1990;10:121–127.

11 CDC: Guidelines for prophylaxis against *Pneumocystis carinii* pneumonia for persons infected with HIV. MMWR 1989;38:(suppl 5):1–9.

12 Volberding PA, Lagakos SW, Koch MA, Pettinelli C, Myers MW, Booth DK: Zidovudine in asymptomatic human immunodeficiency virus infection. N Engl J Med 1990;322:941–949.

CD4 Counts as Surrogate Markers for AIDS

Janossy G, Autran B, Miedema F (eds): Immunodeficiency in HIV Infection and AIDS. EC/FERS/MRC Workshop on Immunodeficiency in HIV-1 Infections, Windsor, Surrey, 1991. Basel, Karger, 1992, pp 1–17

CD4 Counts in Relation to Markers of Immune Activation[1]

Janis V. Giorgi

Department of Medicine and the Jonsson Comprehensive Cancer Center, UCLA School of Medicine, Los Angeles, Calif., USA

HIV infection is accompanied by profound abnormalities in the immune systems of infected persons [1]. CD4 lymphocyte number and function is depressed (see Miedema et al., this volume), and CD8 lymphocyte number is elevated due to an increase in activated and probably immature CD8 cells. B lymphocytes are also activated, and the level of several soluble substances that play a role in immune responses are elevated. NK and T cell cytotoxic functions decline late in disease, and monocyte and dendritic cell functions are impaired. It appears that many of the immune alterations reflect HIV-induced immune deficiency and immune dysfunction while others reflect immune activation. This activation is partly associated with the specific immune responses against HIV-1 but probably also includes immunopathological events which can themselves propagate the disease process. This complex sequence of events unfolds during the clinical stages of HIV disease. These are primary HIV infection, the HIV asymptomatic seropositive period and AIDS. These stages are each characterized by distinctive immune changes, including alterations in the peripheral blood lymphocyte subset profile and elevations in several serologic markers of immune activation. Among the host of immune markers which are associated with HIV disease, only a few appear to have prognostic significance and those with confirmed prognostic value are emphasized in this chapter.

[1] This work was supported by NIH awards AI-7263 and AI-72656.

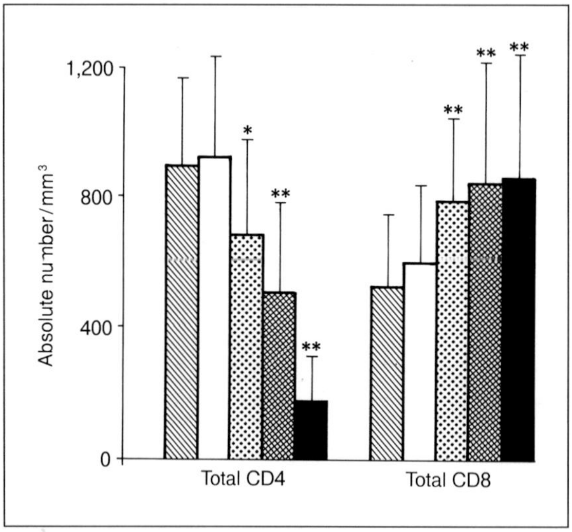

Fig. 1. Absolute value of CD4 and CD8 cells in the peripheral blood of the indicated groups. Mean ± 1 SD; * $p < 0.05$; ** $p < 0.01$. ▧ = Heterosexual controls; □ = seronegative homosexual controls; ▒ = short-term seropositive (< 2 years) homosexuals; ⊠ = long-term seropositive (> 2 years) homosexuals; ■ = AIDS.

CD4 Lymphocyte Decline in HIV-1 Infection

It is well recognized that the alterations of CD4 and CD8 counts, expressed as absolute numbers in the peripheral blood, are associated with how long individuals have been HIV-1 infected and whether AIDS has developed [2–5]. This relationship is shown in figure 1 using data from a cross-sectional study of adult homosexual men most of whom became infected with HIV through sexual exposure. HIV seronegative homosexual and heterosexual control groups both had mean CD4 counts of approximately 1,000/mm^3, while men who had been seropositive for longer periods had progressively lower CD4 numbers. Men recently diagnosed with AIDS-Kaposi's sarcoma generally have relatively high CD4 cell numbers (often higher than 200/mm^3), while those recently diagnosed with AIDS-OI (opportunistic infection) have lower CD4 cell numbers averaging approximately 100/mm^3.

Data from several longitudinal studies indicate that during HIV-1 disease CD4 cell losses occur in four stages [4]. These stages are summarized

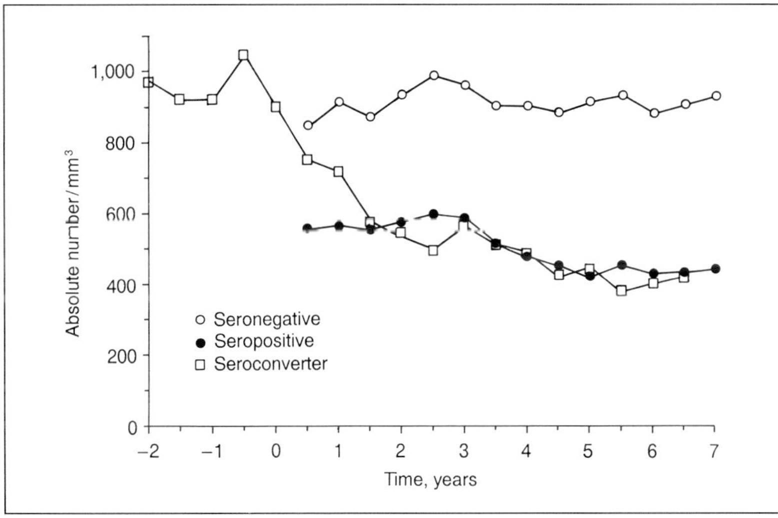

Fig. 2. Changes in the absolute CD4 numbers in three groups of homosexually active men. ● = Seropositive, n = approximately 600 individuals AIDS-free during follow-up; ○ = seronegative, n = approximately 600; □ = seroconverters, n = 78. Time shown is in years. For both the seropositive and the seronegative groups the first observations are placed at the 6-month time point. For the seroconverters, however, time = 0 represents the last time these men were HIV antibody negative and time = 6 months represents when they were first HIV antibody positive.

Table 1. Stages of numerical loss of CD4 lymphocytes

Clinical stage		CD4 number
1	Seroconversion	1,000/mm^3 (normal) drop to 600/mm^3
2	Stable disease	relatively stable or slow decline
3	Development of AIDS-OI	drop to ~ 100/mm^3
4	AIDS	continued drop to 0/mm^3

in table 1. The temporal pattern of loss of CD4 cells can be related to several momentous events. One of these, the event of infection with HIV-1, would be interesting but difficult to study. For this reason, in the discussion here, the CD4 loss is related to the events of seroconversion (fig. 2) and to the development of AIDS (fig. 3). These patterns are shown in a

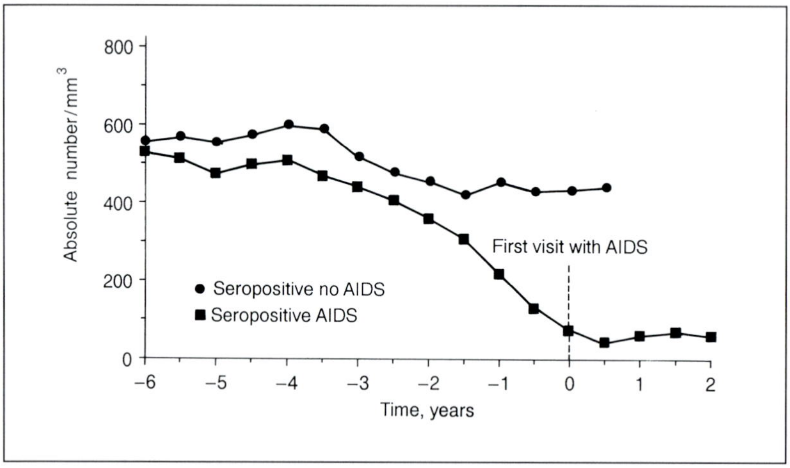

Fig. 3. Changes in the absolute CD4 numbers for the 4 years prior to AIDS development in 137 men who developed AIDS (■) as compared to that in men who did not develop AIDS (●). The latter group includes the approximately 600 individuals from figure 2.

particularly large cohort of people including 600 seropositive men and 78 seroconverters of the UCLA Multicenter AIDS Cohort Study (MACS). *Stage 1* during the first 12–18 months after seroconversion is characterized by a rapid fall in the CD4 cell level, on average from 1,000 to 600/mm^3 (fig. 2). *Stage 2,* which can last several years, is characterized by relatively stable CD4 levels (fig. 2, 3). The length of stage 2 is highly variable. It may even be virtually absent in very rapid progressors, and has now lasted for as long as 12–15 years in some HIV-infected persons. *Stage 3* is again characterized by a period of rapid CD4 cell loss just prior to the development of frank AIDS (fig. 3). *Stage 4,* from AIDS diagnosis to death, is accompanied by a continued decline in the CD4 cell level (fig. 3).

Rapid CD4 cell decline during stages 1, 3 and 4 probably results at least in part from failure to suppress in vivo HIV replication. During stage 2, viral replication appears to be better controlled so less CD4 cell loss occurs. High cytopathicity of certain HIV variants may also contribute to rapid CD4 cell loss [6].

The highly prognostic relationship between low CD4 levels and the development of AIDS is illustrated in figure 4a. CD4 cell numbers are highly prognostic for predicting survival especially during the last two

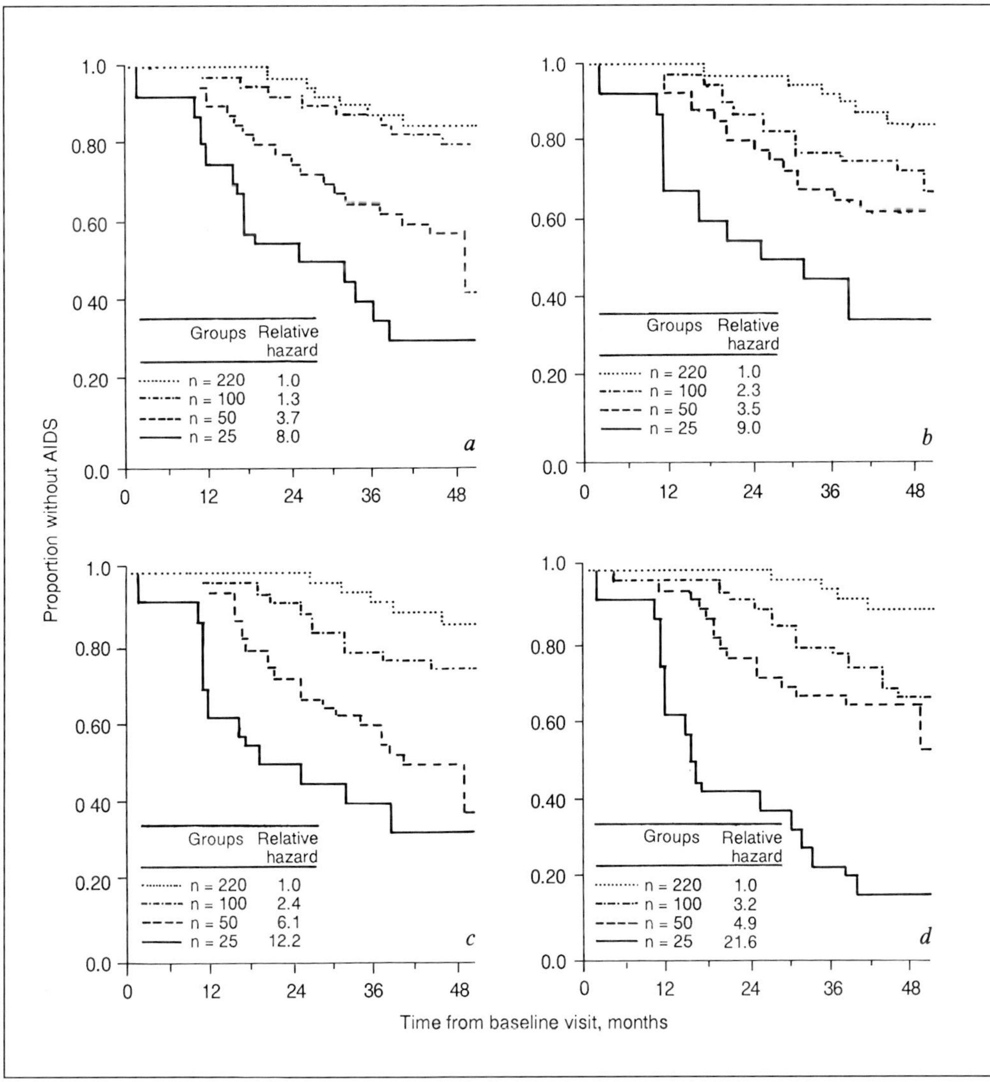

Fig. 4. Kaplan-Meier plots of the proportion without AIDS of 395 homosexual men in whom (*a*) the number of CD4 T cells or (*b*) the neopterin levels were measured; Kaplan-Meier plots for the best linear combination of covariates, combining the number of CD4 T cells and the serum neopterin level (*c*) and the number of CD4 T cells and the serum levels of neopterin, sIL-2R, and p24 antigen (*d*). See Fahey et al. [27] for details.

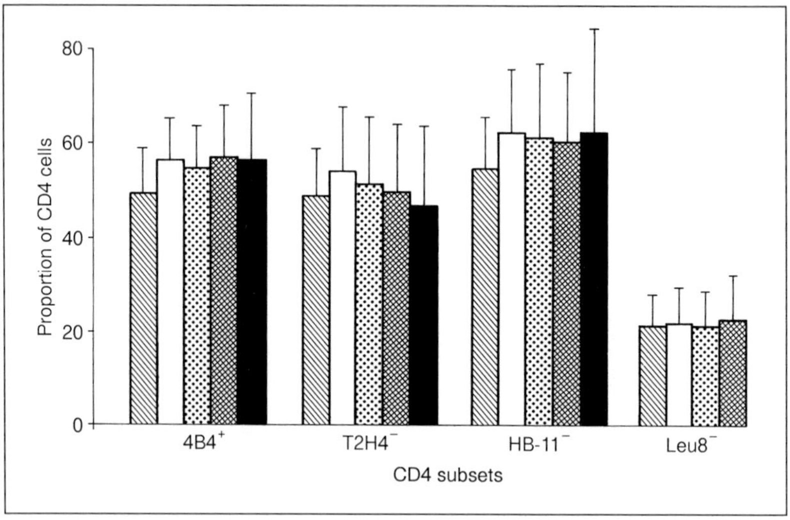

Fig. 5. Proportions of CD4 cells with the phenotypes associated with proliferative responses to soluble antigen, i.e. 4B4+, T2H4−, HB-11−, Leu8−; proportions are the mean ± 1SD of the indicated CD4 subsets for the five groups of men whose CD4 values are shown in figure 1 (see fig. 1 for symbols).

stages, and death occurred within 1 year in 65% of untreated patients whose CD4 cell level had fallen to $<100/mm^3$ [7].

Our dual-color immunofluorescence studies in a large group of HIV-infected individuals indicate that enumeration of additional functional sub-subsets *within* the CD4 population offers no additional prognostic information compared with CD4 measurement alone for evaluating HIV disease progression (fig. 5). This is irrespective whether antibodies such as Leu8 [8], CD29 [9] or CD45RA [10–12] are utilized to identify the particular CD4+ subpopulation. This finding conflicts with earlier reports on smaller study groups [13, 14] that suggested that a selective change in one or another CD4 subset occurs before generalized CD4 cell loss. Our results indicate that the cellular basis for the loss of response to recall antigens during the early stages of HIV infection cannot be attributed to selective loss of CD4 cells exhibiting a memory cell phenotype, detected by the expression of CD45RO or CD29 molecules. This is despite the observation that the CD45RO+ [15] and CD29+ CD4 cells appear to be selectively infected with HIV [16]. A three-color immunofluorescence staining config-

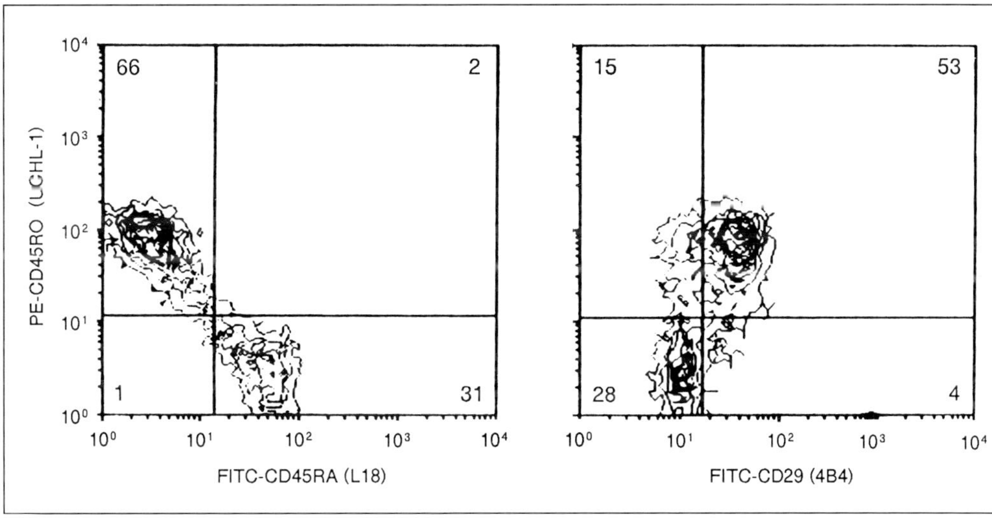

Fig. 6. Three-color immunofluorescence analysis. CD4 lymphocytes were stained with biotinylated CD4 plus avidin-APC (allophycocyanin; red). Staining with anti-CD45RA-FITC (fluorescein; green) versus anti-CD45RO-PE (phycoerythrin; orange) permits objective enumeration of the 'naive' and 'memory' subsets, respectively (left). Staining with anti-CD29-FITC (green) versus anti-CD45RO-PE (orange) is shown for the same person (right).

uration which permits assessment of the bivariant distribution of CD45RO+ vs. CD45RA+ and CD45RO+ vs. CD29+ on CD4 subsets is illustrated in figure 6, and studies are underway in which we will use this three-color immunofluorescence configuration to try to determine the reason for previous conflicting reports on this issue.

In our studies on activated CD4 subsets, CD4 cells expressing CD38 (Leu17) or HLA-DR were observed in untreated AIDS patients (data not shown). These activated CD4 cells may be a response to high levels of HIV replication or coinfections with other pathogens. It is also likely that these activated CD4 cells could support the replication of HIV-1 better than resting CD4 cells. Curiously, IL-2 receptor (CD25) expression, which can be detected at low levels on CD4 cells from normal controls using very sensitive immunofluorescence techniques [17], was not elevated and in fact was expressed at a diminished level on CD4 cells from HIV-infected persons [18]. These observations might be in line with other findings indicating that activated T cells expressing IL-2 receptors may survive very poorly in cultures of blood mononuclear cells taken from asymptomatic individuals with HIV-1 infections (see Gougeon et al., Janossy, Borthwick et al., this volume).

In addition to loss of CD4 cells and low CD4 cell numbers, HIV-1-infected persons have CD4 cells which function poorly. The earliest defects are found in CD4 cell responses to recall antigens [19–21] and alloantigen presented on self-MHC molecules [22]. Later, response to allogeneic stimulator cells and eventually to T cell mitogens including phytohemagglutinin (PHA) is lost [20]. A profound defect in response to anti-CD3, which stimulates both CD4 and CD8 cells, has also been reported [23–25] and may reflect deficient response not only of CD4 cells but interestingly a deficient CD8 cell response as well. These functional investigations are further explored by Miedema et al. and Janossy, Borthwick et al., this volume.

Soluble Markers of Immune Activation

Elevations in several serologic markers of immune activation have been reported. Increased levels of these markers usually occur at the time of seroconversion and persist throughout disease. In general, the increased levels of all the soluble markers are highly correlated with one another and also with low CD4 cell levels. However, they generally add prognostic information to the CD4 level in predicting progression to AIDS. The formulas for calculating the contribution to prognosis for CD4 level plus each of the soluble markers are complex. It further complicates the matters that the best formula to calculate the probability of AIDS development within a certain period of time varies – depending on the stage of disease. Several research groups have proposed writing a computer program utilizing data from natural history studies so that an individual patients' CD4 number and soluble marker values can be combined to predict AIDS. Such a computer program will be essential to allow clinicians to accurately utilize results from multiple laboratory tests to aid prognosis in HIV-infected persons but this tool is not yet available.

Neopterin and γ-Interferon (IFN-γ)

Neopterin is a metabolite of guanosine triphosphate produced by macrophages when they are stimulated by IFN-γ [26]. Kits to measure the level of urinary or serum neopterin are available (Neopterin RIAcid, Henning,

Berlin, FRG) and can be purchased in the United States through DRG International (Mountainside, N.J.) and in Europe through Henning (Berlin, FRG). Normal levels have been reported as 6.6 ± 2.0 nmol/l [27] and the upper limit of normal as 8.7 nmol/l [28].

Urine and serum neopterin levels in seropositive persons are predictive of both progression to AIDS and death. In one study, 45.8% of seropositive individuals who had an initial median serum neopterin level of 24.0 nmol/l (25th–75th percentile, 14.2–41.9 nmol/l) progressed to AIDS in 3 years [28]. As would be expected, serum IFN-γ concentrations correlate with neopterin concentrations in HIV-1 seropositive persons [29]. In fact, it is most likely that high in vivo levels of IFN-γ stimulate neopterin production and are the cause of the elevated serum and urinary neopterin levels in HIV seropositive persons. It has further been hypothesized that the elevation of endogenous IFN-γ levels in HIV seropositive persons is the cause of in vivo hyporeactivity to antigen as reflected by decreased in vitro T cell proliferation and production of lymphokines (including IL-2 and IFN-γ) in response to antigenic stimuli [30].

We have found serum neopterin levels almost as predictive for progression to AIDS [31] as CD4 lymphocyte percentages and absolute numbers [27] (see fig. 4a, b). Elevation of neopterin levels is at least partly independent of changes in CD4 T cells. The combination of CD4 and neopterin measurements is a better indicator of prognosis than either alone [27] (see fig. 4c).

β_2-Microglobulin (β_2-M)

β_2-M is a short polypeptide chain that functions as the light chain of class I histocompatibility molecules. β_2-M is normally present in serum and urine where it can be measured by radioimmunoassay (β_2-M RIA; Pharmacia AB, Uppsala, Sweden). The median levels of β_2-M in serum of heterosexual male adults has been reported to be 1,480 µg/l (25th–75th percentile, 1,005–1,835 µg/l [cf. 32]). β_2-M is expressed during cell turnover, and serum and urine levels are increased in a number of states of immune activation including renal graft rejections, rheumatic diseases, infectious diseases (cytomegalovirus, hepatitis, mononucleosis and HIV infection) and in certain patients with leukemias and lymphomas [reviewed in 32]. Although the marker is not specific for HIV infection, several groups have reported elevated levels of β_2-M in HIV seropositive persons [3, 27, 32, 33]. CD4 numbers and β_2-M are strongly inversely correlated in HIV seropositive homosexual men, and the level of β_2-M in a

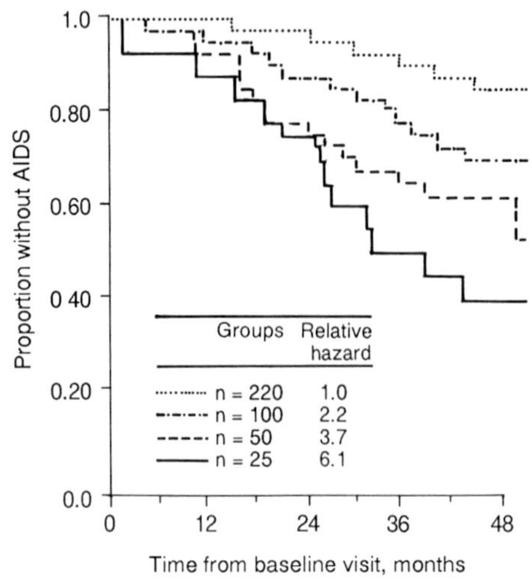

Fig. 7. Kaplan-Meier plots of the proportion without AIDS of those men shown in figure 4 based on the β_2-M levels.

HIV-infected patient is highly predictive either alone (fig. 7) or in conjunction with CD4 levels (data not shown) for progression to AIDS [27, 32].

At seroconversion, β_2-M levels rose in 91% (40/44) of HIV seroconverters at their first seropositive visit [32] and the increase in β_2-M levels could be stratified as low, intermediate or high over the first 6-month period. The magnitude of the initial β_2-M increase predicted the course of disease, and those with the largest increases in β_2-M levels progressed most rapidly. Serum β_2-M level was also found to be a good indicator of AIDS within 3 years in a study of 399 seroprevalent homosexual men (those with $>2{,}800$ μg/l, 46% probability, 1,800–2,800 μg/l, 22% probability; $<1{,}800$ μg/l, 11% probability of AIDS). β_2-M and neopterin levels were highly correlated and had about equal predictive power for AIDS development (fig. 7 and 4b, respectively). β_2-M levels could replace neopterin levels in combination with CD4 levels to provide greater power to predict which seropositive homosexual men would develop AIDS (data not shown). Presumably because both markers reflect CD8 activation, use of both neopterin and β_2-M together with CD4 level did not further increase prognostic power.

Interleukin-2 Receptor (IL-2R) p55 Chain

The high affinity IL-2 receptor is a complex of two chains each of which can independently bind IL-2. The 'low affinity' IL-2R p55 chain is transiently expressed at high levels and shed from the surface of activated mononuclear cells including T lymphocytes and monocytes [34]. Circulating soluble IL-2R p55 chain concentrations are increased during a variety of clinical states which are associated with immune activation such as viral infection, autoimmune diseases or malignancies. Soluble IL-2 receptor levels are elevated in HIV seropositive individuals, and high levels are associated with the increase of other markers of immune activation [35]. Elevated levels have borderline predictive value for development of AIDS [27, 28].

Other Serum Markers: TNF-α, sCD8, IgA

TNF-α induces replication of HIV-1 in latently infected hematopoietic progenitor cells [36]. At least one study has reported increased TNF-α levels in the serum of HIV seropositive individuals [37]. However, this finding is not universal. In another study of HIV seropositive individuals in which elevated serum neopterin and other markers of immune activation were found, TNF-α in serum was not elevated [28]. TNF-α is spontaneously released in vitro from the monocytes of HIV seropositive individuals [38], and it has been suggested that the finding of elevated levels of TNF-α may reflect release in the blood sample after the specimen is collected. The time interval between sample collection and separation of serum for storage should be short (approx. 1 h) to circumvent this artifact. Nevertheless, TNF-α may still play a role in activating HIV production in vivo if high levels are produced then bound to biologically active receptors.

Soluble CD8 is shed from activated CD8 cells and circulating levels are elevated in HIV seropositive individuals [28, 39]. Increased levels are associated with elevations in activated CD8 subsets [39], but the levels of this marker have not yet been shown to be prognostic for AIDS development (but see Janossy, Jiang et al., this volume).

The level of IgA reflects B cell activation. Early in the AIDS epidemic, an increased IgA level was recognized as having prognostic significance for progression to AIDS [7]. Recently, it has been shown to have additional prognostic significance in combination with CD4, serum neopterin (or β_2-M), sIL-2R and p24 antigen (fig. 4d).

p24 Antigen

HIV-1 p24 is the major structural protein of the virus. It has been detected in soluble form in the serum or plasma of 10–20% of HIV-1 seropositive persons. Allain et al. [40], De Wolf et al. [41], and Fahey et al. [27] have noted that the presence of circulating p24 antigen is a poor prognostic sign when it is present. However, a negative p24 test does not indicate that the patient will not develop AIDS and in African studies, few p24-positive individuals were reported [42]. Nevertheless, a positive test is clearly of prognostic value. For example, the relative hazard of developing AIDS was 2.8 times greater among a group of 46 men who were p24 antigen-positive compared with the risk in 349 men who were p24 antigen-negative. Furthermore, a positive p24 test had prognostic value independent of CD4 and soluble markers of immune activation. Consequently, p24 antigen status (positive or negative) was used in the combination of covariates which provided the highest prognostic value for development of AIDS (fig. 4d) [27].

A positive p24 antigen test depends both on the level of circulating anti-p24 antibody in the individual and amount of p24 antigen being produced. This latter factor, i.e., the amount of p24 antigen produced, is presumably dependent on the level of viral replication but other viral factors may play a role as well. p24 antigen tests are most often positive during the early postinfection period before antibodies form, and late in disease when viral replication is high. Some individuals maintain detectable p24 antigen levels for many years without dramatic CD4 cell number decline [42]. It is more frequent, however, that patients with detectable serum p24 lose their CD4 cells more rapidly than patients negative for p24 [43].

McHugh et al. [44] and Nishanian et al. [45] have demonstrated anti-p24/p24 immune complexes in seropositive individuals who test negative on conventional serum p24 assays; studies to determine the prognostic significance of these immune complexes are underway. Infectious virus has been detected in peripheral blood mononuclear cells and plasma of HIV-1 seropositive persons even when p24 antigen tests were negative [46, 47].

Conclusions

The immunology of AIDS encompasses both the immunopathogenesis of HIV infection and specific immunity to HIV. Longitudinal cohort studies of hemophiliacs (see Lee et al., this volume) and homosexually active men have provided detailed information regarding both aspects of the immunology of AIDS in adults. These studies indicate that quantitative measures of CD4 levels including CD4 number and percentage and CD4/CD8 ratio as well as qualitative measures of CD4 cell function are decreased in HIV infection. Knowledge of these parameters provides useful information about the state of an HIV-infected individual's immune system and prognostic information with respect to whether that person will develop AIDS. These studies also indicate that serum levels of neopterin and β_2-M are elevated in HIV infection, and that the levels of these two markers of immune activation are highly associated with an individual's relative risk of developing AIDS.

At this time, laboratory assessment of immune deficiency (reflected in peripheral blood CD4 lymphocyte percentage, CD4 absolute number and CD4/CD8 ratio) and assessment of immune activation (reflected in serum levels of neopterin and β_2-M) provide the most useful prognostic information in HIV-infected adults. Obviously, careful quality control of all laboratory procedures is essential in that laboratory quality control generally leads to more accurate data and subsequently to more accurate prognosis. Discussion of issues related to flow cytometry quality assurance [48, 49] and recommendations for immunophenotyping using flow cytometry [50, 51] can be found in recent publications.

References

1 Lewis DE, Giorgi JV: Immunology of HIV infection. Intern Rev Immunol 1990;7: 1–13.

2 Munoz A, Carey V, Saah AJ, Phair JP, Kingsley LA, Fahey JL, Ginzburg HM, Polk BF, and The Multicenter AIDS Cohort Study: Predictors of decline in CD4 lymphocytes in a cohort of homosexual men infected with human immunodeficiency virus. J AIDS 1988;1:396–404.

3 Moss SR, Bacchetti P, Osmond D, Krampf W, Chaisson RE, Stites D, Wilber J, Allain J-P, Carlson J: Seropositivity for HIV and the development of AIDS or AIDS-related condition: Three-year follow-up of the San Francisco General Hospital cohort. Br Med J 1988;296:745–750.

4 Giorgi JV, Detels R: T-cell subset alterations in HIV-infected homosexual men:

NIAID multicenter AIDS cohort study. Clin Immunol Immunopathol 1989;52:10–18.

5 Giorgi JV, Fahey JL, Smith DC, Hultin LE, Cheng H-L, Mitsuyasu RT, Detels R: Early effects of HIV on CD4 lymphocytes in vivo. J Immunol 1987;138:3725–3730.

6 Tersmette M, de Goede REY, Al BJM, Winkel IN, Gruters RA, Cuypers HTM, Huisman HG, Miedema F: Differential syncytium-inducing capacity of human immunodeficiency virus isolates: Frequent detection of syncytium-inducing isolates in patients with acquired immunodeficiency syndrome (AIDS) and AIDS-related complex. J Virol 1988;62:2026–2032.

7 Taylor J, Afrasiabi R, Fahey JL, Korns E, Weaver M, Mitsuyasu R: Prognostically significant classification of immune changes in AIDS with Kaposi's sarcoma. Blood 1986;67:666–671.

8 Kansas GS, Wood GS, Fishwild DM, Engleman EG: Functional characterization of human T lymphocyte subsets distinguished by monoclonal anti-Leu-8. J Immunol 1985;134:2995–3002.

9 Morimoto C, Letvin NL, Boyd AW, Hagan M, Brown HM, Kornacki MM, Schlossman SF: The isolation and characterization of the human helper inducer T cell subset. J Immunol 1985;134:3762–3769.

10 Morimoto C, Letvin NL, Distaso JA, Aldrich WR, Schlossman SF: The isolation and characterization of the human suppressor inducer T cell subset. J Immunol 1985;134:1508–1515.

11 Tedder TF, Cooper MD, Clement LT: Human lymphocyte differentiation antigens HB-10 and HB-11. II. Differential production of B cell growth and differentiation factors by distinct helper T cell subpopulations. J Immunol 1985;134:2989–2994.

12 Morimoto C, Letvin NL, Distaso JA, Brown HM, Schlossman SF: The cellular basis for the induction of antigen-specific T8 suppressor cells. Eur J Immunol 1986;16:198–204.

13 DeMartini RM, Turner RR, Formenti SC, Boone DC, Bishop PC, Levine AM, Parker JW: Peripheral blood mononuclear cell abnormalities and their relationship to clinical course in homosexual men with HIV infection. Clin Immunol Immunopathol 1988;46:258–271.

14 Nicholson JKA, McDougal JS, Spira TJ: Alterations of functional subsets of T helper and T suppressor cell populations in acquired immunodeficiency syndrome (AIDS) and chronic unexplained lymphadenopathy. J Clin Immunol 1985;5:269–274.

15 Merkenschlager M, Terry L, Edwards R, Beverley PCL: Limiting dilution analysis of proliferative responses in human lymphocyte populations defined by the monoclonal antibody UCHL-1: Implications for differential CD45 expression in T cell memory formation. Eur J Immunol 1988;18:1653–1661.

16 Schnittman SM, Lane HC, Greenhouse J, Justement JS, Baseler M, Fauci AS: Preferential infection of CD4+ memory T cells by human immunodeficiency virus type 1: Evidence for a role in the selective T-cell functional defects observed in infected individuals. Proc Natl Acad Sci USA 1990;87:6058–6062.

17 Jackson AL, Matsumoto H, Janszen M, Maino V, Blidy A, Shye S: Restricted expression of p55 interleukin-2 receptor (CD25) on normal T cells. Clin Immunol Immunopathol 1990;54:126–133.

18 Zola H, Koh LY, Mantzioris BX, Rhodes D: Patients with HIV infection have a reduced proportion of lymphocytes expressing the IL-2 receptor p55 chain (TAC, CD25). Clin Immunol Immunopathol 1991;59:16–25.
19 Shearer GM, Bernstein DC, Tung KS, Via CS, Redfield R, Salahuddin SZ, Gallo RC: A model for the selective loss of major histocompatibility complex self-restricted T cell immune responses during the development of acquired immune deficiency syndrome (AIDS). J Immunol 1986;137:2514–2521.
20 Clerici M, Stocks NI, Zajac RA, Boswell RN, Lucey DR, Via CS, Shearer GM: Detection of three distinct patterns of T helper cell dysfunction in asymptomatic, HIV-seropositive patients: independence of CD4+ cell numbers and clinical staging. J Clin Invest 1989;84:1892–1899.
21 Lane HC, Depper JM, Greene WC, Whalen G, Waldmann TA, Fauci AS: Qualitative analysis of immune function in patients with the acquired immunodeficiency syndrome: Evidence for a selective defect in soluble antigen recognition. N Engl J Med 1985;313:79–84.
22 Clerici M, Via CS, Lucey DR, Roilides E, Pizzo PA, Shearer GM: Functional dichotomy of CD4+ T helper lymphocytes in asymptomatic human immunodeficiency virus infection. Eur J Immunol 1991;21:665–670.
23 Van Noesel CJM, Gruters RA, Terpstra FG, Schellekens PTA, Van Lier RAW, Miedema F: Functional and phenotypic evidence for a selective loss of memory T cells in asymptomatic human immunodeficiency virus-infected men. J Clin Invest 1990;86:293–299.
24 Gruters RA, Terpstra FG, De Jong R, Van Noesel CJM, Van Lier RAW, Miedema F: Selective loss of T cell functions in different stages of HIV infection. Eur J Immunol 1990;20:1039–1044.
25 Fahey JL, Prince H, Weaver M, Groopman J, Visscher B, Schwartz K, Detels R: Quantitative changes in T helper or T suppressor/cytotoxic lymphocyte subsets that distinguish acquired immune deficiency syndrome from other immune subset disorders. Am J Med 1984;76:95–100.
26 Huber C, Batchelor JR, Fuchs D, Hausen A, Lang A, Niederwieser D, Reibnegger G, Swetly P, Troppmair J, Wachter H: Immune response-associated production of neopterin: Release from macrophages primarily under control of interferon-gamma. J Exp Med 1984;160:310–316.
27 Fahey JL, Taylor JMG, Detels R, Hofmann B, Melmed R, Nishanian P, Giorgi JV: The prognostic value of cellular and serologic markers in infection with human immunodeficiency virus type 1. N Engl J Med 1990;322:166–172.
28 Fuchs D, Jager H, Popescu M, Reibnegger G, Werner ER, Dierich MP, Kaboth W, Tilz GP, Wachter H: Immune activation markers to predict AIDS and survival in HIV-1 seropositives. Immunol Lett 1990;26:75–80.
29 Fuchs D, Hausen A, Reibnegger G, Werner ER, Werner-Felmayer G, Dierich MP, Wachter H: Interferon-γ concentrations are increased in sera from individuals infected with human immunodeficiency virus type 1. J AIDS 1989;2:158–162.
30 Fuchs D, Malkovsky M, Reibnegger G, Werner ER, Forni G, Wachter H: Endogenous release of interferon-gamma and diminished response of peripheral blood mononuclear cells to antigenic stimulation. Immunol Lett 1989;23:103–108.
31 Melmed RN, Taylor JMG, Detels R, Bozorgmehri M, Fahey JL: Serum neopterin

changes in HIV-infected subjects: Indicator of significant pathology, CD4 T cell changes, and the development of AIDS. J AIDS 1989;2:70–76.
32 Hofmann B, Wang Y, Cumberland WG, Detels R, Bozorgmehri M, Fahey JL: Serum beta-2-microglobulin level increases in HIV infection: Relation to seroconversion, CD4 T-cell fall and prognosis. AIDS 1990;4:207–214.
33 Zolla-Pazner S, William D, El-Sadr W, Marmor M, Stahl R: Quantitation of β_2-microglobulin and other immune characteristics in a prospective study of men at risk for acquired immune deficiency syndrome. J Am Med Assoc 1984;251:2951–2955.
34 Greene WC, Leonard WJ, Depper JM, Nelson DL, Waldmann TA: The human interleukin-2 receptor: Normal and abnormal expression in T cells and in leukemias induced by the human T-lymphotropic retroviruses. Ann Intern Med 1986;105:560–572.
35 Prince HE, Kleinman S, Williams AE: Soluble IL-2 receptor levels in serum from blood donors seropositive for HIV. J Immunol 1988;140:1139–1141.
36 Folks TM, Justement J, Kinter A, Dinarello CA, Fauci AS: Cytokine-induced expression of HIV-1 in a chronically infected promonocyte cell line. Science 1987; 238:800–802.
37 Lahdevirta J, Maury CPJ, Teppo A-M, Repo H: Elevated levels of circulating cachectin/tumor necrosis factor in patients with acquired immunodeficiency syndrome. Am J Med 1988;85:289–291.
38 Wright SC, Jewett A, Mitsuyasu R, Bonavida B: Spontaneous cytotoxicity and tumor necrosis factor production by peripheral blood monocytes from AIDS patients. J Immunol 1988;141:99–104.
39 Prince HE, Kleinman S, Czaplicki C, John J, Williams AE: Interrelationships between serologic markers of immune activation and T lymphocyte subsets in HIV infection. J AIDS 1990;3:525–530.
40 Allain J-P, Laurian Y, Paul DA, Verroust F, Leuther M, Gazengel C, Senn D, Larrieu M-J, Bosser C: Long-term evaluation of HIV antigen and antibodies to p24 and gp41 in patients with hemophilia. N Engl J Med 1987;317:1114–1121.
41 De Wolf F, Goudsmit J, Paul DA, Lange JMA, Hooijkaas C, Schellekens P, Coutinho RA, Van der Noordaa J: Risk of AIDS-related complex and AIDS in homosexual men with persistent HIV antigenaemia. Br Med J 1987;295:569–572.
42 Lelie PN, Reesink HW, Bakker E, Huisman JG, Ten Veen JH: Clinical importance of HIV antigen and anti-HIV markers in persons infected with HIV. N Engl J Med 1988;318:1204–1205.
43 Phillips AN, Lee CA, Elford J: p24 antigenaemia, CD4 lymphocyte counts and the development of AIDS. AIDS 1991;5:1217–1222.
44 McHugh TM, Stites DP, Busch MP, Krowka JF, Stricker RB, Hollander H: Relation of circulating levels of human immunodeficiency virus (HIV) antigen, antibody to p24, and HIV-containing immune complexes in HIV-infected patients. J Infect Dis 1988;158:1088–1091.
45 Nishanian P, Huskins KR, Stehn S, Detels R, Fahey JL: A simple method for improved assay demonstrates that HIV p24 antigen is present as immune complexes in most sera from HIV-infected individuals. J Infect Dis 1990;162:21–28.
46 Ho DD, Moudgil T, Alam M: Quantitation of human immunodeficiency virus type 1 in the blood of infected persons. N Engl J Med 1989;321:1621–1625.

47 Coombs RW, Collier AC, Allain J-P, Nikora B, Leuther M, Gjerset GF, Corey L: Plasma viremia in human immunodeficiency virus infection. N Engl J Med 1989; 321:1626–1631.
48 Giorgi JV, Cheng H-L, Margolick JB, Ferbas J, Waxdal M, Schmid I, Hultin LE, Jackson AL, Park L, Taylor JMG, and The Multicenter AIDS Cohort Study Group: Quality control in the flow cytometric measurement of T-lymphocyte subsets: The multicenter AIDS cohort study experience. Clin Immunol Immunopathol 1990;55: 173–186.
49 Parker JW, Adelsberg B, Axen SP, Boone D, Fletcher MA, Gjerset GF, Hassett J, Kaplan J, Niland JC, Odom-Maryon T, Operskalski EA, Prince H, Scott D, Stites DP, Mosley JW, and The Transfusion Safety Study Group: Leukocyte immunophenotyping by flow cytometry in a multisite study: Standardization, quality control, and normal values in the transfusion safety study. Clin Immunol Immunopathol 1990;55:187–220.
50 Giorgi JV, Hultin LE: Lymphocyte subset alterations and immunophenotyping by flow cytometry in HIV disease. Clin Immunol Newslett 1990;10:55–61.
51 Landay A, Ohlsson-Wilhelm B, Giorgi JV: Application of flow cytometry to the study of HIV infection. AIDS 1990;4:479–497.

Janis V. Giorgi, PhD, UCLA School of Medicine,
Department of Medicine/CIA/CIC, 10833 Le Conte Avenue,
Los Angeles, CA 90024–1745 (USA)

Janossy G, Autran B, Miedema F (eds): Immunodeficiency in HIV Infection and AIDS.
EC/FERS/MRC Workshop on Immunodeficiency in HIV-1 Infections, Windsor, Surrey, 1991.
Basel, Karger, 1992, pp 18–31

Normal Control Values for CD4 and CD8 T Lymphocytes and Their Relevance for HIV-1 Infections

Margarita Bofill[a], *George Janossy*[a], *Andrew N. Phillips*[b], *Anthony Timms*[a], *David MacDonald-Burns*[c], *Margaret A. Johnson*[d], *Christine A. Lee*[e]

Departments of [a]Clinical Immunology, [b]Public Health and Primary Care, [c]Genito-Urinary Medicine, [d]Thoracic Medicine, and [e]Haemophilia Centre, Royal Free Hospital and School of Medicine, Hampstead, London, UK

During infection with human immunodeficiency virus (HIV-1) CD4 lymphocytes are selectively lost [1, 2] and the depletion of CD4 cells in the blood is one of the best markers to predict progression to AIDS (see Lee et al., and Giorgi, this volume) [2–5]. The CD4 counts and morbidity are related. People with $>200/mm^3$ CD4 levels very infrequently develop AIDS except with Kaposi sarcoma, while about half of patients develop the disease before falling to a CD4 count of $50/mm^3$ [2, 6].

The severity of depletion is reflected by three measures: the absolute CD4 cell numbers per blood volume, the percentages of CD4 cells (CD4%) among lymphocytes and the CD4/CD8 ratio [7]. These measurements closely correlate with each other and all show prognostic significance [2, 7, 8]. Nevertheless, they also have different roles in immunodiagnosis: inverted CD4/CD8 ratios are seen in a host of viral infections [1, 9–13] and complement absolute CD4 counts which are more characteristic, albeit not specific, for HIV-1 involvement. A great disadvantage of absolute CD4 counts is that when these values are calculated the errors which are inherent in WBC counting are also included. Some of these problems may be remedied by the arrival of flow cytometers with absolute counting facility. Many investigators favour absolute CD4 counts despite these problems because the values also take into account lymphopenia which is thought to be part of disease progression in HIV-1 infections. For these reasons, CD4 levels are regularly determined during follow-up, and low absolute CD4 count is an important indication to initiate antiviral drugs and primary prophylaxis [14, 15].

The precision of flow cytometry has recently been greatly increased and the T cell subset reference ranges have been defined [16–21] by two criteria. Firstly, the mean values have been documented in infants, children and adults [18–21] which help assess the degree of changes required for developing symptoms of immunodeficiency. Secondly, the lower normal limits of CD4 counts are also known for the interpretation of putative HIV-1-associated changes. These investigations include volunteers [18–21], patients with coagulation disorders but with no lymphocyte abnormalities [21] as well as HIV-1-negative homosexual males [17, 21]. The latter group may be regarded as the most appropriate controls for HIV-1-positive homosexuals as exposures to factors other than HIV-1 may also alter T lymphocyte subsets, and these are similar in the seropositive and seronegative homosexual people. These normal values are reviewed below.

Methods

Since the introduction of whole blood methods by Hoffman et al. [16] the superiority of this technique over the analysis of Ficoll-Hypaque separated mononuclear cells has been fully accepted [2, 17, 18, 21–24]. The whole blood method together with dual fluorescence utilizing pairs of antibodies directly conjugated with fluorescein (FITC; green) and phycoerythrin (PE; orange), respectively, elicited with a single laser light source offers both the advantages of speed and precision [18]. These IF analyses on flow cytometers show <5% coefficients of variations 'within-sample' and <12% 'within-person' [17, 18, 22]. Gating of lymphocytes and monocytes has been aided by using CD45 and CD14 antibodies in conjunction with 90° (side) and forward angle laser light scatter [18]. In the routine laboratory the regular use of CD3/CD4 and CD3/CD8 doubles and perhaps an additional CD4/CD8 double are now among the essential requirements for determining CD4% and CD8% values accurately within the lymphocyte gate as well as within the CD3+ population. These results also yield exact CD4/CD8 ratios [17, 18, 21]. Pitfalls of T cell counting are readily avoided. These would be inclusion of CD4+ monocytes in CD4+ counts, and lack of information about CD3–, CD8+ cells. The non-T CD8+ cells are, however, mostly present in low proportions (<10% of all CD8+ cells), and are included in the CD8 counts. Finally, it is also common practice that the sum of CD3% (T cells), CD19% (B cells) and CD16% (NK-like cells) is required to be around 100% on cells within the lymphocyte gate – a further simple check on quality [18, 24].

Table 1. Changes of lymphocyte subsets in different age groups[a]

Cells	Children					Adolescents and adults, 11–79 years			
	1–2 years	3–4 years	5–6 years	7–8 years	9–10 years	median	mean ± SD	SEM	range[b]
Absolute number of lymphocytes/mm³									
Lymphocytes	5,390 ± 1,140	3,510 ± 971	3,290 ± 812	2,840 ± 673	2,440 ± 586	1,800	1,900 ± 549	22	*1,000–3,200*
CD3	4,050 ± 955	2,740 ± 889	2,470 ± 650	2,150 ± 466	1,820 ± 486	1,400	1,450 ± 458	19	*780–249*
CD4	2,530 ± 648	1,560 ± 528	1,460 ± 487	1,140 ± 289	980 ± 243	790	830 ± 288	12	
Range[c]	*1,600 – 3,600*	*1,040 – 2,700*	*980 – 2,100*	*880 – 1,400*	*720 – 1,330*				*410–1,540*
Lowest value[b]	*1,400*	*1,020*	*960*	*540*	*710*				*220*
CD8	1,390 ± 363	1,050 ± 371	870 ± 331	890 ± 229	720 ± 225	520	520 ± 231	9	*230–1,090*
Ratios									
CD4/CD8	1.83 ÷ 1.31	1.50 ÷ 1.33	1.72 ÷ 1.48	1.28 ÷ 1.21	1.40 ÷ 1.32	1.50	1.51 ÷ 1.50		*0.66–3.52*
Percentages among lymphocytes									
%CD3	75.6 ± 8.84	77.3 ± 6.85	75.2 ± 8.22	76.6 ± 7.01	75.0 ± 8.81	77.8	76.3 ± 9.23	0.38	*58–91*
%CD4	47.2 ± 7.79	44.5 ± 7.97	44.7 ± 9.92	40.3 ± 5.23	41.1 ± 7.90	43.0	43.6 ± 8.88	0.36	
Range[c]	*37 – 58*	*34 – 54*	*34 – 57*	*33 – 48*	*27 – 54*				*27–61*
Lowest value[b]	*36*	*33*	*32*	*29*	*25*				*22*
%CD8	25.9 ± 4.78	30.0 ± 7.46	25.9 ± 5.13	31.5 ± 5.07	29.2 ± 4.79	29.0	29.5 ± 8.23	0.34	*14–46*
n[d]	22	22	15	16	25	600			

[a] From Bofill et al. [21] with permission.
[b] Outlier extreme values for CD4 cells are separately shown in italics.
[c] Ranges (shown in italics) are truncated to include 95% of data.
[d] Number of tests in each group.

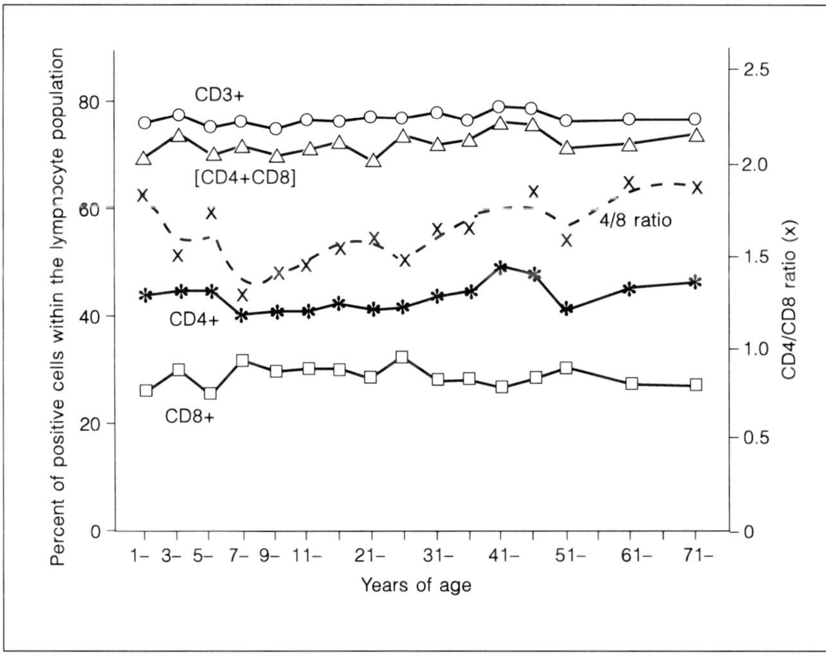

Fig. 1. The stability of the percentage values of lymphocyte subsets in the different age groups [from 21, with permission]. The percentages of subsets within the lymphocyte population and the CD4/CD8 ratios are shown. ○ = CD3%; △ = CD4 + CD8 (calculated); * = CD4%; □ = CD8%; × = CD4/CD8 ratio. The CD4% values are consistently > 25% in normal individuals, and in children born to HIV-1-positive mothers values of CD4% < 25% may indicate HIV-1 infection.

Changes of T Cell Subsets with Age

CD3+, CD4+ and CD8+ lymphocytes are present in stable proportions in the different age groups (fig. 1). In both children and adults the mean values for CD3% are 76 ± 8% (SD), for CD4% 43.5 ± 8%, and for CD8% 29 ± 7 (table 1). The lowest CD4% values are invariably >25% in children and are also >25% in 99% of adults. The sums of [CD4% + CD8%] are around 72%. These values and the fact that some CD8±, CD3– lymphocytes (2.5% of total) are included in the CD8 counts indicate that about 6% of CD3+ T cells have undetectable amounts of CD4 and CD8 (CD3+, CD4–, CD8–). Some of these carry TCRγδ.

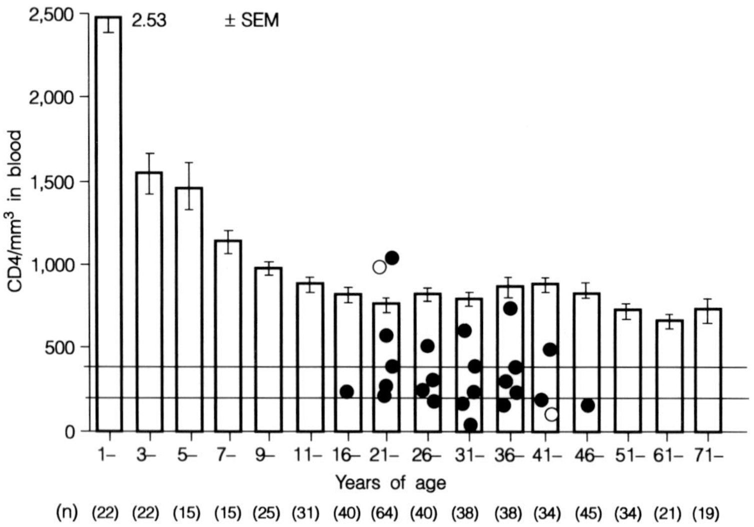

Fig. 2. The absolute CD4 counts in the blood of 25 unselected HIV-1-positive patients attending an Open Access Testing Clinic. Only 2 of these individuals had CD4/CD8 ratios above unity (○) and 23 had inverted CD4/CD8 ratios (●). The values are projected onto the background of normal CD4 counts shown at the different age groups (see also table 1). The horizontal lines are 400/mm³ and 200/mm³ CD4 levels.

The means for the CD4/CD8 ratios also remains stable with age, with an average value of 1.51 ÷ 1.50 (geometrical mean) in adults (table 1). Nevertheless, cord blood samples have high CD4/CD8 ratios (3.3 ÷ 1.2) and babies are born with high CD4%. These CD4/CD8 ratios reach normal values within the first year of life [19, 20, 25].

The age-related changes of CD4 and CD8 counts are governed by the alterations of absolute lymphocyte counts. In babies the counts are 2.5–3 times higher, and in infants about 2 times higher than in adults. These counts gradually decrease between years 1 and 11 (fig. 2). In adolescents (11–16 years), however, absolute counts are no longer significantly above the adult values. The means of lymphocyte counts/mm³ for adults are: lymphocytes 1,910 ± 560 (SD), CD3+ 1,450 ± 470, CD4+ cells 810 ± 280 and CD8+ lymphocytes 570 ± 240 [21].

Table 2. Absolute numbers and percentage values of T cell populations in the blood of HIV-1-seronegative adults within the different control groups[a, b]

	Heterosexual males	Heterosexual females	Haemophilia B	von Willebrand disease	Haemophilia A	Homosexual males
Absolute number of lymphocytes/mm³						
Lymphocytes	2,010 ± 525	2,120 ± 572	1,970 ± 586	1,780 ± 454	1,820 ± 526	1,980 ± 604
CD3	1,490 ± 453	1,690 ± 479	1,500 ± 504	1,360 ± 375	1,410 ± 440	1,470 ± 488
CD4	840 ± 285	1,050 ± 377	840 ± 259	810 ± 228	790 ± 263	800 ± 324
CD8	550 ± 227	610 ± 213	600 ± 263	480 ± 176	540 ± 226	650 ± 245
Ratios						
CD4/CD8	1.56÷1.50	1.70÷1.55	1.45÷1.38	1.72÷1.42	1.51÷1.47	1.25÷1.62
n[c]	50	50	86	112	202	100

[a] From Bofill et al. [21] with permission.
[b] Results shown are means (x) ± SD.
[c] Numbers of individuals tested.

The lymphoid cells in the non-T CD3-negative category are represented by an equal number of CD19+ B lymphocytes and CD16+ NK cells. The CD19% and CD16% values are 12% amounting to 230/mm³ circulating cells [18].

T Cell Subpopulations in the Different Control Groups

In a recent study from our laboratory the absolute numbers and percentages of T cell populations have been compared in six different control groups [21]. These included a total of 600 individuals within the categories of heterosexual males, heterosexual females, patients with haemophilia A, haemophilia B and von Willebrand disease, and homosexual males (table 2). The homosexual males had slightly increased CD8% (32.7 ± 10.1%) and CD8 absolute counts (640 ± 250/mm³) when compared to the rest of controls, and the increased CD8 contribution was also reflected in the high proportions of homosexual males (27%) with inverted CD4/CD8 ratios. In female heterosexuals, as a result of cumulative increases in lym-

Table 3. Extreme low CD4%, absolute CD4 counts and inverted CD4/CD8 ratios amongst HIV-1-seronegative controls

Groups	Number tested in the group	Proportion of individuals (%) with					
		CD4% <25% A	abs. CD4 (<400/mm³) B	inverted CD4/8 ratio C	B+C[a]	A+B+C[b]	abs. CD4 <200/mm³
HIV-1 seronegative groups[c]							
Heterosexual males	50	2.0	2.0 (380)[d]	10.0	2.0 (380)[d]	2.0	0
Heterosexual females	50	2.0	4.0 (290 340)	12.0	2.0 (290)	2.0	0
Haemophilia B	86	0	2.3 (270 390)	9.3	0	0	0
von Willebrand disease	112	0	3.6 (270 290 320 320 390)	4.5	2.0 (320 390)	0	0
Haemophilia A	202	0.5	2.6 (280 320 340 340 360)	12.4	1.5 (280 340 340)	0.5	0
Homosexual males	100	3.0	4.0 (210 230 350 390)	27.0	3.0 (210 230 350)	3.0	0
HIV-1 seropositive group							
At an Open Clinic[e]	25	56.0	68 (a range of 3–390)	92.0	64 (3–390)	52.0	20.0

[a] People with <400/mm³ CD4 count *and* inverted ratio.
[b] People with <25% CD4% + <400/mm³ CD4 + inverted ratio.
[c] People aged between 11 and 79 years.
[d] Numbers in parentheses refer to absolute CD4 counts detected in each case.
[e] See Squire et al. [28].

phocyte counts and CD4%, the absolute CD3 and CD4 counts were 16–28% higher than the normal ranges. As these samples were frequently drawn in the afternoon the increase may reflect diurnal factors (see below).

Previously low lymphocyte counts, CD4 lymphopenia, increased CD4/CD8 ratios and hypergammaglobulinaemia have been reported to occur in patients with coagulation disorders [26, 27], which we could no longer confirm. The probable explanations for the changes are twofold. In the early studies patients developing HIV-1 seropositivity were still included [26, 27], and recently made factor VIII cryoprecipitates are cleaner and have lower immunogenicity.

Extreme Low CD4 Values and Related Parameters in Control Groups

From a study of 600 normal people, individuals with low CD4% (<25%), low absolute CD4 counts (<400 mm^3) and inverted CD4/CD8 ratios in the various control groups are listed in table 3. About 2.8% of people showed CD4 lymphopenia and 9–10% had inverted ratios in the control and 27% had inverted ratios in the HIV-1-negative homosexual group. These were mostly different individuals; only 7 out of 500 controls (1.4%) and 3 out of 100 HIV-1-seronegative homosexual males (3%) exhibited <400 mm^3 CD4 plus inverted CD4/CD8 ratio. Only 0.6% of 500 controls had all three features.

The same parameters were also analysed in 25 HIV-1-seropositive patients attending an open access clinic [28]. In 23 (92%) inverted CD4/CD8 ratios were seen, and 17 (68%) had CD4 lymphopenia <400/mm^3 with 64% coincidence (fig. 2). The majority of patients (52%) showed all three features including <25% CD4%. Five (20%) had their CD4 counts below 200/mm^3 (table 3).

The Significance of CD4 Counts in Assessing HIV-1-Related Changes

The observations about the normal values shown in table 1 may assist the diagnosis of HIV-1 infection in children born to HIV-1-seropositive mothers where serology is unhelpful and in a proportion of cases both HIV-1 isolation and polymerase chain reaction can yield false-negative results [29]. Normal children had >25% CD4%. Thus, values with <25%

Table 4. Normal lymphocyte values in the blood[a]

	Our study[a]	Diurnal changes[b] blood taken at		Multicentre AIDS Cohort Study (4 centres)[c]	Becton-Dickinson Clinical Monograph (7 centres)[d]
		8 a.m.	10 p.m.		
Absolute numbers of lymphocytes/mm³					
Total	1,900 ± 550	1,900	2,810	2,250 ± 570	1,870 ± n.a.
CD3	1,450 ± 460			1,640 ± 480	1,370 ± n.a.
CD4	830 ± 290	820	1,320	1,020 ± 330	(830) ± n.a.[e]
CD8	560 ± 230			610 ± 230	(620) ± n.a.
Ratios					
CD4/8	1.51÷1.50	1.56	1.84	n.a.	1.4 ± 0.6
Percentages within lymphocytes					
%CD3	76.3 ± 9.2			71.1 ± 9.0	73 ± 6.5
%CD4	43.6 ± 8.9	43.2	47.7	44.5 ± 7.5	44 ± 7.6
%CD8	29.5 ± 8.2	29.2	27.3	26.8 ± 7.2	33 ± 7.4
n[f]	600	small study		2,700	304

n.a. = Not available.
[a] See Bofill et al. [21].
[b] From Malone et al. [31] to indicate sources of variability (e.g. diurnal changes).
[c] From Giorgi et al. [17] (UCLA MACS Group) including male homosexuals.
[d] From Reichert et al. [18] compiled from seven European centres including normal controls.
[e] The values in parentheses are calculated.
[f] The number of observations in the publication.

CD4% and below the lowest absolute counts shown in table 1 may suggest underlying viral, including HIV-1, involvement.

Apart from the age-related long-term changes, many factors contribute to the variability of results. First, in different individuals the absolute CD4 counts tend to be fixed at different levels for a long span of time: in our study for at least 5 years [21]. Secondly, the diurnal changes of lymphocyte counts are known [30, 31] and blood samples for CD4 counts should therefore be taken at a standard time of the day [31]. From low values in the morning, both lymphocyte absolute counts and CD4% values increase by as much as 48 and 10%, respectively, to peak values at 10 p.m. [31]. The increment in CD4 absolute count is therefore a cumulative 60% – from 820 to 1,320/mm^3 (table 4). Most people in our study [21] donated blood during the morning, as reflected in our low ranges (table 4). The female control samples were, however, mostly taken in the afternoon in line with their significantly higher absolute CD4 counts (1,050/mm^3). The largest American multicentre study also shows moderately higher absolute values [17], and these may be due to diurnal factors and/or minor differences in setting up WBC counting.

The observations above are relevant for assessing HIV-1-related diagnostic parameters. CD4 lymphopenia <400/mm^3 is rare in normals but in some the CD4 levels are set at such a low value [21]. In these people serial CD4 counts are stable, unlike in HIV-1 infection where negative slopes with a median value of 0.08×10^9/litre/year are observed [32]. Among diseases tuberculosis [33], early acute phases of viral infections such as influenza [34] and Dengue fever [35] can also cause CD4 depletion.

Inverted CD4/CD8 ratios among the nonhospitalized controls are seen at 10% frequency, and even more frequently among seronegative homosexual men (27%) probably reflecting an increased exposure to chronic vial infections. High CD8% and inverted ratios are also reported in disease conditions with antigen overload such as frequently transfused patients with thalassaemia [36].

Nevertheless, when both the parameters of <400/mm^3 CD4 lymphopenia and the inverted CD4/CD8 ratio are evaluated, 1.4–3% of individuals in the normal and homosexual control groups exhibit both features. As a contrast, 64% of the HIV-1-positive group from an Open Access Clinic fully satisfy these criteria (table 3). The powerful discrimination between the HIV-1-negative and -positive groups by these combined parameters warrants a careful follow-up on all patients who show both abnormal parameters – including both the repeats of subset analysis and serology, and PCR for HIV-1. These subset aberrations are only suggestive of chronic viral disease because in normal lymphopenic people acute viral infections other than HIV-1 may also cause CD8 lymphocytosis which, together with their low CD4 values, mimic HIV-1-associated cellular parameters [13, 37]. But such changes induced by other viruses are transient rather than progressive. Chronic CD4 depletion with consistently inverted CD4/CD8 ratios are almost exclusively seen in HIV-1 infections apart from cases of congenital CD4 deficiency and common variable hypogammaglobulinaemia [38], or from defects of immune reconstitution following bone marrow transplantation [39].

Conclusions

With the advent of a well-standardized method using whole blood lysis, double colour fluorescence and flow cytometry the normal ranges of CD3, CD4 and CD8 T lymphocyte subsets can be established in the routine laboratory. These values assist the definition of HIV-1-related deviations. In children CD4% values <25% are abnormal. In male homosexual patients higher CD8 levels were seen compared to heterosexual males and 27% had an inverted CD4/CD8 ratio, mostly without CD4 lymphopenia (CD4 <400/mm^3). Others may have had low CD4 counts <400/mm^3 with normal CD4/CD8 ratios. Only 3 and 1.4% of these seronegative normals had inverted CD4/CD8 ratios together with <400/mm^3 CD4 counts. Amongst the HIV-1-seropositive asymptomatic patients, however, >60% had such characteristic combined changes. Consequently, individuals who have the laboratory findings of $<0.4 \times 10^9$/litre CD4 count and a CD4/CD8 ratio of less than unity should not be lost from follow-up investigations and need to be investigated further by serology, subset analysis and even molecular probes to exclude or confirm chronic infections with viruses such as HIV-1 and other causes of immunodeficiency.

Acknowledgements

We are particularly grateful to Dr. E. Goldman of Haemophilia Centre and Dr. S.B. Squire of Department of Thoracic Medicine, RFH, for data on patients included, as well as to Prof. P. Griffiths and Dr. A. Webster, Department of Virology, RFHSM, for the results on HIV-1 serology. Caroline Sabine assisted in the statistical calculations. This study was supported by a grant from the MRC AIDS Directed Programme to Prof. G. Janossy.

References

1 Gottlieb MS, Schroff R, Schanker HM, Weisman JD, Fan PT, Wolf RA, Saxon A: *Pneumocystis carinii* pneumonia and mucosal candidiasis in previously healthy homosexual men: evidence of a new acquired cellular immunodeficiency. N Engl J Med 1981;305:1425–1431.

2 Fahey JL, Taylor JM, Detels R, Hofmann B, Melmed R, Nishanian P, Giorgi JV: The prognostic value of cellular and serologic markers in infection with human immunodeficiency virus type 1. N Engl J Med 1990;322:166–172.

3 Tindall B, Cooper DA, Donovan B, Barnes T, Philpot CR, Gold J, Penny R: The Sydney AIDS Project: development of acquired immunodeficiency syndrome in a group of HIV seropositive homosexual men. Aust NZ J Med 1988;18:8–15.

4 Phillips A, Lee CA, Elford J, Janossy G, Bofill M, Timms A, Kernoff PB: Prediction of progression to AIDS by analysis of CD4 lymphocyte counts in a haemophilic cohort. AIDS 1989;3:737–741.
5 Stites DP, Moss AR, Bacchetti P, Osmond D, McHugh TM, Wang YJ, Hebert S, Colfer B: Lymphocyte subset analysis to predict progression to AIDS in a cohort of homosexual men in San Francisco. Clin Immunol Immunopathol 1989;52:96–103.
6 Detels R, English PA, Giorgi JV, Visscher BR, Fahey JL, Taylor JM: Patterns of CD4+ cell changes after HIV-1 infection indicate the existence of a codeterminant of AIDS. J Acquir Immune Defic Syndr 1988;1:390–395.
7 Taylor JM, Fahey JL, Detels R, Giorgi JV: CD4 percentage, CD4 number, and CD4:CD8 ratio in HIV infection: which to choose and how to use. J Acquir Immune Defic Syndr 1989;2:114–124.
8 Fahey JL, Prince H, Weaver M, Groopman J, Visscher B, Schwarz K, Detels R: Quantitative changes in T helper or T suppressor/cytotoxic lymphocyte subsets that distinguish acquired immune deficiency syndrome from other immune subset disorders. Am J Med 1984;76:95–100.
9 Reinherz EL, O'Brien C, Rosenthal P: The cellular basis for viral induced immunodeficiency – analysis by monoclonal antibodies. J Immunol 1980;175:1269–1276.
10 Maher P, O'Toole CM, Wreghitt TG, Spiegelhalter DJ, English TA: Cytomegalovirus infection in cardiac transplant recipients associated with chronic T cell subset ratio inversion with expansion of a Leu-7+ TS-C+ subset. Clin Exp Immunol 1985;62:515–524.
11 Crawford DH, Brickell P, Tidman N, McConnell T, Hoffbrand AV, Janossy G: Increased numbers of cells with suppressor T cell phenotype in the peripheral blood of patients with infectious mononucleosis. Clin Exp Immunol 1981;43:291–297.
12 Thomas HC, Brown D, Routhier G, Janossy G, Kung PC, Goldstein G, Sherlock S: Inducer and suppressor T cells in hepatitis B virus-induced liver disease. Hepatology 1982;2:202–204.
13 Tindall B, Barker S, Donovan B, Barnes T, Roberts J, Kronenberg C, Gold J, Penny R, Cooper D: Characterization of the acute clinical illness associated with human immunodeficiency virus infection. Arch Intern Med 1988;148:945–949.
14 Volberding PA, Lagakos SW, Koch MA, Pettinelli C, Myers MW, Booth DK: Zidovudine in asymptomatic human immunodeficiency virus infection. N Engl J Med 1990;322:941–949.
15 Hirschel B, Lazzarin A, Chopard P, Opravil M, Furrer HJ, Ruttmann S: A controlled study of inhaled pentamidine for primary prevention of *Pneumocystis carinii* pneumonia. N Engl J Med 1991;324:1079–1083.
16 Hoffman RA, Kung PC, Hansen WP, Goldstein G: Simple and rapid measurement of human T lymphocytes and their subclasses in peripheral blood. Proc Natl Acad Sci USA 1980;77:4914–4917.
17 Giorgi JV, Cheng HL, Margolick JB, Bauer KD, Ferbas J, Waxdal M: Quality control in the flow cytometric measurement of T-lymphocyte subsets: the multicenter AIDS cohort study experience. The Multicenter AIDS Cohort Study Group. Clin Immunol Immunopathol 1990;55:173–186.

18 Reichert T, De Bruyere M, Deneys V, et al: Lymphocyte subset references ranges in adult caucasians. Clin Immunol Immunopathol 1991;60:190–208.
19 Falcao RP, Ismael SJ, Donadi EA: Age-associated changes of T lymphocyte subsets. Diagn Clin Immunol 1987;5:205–208.
20 Goto M, Nishioka K: Age- and sex-related changes of the lymphocyte subsets in healthy individuals: an analysis by two-dimensional flow cytometry. J Gerontol 1989;44:51–56.
21 Bofill M, Janossy G, Lee CA, MacDonald-Burns D, Phillips AN, Sabin C, Timms A, Johnson MA, Kernoff PBA: Laboratory control values for CD4 and CD8 T lymphocytes – implications for HIV-1 diagnosis. Clin Exp Immunol 1992;88:243–252.
22 Parker JW, Adelsberg B, Axen SP, Boone B, Fletcher MA, Gjerset GF, Hassett J, Kaplan J, Niland JC, Odom-Maryon T, et al, and The Transfusion Safety Study Group: Leukocte immunophenotyping by flow cytometry in a multisite study: Standardization, quality control, and normal values in the transfusion safety study. Clin Immunol Immunopathol 1990;55:187–220.
23 Renzi P, Ginns LC: Analysis of T cell subsets in normal adults. Comparison of whole blood lysis technique to Ficoll-Hypaque separation by flow cytometry. J Immunol Methods 1987;98:53–55.
24 Landay A, Ohlsson-Wilhelm B, Giorgi JV: Application of flow cytometry to the study of HIV infection. AIDS 1990;4:479–497.
25 Wilson M, Rosen FS, Schlossman SF, Reinherz EL: Ontogeny of human T and B lymphocytes during stressed and normal gestation: phenotypic analysis of umbilical cord lymphocytes from term infants. Clin Immunol Immunopathol 1985;37:1–12.
26 Lee CA, Bofill M, Janossy G, Thomas HC, Rizza CR, Kernoff PBA: Relationships between blood product exposure and immunological abnormalities in English haemophiliacs. Br J Haematol 1985;60:161–172.
27 Beddall AC, Al Rubei K, Williams MD, Hill FG: Lymphocyte subset ratios and factor VIII usage in haemophilia. Arch Dis Child 1985;60:530–536.
28 Squire SB, Elford J, Bor R, Tilsed G, Salt H, Bagdades EK, Janossy G, Griffiths PD, Johnson MA: Open access clinic providing HIV-1 antibody results on day of testing: the first twelve months. Br Med J 1991;302:1383–1386.
29 Blanche S, Rouzioux C, Moscato ML, Veber F, Mayaux MJ, Jacomet C, Tricoire J, Deville A, Vial M, Firtion G: A prospective study of infants born to women seropositive for human immunodeficiency virus type 1. HIV Infection in Newborns French Collaborative Study Group. N Engl J Med 1989;320:1643–1648.
30 Levi FA, Canon C, Blum JP, Mechkouri M, Reinberg A, Mathe G: Circadian and circahemidan rhythms in 9 lymphocyte related variables from peripheral blood of healthy subjects. J Immunol 1985;134:217–222.
31 Malone JL, Simms TE, Gray GC, Wagner KF, Burge JR, Burke DS: Sources of variability in repeated T-helper lymphocyte counts from human immunodeficiency virus type 1-infected patients: total lymphocyte count fluctuations and diurnal cycle are important. J Acquir Immune Defic Syndr 1990;3:144–151.
32 Phillips AN, Lee CA, Elford J, Janossy G, Timms A, Bofill M, Kernoff PB: Serial CD4 lymphocyte counts and development of AIDS. Lancet 1991;337:389–392.
33 Beck JS, Potts RC, Kardjito T, Grange JM: T4 lymphopenia in patients with active pulmonary tuberculosis. Clin Exp Immunol 1985;60:49–54.

34 Lewis DE, Gilbert BE, Knight V: Influenza virus infection induces functional alterations in peripheral blood lymphocytes. J Immunol 1986;137:3777–3781.
35 Sarasombath S, Suvatte V, Homchampa P: Kinetics of lymphocyte subpopulations in dengue hemorrhagic fever/dengue shock syndrome. Southeast Asian J Trop Med Public Health 1988;19:649–656.
36 Grady RW, Akbar AN, Giardina PJ, Hilgartner MW, De Sousa M: Disproportionate lymphoid cell subsets in thalassemia major: the relative contribution of transfusion and splenectomy. Br J Haematol 1985;59:713–723.
37 Pedersen C, Dickmeiss E, Gaub J, Ryder LP, Platz P, Lindhardt BO, Lundgren JD: T-cell subset alterations and lymphocyte responsiveness to mitogens and antigen during severe primary infection with HIV: a case series of seven consecutive HIV seroconverters. AIDS 1990;4:523–526.
38 Seligmann M, Aractingi S, Oksenhendler E, Rabian C, Ferchal F, Gonnot G: CD4+ lymphocytopenia without HIV in patients with cryptococcal disease. Lancet 1991; 337:57–58.
39 Janossy G, Prentice HG, Grob JP, Ivory K, Tidman N, Grundy J, Favrot M, Brenner M, Campana D, Blacklock HA, Gilmore MJML, Patterson J, Griffiths P, Hoffbrand AV: T lymphocyte regeneration after transplantation of T cell depleted allogeneic bone marrow. Clin Exp Immunol 1986;63:577–586.

Dr. Margarita Bofill, Department of Clinical Immunology, Royal Free Hospital and School of Medicine, Hampstead, GB–London NW3 2QG (UK)

Janossy G, Autran B, Miedema F (eds): Immunodeficiency in HIV Infection and AIDS.
EC/FERS/MRC Workshop on Immunodeficiency in HIV-1 Infections, Windsor, Surrey, 1991.
Basel, Karger, 1992, pp 32–45

Applications of CD4 Counts in a Cohort of HIV-1 Seropositive Patients with Haemophilia[1]

Christine A. Lee[a], *Andrew N. Phillips*[b,f], *Jonathan Elford*[b], *Paul Griffiths*[c], *Margarita Bofill*[d], *Leonie Swaden*[e], *George Janossy*[d], *Peter B.A. Kernoff*[a]

[a] Haemophilia Centre and Academic Departments of Haematology,
[b] Public Health and Primary Care, [c] Virology, [d] Clinical Immunology and
[e] Pharmaceutics, Royal Free Hospital and School of Medicine, and
[f] Department of Genito-Urinary Medicine,
University College and Middlesex School of Medicine, London, UK

Although in Western countries by the end of this century AIDS is likely to be a predominantly heterosexual disease, the best studied oldest HIV-1-infected cohorts are those of homosexual men and haemophiliacs. The natural history of HIV-1 infection in these groups therefore presents a set of invaluable information for planning decisions in the management of this disease [1]. The progressive loss of the CD4 lymphocyte population has been found to fall at a rate of 85 cells/mm^3/year amongst homosexuals [2]. The advantage of cohorts of haemophiliacs and blood transfusion recipients is that the date of infection with HIV is well defined. This has enabled the observation to be made that the size of the HIV inoculum and strain variation may be cofactors that influence the rate of progression to AIDS [3]. In addition, haemophiliacs have a wider age range at the time of seroconversion and thus it has been possible to show that there is an increased risk for progression to AIDS associated with increasing age at seroconversion [4–6]. The study of the natural history of HIV infection continues to be of primary importance not only to provide information

[1] This work is dedicated to Peter Kernoff as a brilliant physician, scientist and colleague.

The study was supported by the MRC AIDS Directed Programme of the United Kingdom.

about the demand for clinical services by large numbers of infected individuals progressing to AIDS, but also to establish the time for appropriate therapeutic intervention for HIV-infected individuals.

Haemophilic Cohort at the Royal Free Hospital

In the United Kingdom, at least 1,206 haemophiliacs are known to have been infected with HIV and of these a cohort of 111 haemophiliacs has been followed at the Royal Free Hospital Haemophilia Centre [7]. These patients have been assessed at least every 6 months. The majority of these patients, 101 individuals, had severe haemophilia A (factor VIII <2 U/dl), 7 had moderate/mild haemophilia A (factor VIII >2 U/dl), 1 had severe haemophilia B (factor IX <2 U/dl) and 2 had von Willebrand's disease. The majority (109) of these patients were treated with unheated factor VIII concentrates, 1 patient received unheated factor IX concentrate and 1 patient with von Willebrand's disease had other risk factors. The dates of seroconversion between 1979 and 1985 were known for 63 patients and were estimated for the rest. The first patient seroconverted in October 1979 following infusion in August 1979. He is still asymptomatic and fathered a child who is HIV-1 seronegative [8]. The median age at time of initial seropositivity was 24 years (range 2–77 years). Sterilized clotting factor concentrate became available in late 1984 and early 1985, and the last seroconversion occurred in early 1985. AIDS was classified according to the Centres for Diseases Control [9]. The HIV-related clinical symptoms that were assessed included seborrhoeic dermatitis, oral candidiasis, bacterial infections, herpes zoster, thrombocytopenia (platelet count $<50 \times 10^9$/l) and lymphadenopathy.

The investigation of T-lymphocyte subsets started in November 1982. Between 1982 and 1985 the subsets were determined on Ficoll-Hypaque-separated mononuclear cells. Since then a whole blood method has been introduced. The absolute lymphocyte counts of CD4 and CD8 subsets were calculated as described by Bofill et al., this volume.

CD4 Counts Are Related to Disease Progression

The relationship between the CD4 count and time since seroconversion is shown in figure 1. When a linear regression model was fitted for each patient the median slope for the whole population was -85 mm^3/year

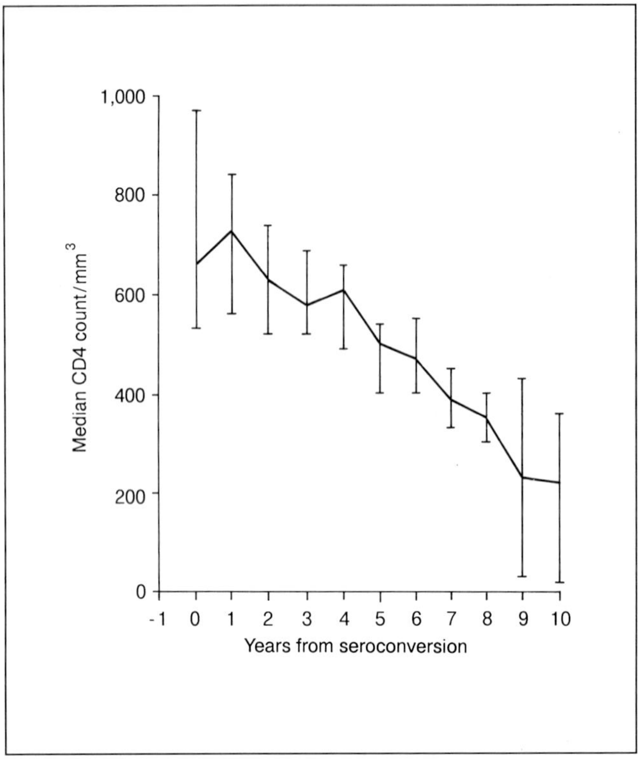

Fig. 1. The relationship between the CD4 count and time since seroconversion.

[see also below and 10]. The cumulative probability of developing HIV-related clinical symptoms or AIDS before falling to a given CD4 count was calculated (fig. 2). For a given CD4 count the numerator for the probability is the number of patients diagnosed as having symptoms or AIDS while having a CD4 count above this level. The denominator consists of all those in the numerator plus the number of patients who reached the CD4 cut-off value symptom- or AIDS-free. In this way it was shown that there was only a 50% chance of being symptom-free at a CD4 lymphocyte count of 200/mm^3 and AIDS-free at a CD4 count of 50/mm^3. The observation that there was an increasing probability of AIDS occurring at a CD4 count of 200/mm^3 supported a treatment policy of introduction of antiviral and prophylactic therapy at this level.

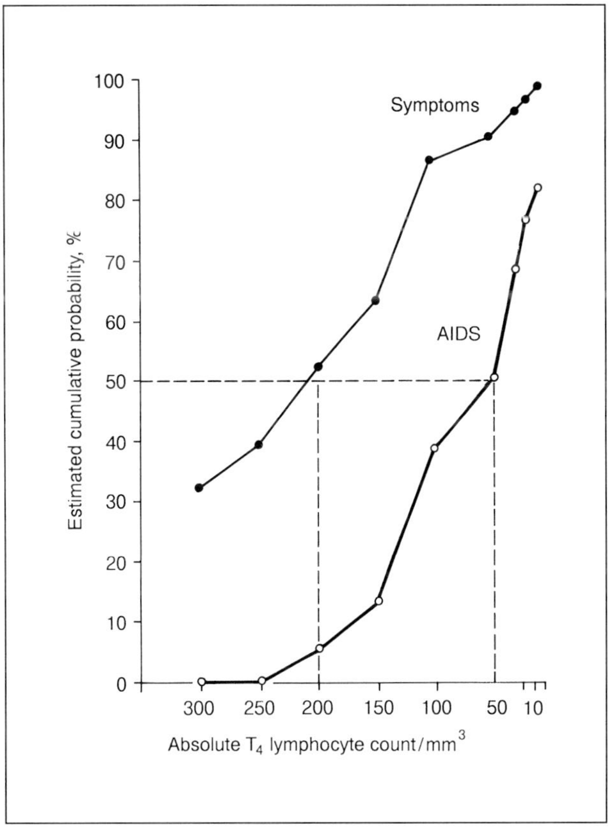

Fig. 2. The cumulative probability of developing HIV-related clinical symptoms or AIDS before falling to a given CD4 count.

The Slope of Decline in CD4 Counts

As multiple CD4 determinations have been available in each patient from the time of seroconversion the rate of CD4 decline could be calculated (table 1). The percentages of patients with slopes within a given range are shown in figure 3. Only 6% of patients have slopes around zero and none of them has developed AIDS. Most patients (94%) have negative slopes. AIDS is more common among patients with steeper negative slopes (fig. 4a, b) [10].

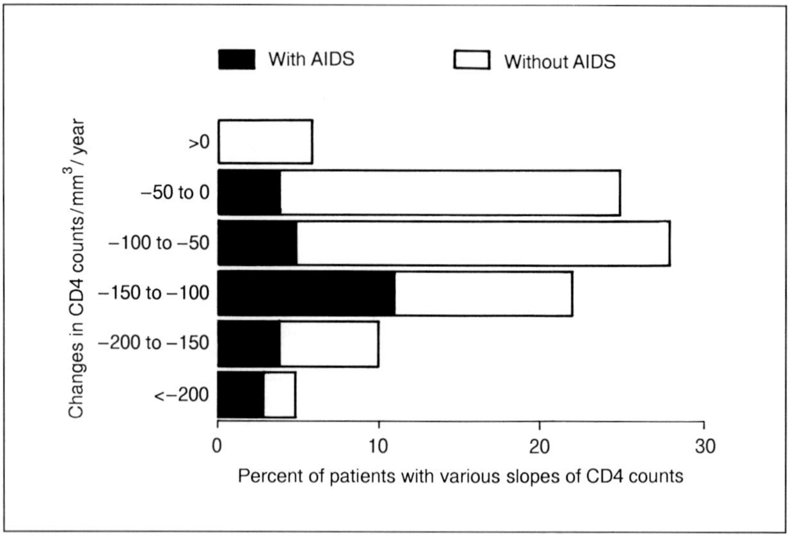

Fig. 3. The percentages of patients with CD4 slope within a given range.

Table 1. Slopes of change in CD4 counts with time in the different groups of patients

	Median CD4/mm^3/year ± SEM
HIV-1 seropositive, all patients	−85 ± 6
AIDS-free by Jan. 1, 1990	−69 ± 7[a]
Developed AIDS by Jan. 1, 1990	−121 ± 13[a]
p24 antigen-free	−73 ± 8[b]
p24 antigen detected	−113 ± 10[b]
Age group	
10–19 years	−83 ± 11[c]
20–29 years	−77 ± 12[c]
> 30 years	−95 ± 10[c]
CMV-negative	−82 ± 10[d]
CMV-positive	−87 ± 8[d]

[a] The time period of observations is from seroconversion to Jan. 1, 1990. From Phillips et al. [10].
[b] From Phillips et al. [12].
[c] From Phillips et al. [15].
[d] From Webster et al. [16].

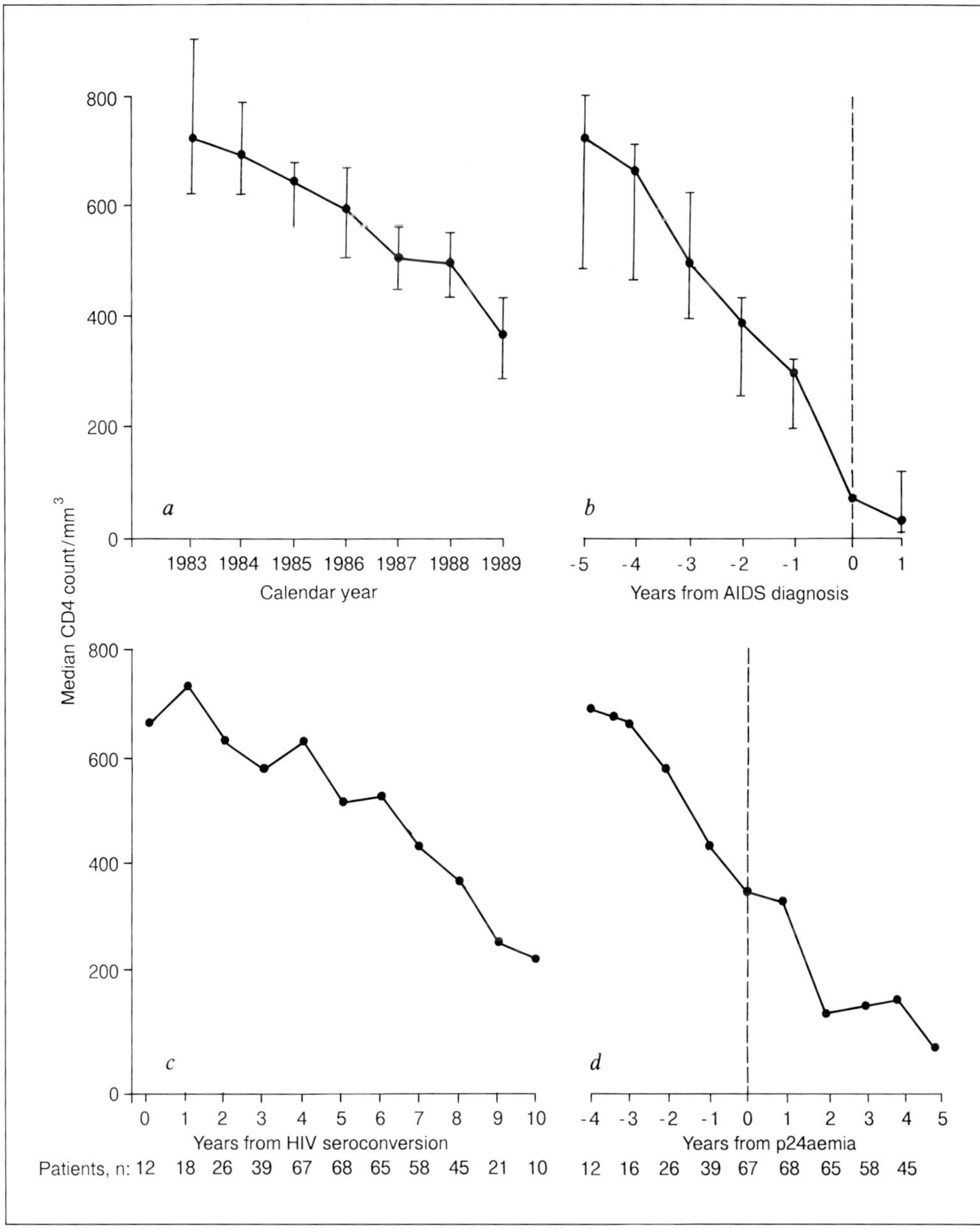

Fig. 4. The slope of CD4 decline in patients without *(a)* and with AIDS *(b)*, and without *(c)* and with p24 antigenaemia *(d)*.

To assess the strength of the relation between the CD4 count and progression to AIDS, a simple model was proposed for the natural history of HIV infection. Since it has been observed that AIDS develops in 50% of untreated patients before the CD4 count falls to 50/mm^3, in this model it was assumed that the decline in CD4 count is near-linear and that when a count of 50/mm^3 is reached the patient manifests AIDS [10]. By fitting a straight line to the CD4 counts of each patient and calculating when the line will cross the count of 50/mm^3 it is estimated when the patient will get AIDS. This model was used to estimate that 30 of 96 patients would develop AIDS by January 1, 1990 and 66 would not. The actual figures were 27 with AIDS and 69 without. Thus, 81 (84%) of patients were correctly classified ($p < 0.0001$; χ^2).

Although the presence of p24 antigenaemia and low CD4 counts in HIV-infected individuals has been shown to be related to increased risk of AIDS, the way in which these factors interrelate in exerting an effect on the risk of AIDS has not been fully evaluated [11]. We observed p24 antigen to be present in the serum of 27 of our 111 patients at least once after seroconversion to anti-HIV and before a diagnosis of AIDS, and in a few patients p24 antigenaemia was consistently detected [7]. For patients in whom p24 antigenaemia appeared, the median CD4 count at the time of the first p24 antigen-positive serum was 380/mm^3 (range 80–510). CD4 counts in those who showed p24 antigenaemia tended to decline at a more rapid rate (–114/mm^3/year) than counts in patients persistently p24 negative (–75/mm^3/year; $p = 0.007$ with t test in fig. 4c, d). Thus the association between p24 antigenaemia and the rate of progression of AIDS can be largely explained by a more rapid decline in CD4 count among patients with p24 antigenaemia than those without [12]. The major pathological effects of persisting viral replication as detected by p24 antigenaemia appear to act via progressive CD4 depletion.

The Influence of Age and CMV Infection

It has been shown that older homosexuals [1], haemophiliacs [6, 7, 13] and blood transfusion recipients [14] progress more rapidly to AIDS than younger individuals. This has been confirmed in our study (fig. 5a). The AIDS case rate per patient year was lowest in the young '10–19 year' age group (2.2 cases per 10 patient-years), intermediate in the '20–29 year' age group (2.5 cases per 10 years) and high in the older group (4.4 cases per 10

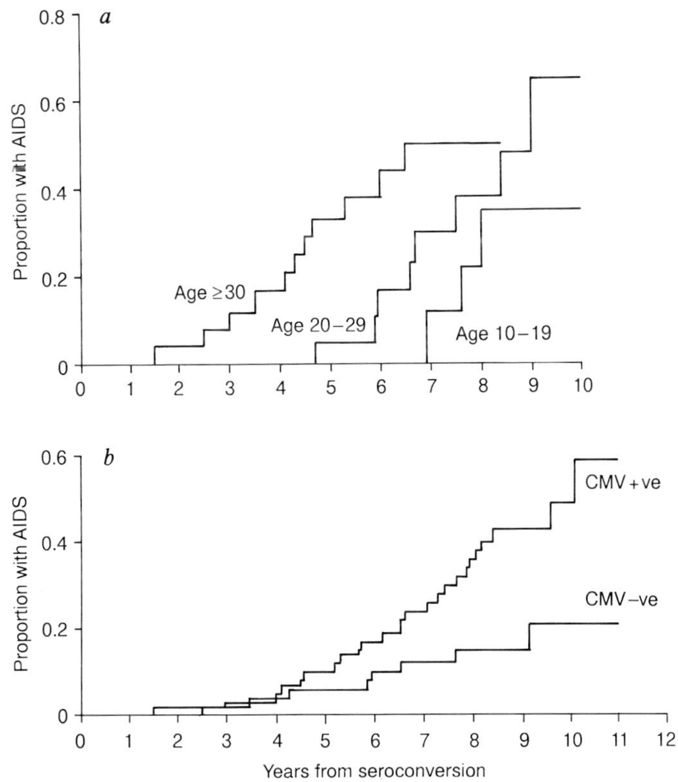

Fig. 5. Progression to AIDS in relation to age *(a)* and to CMV infections *(b)*.

years) [15]. We then used data from our haemophilic cohort to investigate two possible pathways for the more rapid progression rate to AIDS in older people: the CD4 count may fall to a low level more rapidly in older than younger people or there may be a greater risk of developing AIDS among older people at a given low CD4 count.

The age-matched control values and the physiological decline in CD4 counts of HIV-1 seronegative patients are documented by Bofill et al., this volume. These control observations also include HIV-1-negative haemophilic patients and show that from 10 years of age onwards CD4 counts remain at a stable level around 800/mm^3. In our study, there was no suggestion that younger HIV-1 seropositive patients in the 10–19 year age group reach low CD4 levels significantly more slowly than the older patients. The mean decline in CD4 counts in the different age groups was similar (table 1) [see also 15].

We have previously shown that patients seropositive for CMV progress to CDC IV disease quickly and that this effect is independent of age [16]. More recently we have confirmed these observations showing an estimated rate of progression to AIDS at 11 years of 59% (CI 39–81%) in CMV-positive patients, compared with 21% (CI 6–36%) in CMV-negative patients (fig. 5b) [13]. The relative risk is 3.0 (CI 1.4–6.8; $p = 0.006$). In order to prevent this accelerating influence of CMV, trials of acyclovir are being conducted in CMV-positive HIV-infected individuals. However, CMV-positive patients tend to be older than CMV-negative patients and after adjustment for age the significance falls to $p = 0.03$. Again, the slopes of the CD4 count against time since HIV-1 seroconversion in CMV-positive patients was similar to that seen in CMV-negative patients (table 1). From these studies it can be concluded that both the influences of age and CMV infection manifest themselves through factors other than CD4 counts in facilitating the development of symptoms in HIV-1 seropositive patients. In these patients lymphocyte performance assays are being conducted.

Effects of Antiviral Therapy

In our centre, zidovudine was first used to treat patients with AIDS in August 1987 [17] when efficacy had been reported in the treatment of AIDS and AIDS-related complex [18]. Asymptomatic patients were first treated with zidovudine (1,000 mg) when entry into the MRC Concorde trial began in November 1988. In January 1990, treatment with zidovudine (500 mg) at a CD4 count of 200/mm^3 began. In addition, since February 1989, 300 mg nebulized pentamidine was used fortnightly for secondary prophylaxis and monthly for primary prophylaxis against pneumocystis pneumoniae (pcp) [19, 20]. The prophylaxis was given to patients at a CD4 count of $<$200/mm^3. In order to assess whether the risk of AIDS has changed we have considered the total patient-years of AIDS-free experience after the CD4 count has fallen below certain low cut-off levels and the number of AIDS cases that have occurred (table 2) [17]. It can be seen

Table 2. Patient-years AIDS-free at certain CD4 counts before and after November 1988

After CD4 count has reached		Pre-November 88		Post-November 88		Relative risk	Confidence intervals	p value
$\times 10^9/l$	$/mm^3$	patient-years AIDS-free	AIDS cases	patient-years AIDS-free	AIDS cases			
0.05	50	3.6	8	12.6	7	4.0	1.45–11.03	<0.01
0.10	100	10.2	11	18.9	7	2.91	1.13–7.51	=0.03
0.15	150	28.3	16	30.5	7	2.46	1.01–5.98	<0.05
0.20	200	58.2	19	47.6	8	1.94	0.84–4.43	=0.1

that in the 50–150/mm^3 CD4 range there was a lower risk of progression to AIDS after November 1988, i.e. following the introduction of prophylactic therapy.

In a further analysis, 29 patients in whom the CD4 count was known at the time that AIDS developed by November 1988 [17] were compared to those who developed AIDS more recently. Patients undergoing prophylaxis had significantly lower CD4 counts at the time of AIDS development than the previous group before November 1988. The geometric means were 34 and 110/mm^3, respectively ($p = 0.003$). This difference remained virtually unchanged after adjustment for age and CMV serostatus, as well as after exclusion of occasional patients who developed AIDS with a CD4 count above 200/mm^3.

There is also another way the effect of treatment can be analysed. We have previously reported a method for predicting future numbers of AIDS cases based on a linear decline in CD4 counts [21]. This was achieved by fitting a straight line through the plot of linear regression, and noting the intersection of that line with the CD4 count 50/mm^3 point. This was the median value at which AIDS occurred and was thus chosen as the point where AIDS would be expected to occur in AIDS-free patients. Such a simple model has limited application for prediction in an individual but was found to be useful for groups of patients. Using data up to November 1988 in our cohort we were able to predict the number of patients who would progress to AIDS by each calendar year. Such a prediction could be made for 81 patients for whom five or more CD4 counts were then available.

We have now been able to compare subsequently observed numbers of cases of AIDS with those expected on the basis of the model. The observed

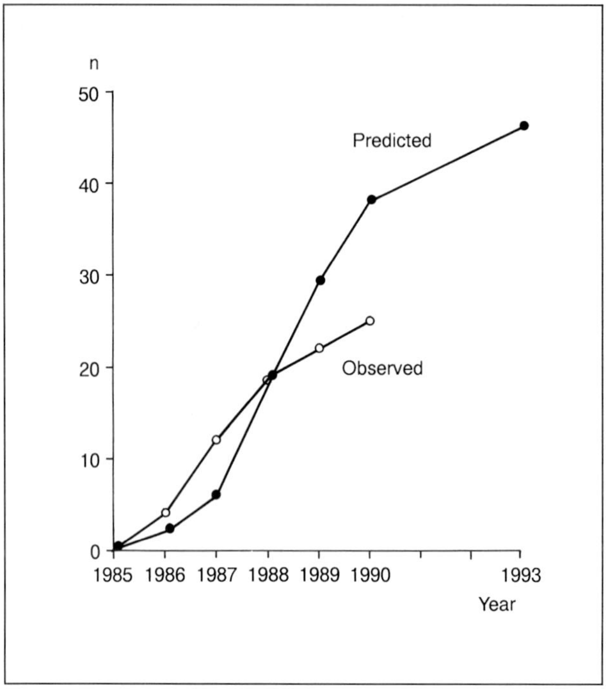

Fig. 6. Predicted and observed numbers of patients progressing to AIDS by calendar year.

Table 3. Relationship between CD4 count and costs of treatment in HIV-1 disease

Mean CD4 count/mm³	AIDS	Mean cost/ patient-year	Total patient cost/group	Cost/patient-year for most expensive 10%
< 200[a]	some	£ 1,981	£ 53,500 (92.3%)	> £ 4,763
< 200[b]	yes	£ 2,869		> £ 4,763
< 200[b]	no	£ 1,020		£ 3,527
201–500	no	£ 120	£ 3,900 (6.7%)	£ 166
> 500	no	£ 34	£ 900 (1.0%)	£ 50

[a] All patients in this category are shown including patients with or without AIDS.
[b] The observations reveal substantial costs in both groups of patients with AIDS and without AIDS. In the latter group preventive treatment is given (see figs. 6 and 7 for details).

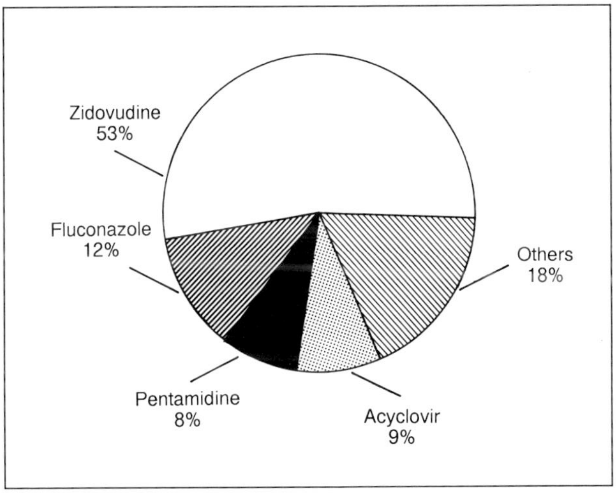

Fig. 7. Relative costs of drugs utilized for treating patients with HIV-1 infection.

numbers of patients progressing to AIDS in our cohort by calendar year begin to diverge from the curve of predicted numbers of patients at the end of 1988, the time at which treatment for patients without AIDS was first introduced (fig. 6) [17]. The difference between the observed and predicted numbers for 1990 are significant ($p = 0.03$).

Predicting the Expenses of Preventive Therapy

There is little published information about the cost of drug treatment for HIV seropositive patients. In order to provide a useful measure which would enable prediction of future drug costs not only in our own cohort but also in other anti-HIV-positive groups, we related drug costs to CD4 counts (table 3; presented by Swaden et al. at the VIIth World Congress on AIDS, Florence 1991). The practice of introducing antiviral treatment and prophylaxis at a CD4 count of 200/mm^3 will result in increasing costs at this level.

The total costs of all 76 patients for the year July 1989–June 1990 was £58,500 and this was accounted for by £31,000 (53%) for zidovudine, £6,900 (12%) for fluconazole and £4,700 (8%) for pentamidine (fig. 7).

The remaining £15,900 (27%) drug costs covered the provision of antiviral and antibacterial therapies.

The use of therapeutic protocols to be introduced at specific CD4+ counts together with forecasts of progression of HIV disease using CD4+ counts will enable planning and budgeting to be more reliable.

Conclusions

The serial CD4+ measurements in this haemophilic cohort together with the detailed clinical follow-up has enabled a number of observations to be made on factors influencing the progression of HIV disease, the impact of therapy and the economics of HIV disease. Serial CD4 lymphocyte counts will continue to be useful for monitoring patients for assessing the time at which to introduce prophylactic therapy and for predicting resource requirements.

References

1 Moss AR, Bacchetti P: Natural history of HIV infection. AIDS 1989;3:55–61.
2 Moss AR, Bacchetti P, Osmond D, et al: Seropositivity for HIV and the development of AIDS-related conditions: Three-year follow-up of the San Francisco General Hospital cohort. Br Med J 1988;296:745–750.
3 Ward JW, Bush TJ, Perkins HA, et al: The natural history of transfusion-association infection with human immunodeficiency virus. N Engl J Med 1989;321:947–952.
4 Eyster ME, Mitchell GH, Ballard JO, et al: Natural history of human immunodeficiency virus infections in haemophiliacs: Effects of T-cell subsets, platelet counts and age. Ann Intern Med 1987;107:1–6.
5 Darby SC, Rizza CR, Doll R, et al: Incidence of AIDS and excess of mortality associated with HIV in haemophiliacs in the UK: Report on behalf of the directors of haemophilia centres in the UK. Br Med J 1989;298:1064–1068.
6 Goedert JJ, Kessler CM, Aledort LM, et al: A prospective study of human immunodeficiency virus type I infection and the development of AIDS in people with haemophilia. N Engl J Med 1989;321:1141–1148.
7 Lee CA, Phillips AN, Elford J, et al: The natural history of human immunodeficiency virus infection in a haemophilic cohort. Br J Haematol 1989;73:228–234.
8 Lee CA, Webster A, Griffiths PD, Kernoff PBA: Symptomless HIV infection after more than ten years. Lancet 1990;335:425–426.
9 CDC Revision of the CDC Surveillance Case Definition for Acquired Immunodeficiency Syndrome. MMWR 1987;36(suppl 15):1–157.
10 Phillips AN, Lee CA, Elford J, et al: Serial CD4 lymphocyte counts and development of AIDS. Lancet 1991;337:389–392.

11 Pedersen C, Nielsen CM, Vestergaard BF, et al: Temporal relation of antigenaemia and loss of antibodies to core antigens to development of clinical disease in HIV infection. Br Med J 1987;295:567–569.
12 Phillips AN, Lee CA, Elford J, et al: p24 antigenaemia, CD4 lymphocyte counts and the development of AIDS. AIDS 1991;5:1217–1222.
13 Ragni MV, Kingsley LA: Cumulative risk for AIDS and other outcomes in a cohort of haemophiliacs in Western Pennsylvania. J Acquir Immune Defic Syndr 1990;3: 708–713.
14 Blaxhult A, Granath F, Lidman K, Giesecke J: The influence of age on the latency of AIDS in people infected by HIV through blood transfusion. AIDS 1990;4:125–129.
15 Phillips AN, Lee CA, Elford J, et al: More rapid progression to AIDS in older HIV-infected people: the role of CD4+ T-cell counts. J Acquir Immune Defic Syndr 1991; 4:970–975.
16 Webster A, Lee CA, Cook DG, et al: Cytomegalovirus infection and progression towards AIDS in haemophiliacs with human immunodeficiency virus infection. Lancet 1989;ii:63–66.
17 Lee CA, Phillips AN, Elford J, et al: Progression of HIV disease in a haemophilic cohort followed for eleven years and the effect of treatment. Br Med J 1991;in press.
18 Fischl MA, Richman DD, Grieco MH, et al: The efficacy of azidothymidine in the treatment of patients with AIDS and AIDS-related complex: A double-blind, placebo-controlled trial. N Engl J Med 1987;317:185–191.
19 Simonds AK, Newman SP, Johnson MA, et al: Simple nebuliser modification to enhance alveolar deposition of pentamidine. Lancet 1989;ii:953.
20 Simonds AK, Newman SP, Johnson MA: Alveolar targeting of aerosol pentamidine. Am Rev Respir Dis 1990;141:827–829.
21 Phillips AN, Lee CA, Elford J, et al: Prediction of progression to AIDS by analysis of CD4 lymphocyte counts in a haemophilic cohort. AIDS 1989;3:743–749.

Dr. Christine A. Lee,
Haemophilia Centre and Academic Departments of Haematology,
Royal Free Hospital, Pond Street, GB–London NW3 2QG (UK)

Janossy G, Autran B, Miedema F (eds): Immunodeficiency in HIV Infection and AIDS.
EC/FERS/MRC Workshop on Immunodeficiency in HIV-1 Infections, Windsor, Surrey, 1991.
Basel, Karger, 1992, pp 46–53

Biphasic Decline of CD4+ T Cells during Progression to AIDS: Correlates with the Phenotype of HIV-1 Isolate[1]

Peter Th.A. Schellekens[a,b], *Matthijs Tersmette*[a],
Marijke Th.L. Roos[a], *Frank Miedema*[a]

[a]Department of Clinical Immunovirology, Central Laboratory of The Netherlands Red Cross Blood Transfusion Service, incorporating the Laboratory of Experimental and Clinical Immunology; [b]Department of Internal Medicine, Academic Medical Centre, University of Amsterdam, The Netherlands

There is ample evidence that in HIV-1-infected individuals the number of CD4+ lymphocytes in the peripheral blood is the best prognostic laboratory marker for progression to AIDS [1]. Longitudinal data on the rate of decline of CD4+ lymphocyte numbers in individuals who remained asymptomatic and who progressed to AIDS are, however, relatively sparse [1–4]. These studies show a low and constant rate of decline of CD4+ cells in groups of asymptomatic HIV-infected persons and a constant but more rapid loss of CD4+ cells in groups of persons who finally progress to AIDS. In one of these investigations Phillips et al. [3] concluded that differences in the rate of decline of CD4+ lymphocytes, together with differences in CD4+ cell numbers at the time of infection, may account for the variation observed in the duration of the asymptomatic period. These authors utilize the known prognostic value of CD4 counts to predict the expenses of medical care in order to give appropriate allocations for antiviral drugs and primary prophylaxis (see also Lee et al., this volume). Despite the caution expressed in the paper by Phillips et al. [3], the investigations of the rate of

[1] This study was supported by grants from the Netherlands Foundation for Preventive Medicine and the Ministry of Health. F. Miedema is a senior fellow of the Royal Netherlands Academy of Arts and Sciences.

CD4 decline might be extended to follow individual patients as well, in order to identify progressors from nonprogressors. This suggestion presupposes that the rate of decline is relatively constant over a long period of time, which is obviously not necessarily the case. Indeed, in a preliminary study we have obtained evidence that different rates of CD4 decline may be associated with the presence of different biological variants of HIV-1 [5]. In particular, rapid CD4+ cell decline was observed in individuals with syncytium-inducing (SI) isolates that have a broad host range and high replication rates in culture. Thus, these variants have characteristics in vitro that would make them highly virulent and disease-inducing in vivo [5–8]. These observations might be important for the following reasons. These isolates are absent from seroconversion but in 50% of the patients they emerge during the course of HIV-1 infection [7]. One would predict that the appearance of the more lymphotropic aggressive SI variants may induce a change in the rate of CD4+ cell decline. This relationship might then be documented in individual patients by the correlation of the appearance of SI variants and the steeper CD4 slopes.

To address these points we have prospectively followed the rate of CD4+ cell decline in 187 HIV-1-infected homosexual men over a period of 5 years but with particular emphasis on the time period just before the diagnosis of AIDS. In addition, in 43 individuals of this cohort the results on the virological investigations in respect of tropism and growth pattern of the HIV-1 isolates have also been available. In these individuals, the CD4+ lymphocyte decline and the phenotype of HIV-1 isolates were correlated.

Study Population and Methods

From October 1984 to March 1986, 961 homosexual men were enroled into a prospective longitudinal study on risk factors for HIV infection and AIDS as previously described. In 238 men antibodies to HIV-1 were present at the onset of the study. A series of at least 3 CD4+ lymphocyte counts was available for 198 of these 238 men. By January 1, 1990 AIDS had developed in 56 of the 198 men with at least 3 CD4+ cell counts. Kaposi's sarcoma which may manifest at higher CD4+ lymphocyte counts than other AIDS-defining conditions (n = 11), were excluded from the analysis. Of the 187 HIV-1-seropositive individuals left in the study 45 developed AIDS during the observation period.

The phenotype of HIV-1 isolates was determined in 43 men, 23 of whom were randomly chosen from the group who was still asymptomatic at the end of the study and 20 randomly selected from the group who progressed to AIDS. As controls, 44 HIV-1 seronegative homosexual men were also followed longitudinally during the same observation period.

At three-monthly intervals peripheral blood mononuclear cells (PBMC) were isolated from heparinized venous blood by density gradient centrifugation on Ficoll-Paque. The proportions of CD4+ lymphocytes were determined by flow cytometry and the absolute numbers in blood were calculated.

HIV-1 was recovered from cryopreserved, sequentially sampled, peripheral blood mononuclear cells (PBMC) as described before [8]. Briefly, patient cells were repeatedly cocultivated with phytohaemagglutinin (PHA) prestimulated PBMC from seronegative individuals in the presence of polybrene and IL-2, and cultures were screened for syncytium formation. Virus replication was detected by the testing of culture supernatants for HIV-1 p24 core antigen in a capture assay [8]. According to the in vitro properties the isolates could be divided into three groups: (i) syncytium-inducing isolates, referred to as SI, with high replication rates and producing detectable viral replication after 4–7 days of culture, which could be transmitted to the H9 T-cell line; (ii) highly replicating but non-syncytium-inducing isolates, referred to as Fast-NSI, which could not be transmitted to the H9 cell line, and (iii) low replicating isolates, referred to as Slow-NSI, producing detectable replication after 9–28 days of culture and unable to induce syntitium formation or to transmit to the H9 cell line. In previous studies we have demonstrated that this division of HIV variants into categories can be done objectively and is highly reproducible [7, 8]. When an SI variant was found in any patient, retrospective analysis of frozen samples stored one and two years before the final sample was also carried out to establish the approximate time when the new aggressive mutant emerged.

Kinetics of CD4 Cell Decline in All Patients Studied

Individuals who had lately become known as progressors to AIDS (45 patients) and nonprogressors (142 individuals) were placed into two groups, and the kinetics of CD4+ cell decline was analyzed and compared in these two groups during a 5-year follow-up period. During the follow-up period, in nonprogressors a continuous and steady decline of the mean number of CD4+ cells occurred, from a mean of 700/mm^3 at 9 months to 400/mm^3 at 54 months. In the same observation period the CD4+ cell numbers in HIV-1-seronegative controls remained stable without a statistically significant decline. In progressors, until 18 months before AIDS diagnosis, the decline of CD4+ lymphocytes was slow and steady as in the nonprogressor group. Nevertheless, from 18 months onward, when mean CD4+ cell numbers were around 400/mm^3, the decline was faster. At the time of diagnosis of AIDS mean CD4+ cell numbers were 140/mm^3 (10th–90th percentiles: 30–310/mm^3).

To determine the rate of CD4+ lymphocyte decline, linear trend regression analysis was performed. In nonprogressors, the linear trend

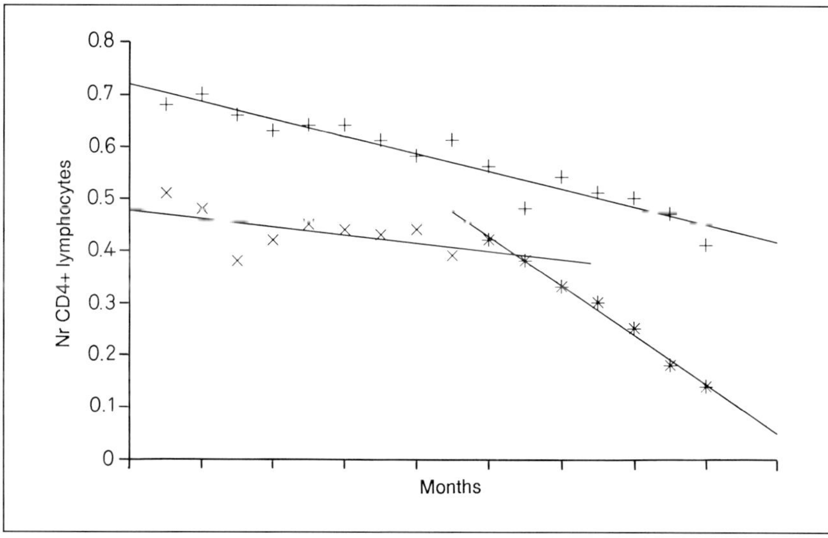

Fig. 1. Linear trend lines of CD4+ cell decline in HIV-1-infected homosexual men. + = Nonprogressors; X-axis indicates months follow-up. Progressors divided into two periods: × = more than 18 months preceding AIDS diagnosis; * = 18 months or less before diagnosis; X-axis indicates months before AIDS. Datapoints are at 3-month intervals.

Table 1. Rate of CD4 decline in HIV-positive men

		CD4 loss $\times 10^6$/l/month	SE	r-squared
Nonprogressors		5.6	0.5	0.92
Progressors	> 18 months pre-AIDS	3.3	1.2	0.47
	< 18 months pre-AIDS	15.7	1.0	0.98

lines fitted through the data points of CD4 counts (fig. 1). The regression coefficients of these trend lines showed that nonprogressors lose their CD4+ cells at a rate between 4.9 $\times$ 10^6/l/month and 6.9 $\times$ 10^6/l/month (table 1). In progressors, however, the linear trend lines could be divided into periods before and after the 18-months time point preceding AIDS.

CD4+ cell decline in the progressors in the period more than 18 months before diagnosis of AIDS was low (3.3 $\times$ 10^6/l/month) while in the last 18 months preceding diagnosis the decline has accelerated to about 5 times faster rate (15.7×10^6/l/month) (table 1). The regression coefficients were determined for each single person in the three groups: i.e. for nonprogressors and for progressors before and after the 18-months time point, respectively. The rate of CD4+ cell loss in nonprogressors and progressors more than 18 months before AIDS were similar. The median of CD4+ cell decline in nonprogressors was 5×10^6/l/month, and the progressors in the early period ($>$ 18 months preceding AIDS diagnosis) lost 4×10^6 CD4+ cells/l/month. In the late period ($<$ 18 months) the median CD4+ decline was 17×10^6/l/month. As expected, these cell losses differ only slightly from those calculated on the lines fitted to the means of the CD4+ cells per group (table 1).

When one pays attention to the variability in each group, it is apparent that, as reflected in positive regression coefficients, CD4+ cell numbers remained perfectly stable in about 15–20% of the nonprogressors, as also seen in other studies [3]. Among the progressors approximately 25% had only a mild decline in the period $<$ 18 months before AIDS diagnosis, similarly to the regression rates regularly seen in the asymptomatic group or in the progressors $>$ 18 months preceding diagnosis. Less than 10% of the progressors did not show any decline at all in the period less than 18 months preceding AIDS, but these men had already lost their CD4 cells earlier as could be seen from their extremely low CD4 count (data not shown).

Association of CD4+ Cell Decline and HIV-1 Subtypes

Figure 2 shows box plots of the rates of CD4+ lymphocyte decline in individuals divided into groups according to biological phenotypes of HIV-1 isolates. The median and range of CD4+ decline in the SI group was lower than in the Fast-NSI or Slow-NSI group with a highly significant difference ($p < 0.0005$). It appeared that median and range of CD4+ cell loss in individuals with SI variants were almost identical with those in progressors in the 18-month period before diagnosis. The median of the Fast-NSI and Slow-NSI groups were in the range of nonprogressors and of the progressors more than 18 months before diagnosis. When an SI type was present there was always a loss of CD4+ cells, while a substantial part of individuals with Slow-NSI and Fast-NSI types did not lose these cells during the observation period.

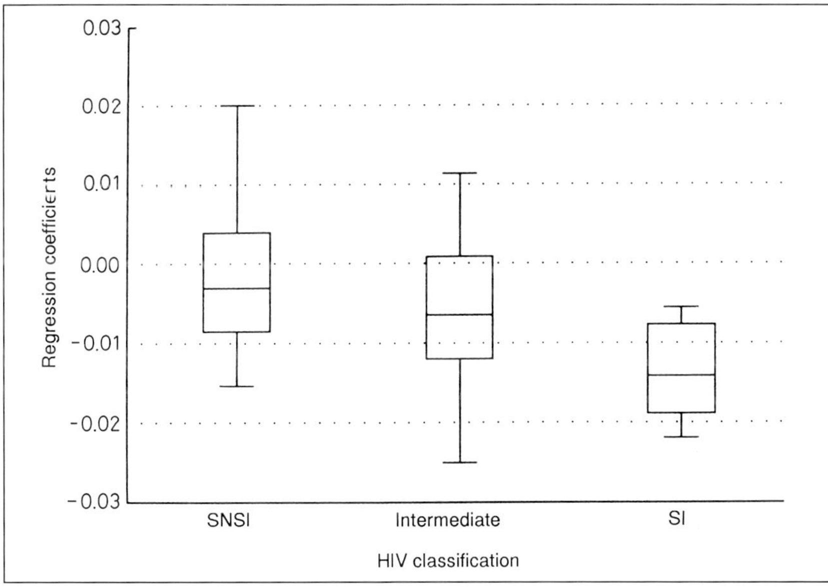

Fig. 2. Box plots of regression coefficients of the linear trend lines fitted through the CD4+ cell numbers of each individual, in which the time periods were classified according to the presence of different HIV isolates. SNSI = Time period during which slow growing nonsyncytium-inducing isolates were present (n = 16); Intermediate (FAST-NSI) = time period during which fast-growing nonsyncytium-inducing isolates were present (n = 21); SI = time period during which syncytium-inducing isolates were present (n = 14).

The Significance of Rapid CD4+ Decline and Isolation of SI HIV-1 Variants

In this paper investigations carried out during a period of 4–5 years are described. We have observed that in nonprogressing individuals the constant rate of decline was about 5×10^6 lymphocytes per month, which is the same rate as reported in the literature [3]. However, when analyzing CD4+ cell depletion in persons who progressed to AIDS, at about 18 months before AIDS in about 50% of the persons in this group a 'critical event' seems to occur leading to a 3–5 times more rapid decline of CD4+ lymphocytes. Because around this point of time neither a change in the clinical condition of the individuals was observed nor indications were present for the occurrence of infections other than HIV-1 we hypothesized that a change in the behavior of HIV-1 might be responsible. This hypothesis was substantiated by the demonstration of a significant correlation between the fast decline of CD4+ cells and the isolation of more virulent HIV-1 subtypes from these patients. These results, obtained in a large cohort of infected people, are compatible with a role for HIV biological variants in AIDS pathogenesis postulated earlier on the basis of smaller numbers of observations [5, 8].

Phillips et al. [3] and Lee et al. (this volume) reported a much faster decline in individuals that progressed to AIDS over the total period of HIV infection compared to those who over that same period of time remained asymptomatic. The issue here is that in approximately 50% of the people the patterns change, demonstrably, within a time interval of 18 months before AIDS is diagnosed. The apparent discrepancy may well be within the limits of sampling errors and be the consequence of the aim of the different analyses. The data presented by Phillips et al. [3] do in fact show a somewhat faster decline in the later stages of infection just before AIDS. We regard the correlation between CD4 decline patterns and the results of assays for the isolation of SI HIV-1 strains important. These two parameters together have a better likelihood of selecting progressors early.

These observations, taken together, establish that in addition to the absolute numbers of CD4+ cells the rate of CD4+ cell decline is a powerful prognostic marker for progression to AIDS.

Fast declining CD4+ cell numbers might thus be considered as an additional treatment criterion for early therapy. However, a considerable overlap exists between the rate of decline of the group that will progress to AIDS in less than 18 months and of both groups of nonprogressors and progressors who were more than 18 months before AIDS. It appeared that using CD4+ cell decline progressors and nonprogressors can be discriminated only within a 18-month period before diagnosis of AIDS. To accurately assess the rate of decline, at least 3 time points over a period of 6 months should be used. Calculation of decline over such a short period will result in even more overlap of rates of decline between groups of progressors and nonprogressors than in the current study. Besides, the period of time preceding AIDS will be reduced to a mere 9–12 months, which may prove to be late for the installation of antiretroviral treatment, e.g. with zidovudine. Once more evidence for the predictive value of SI variants for subsequent rapid CD4 cell depletion has been produced, this virological marker can be used to discriminate progressors from nonprogressors and with the practical advantage of giving a direct test result on a sample taken at any point in time.

The novel implication of this report is that biological phenotyping of HIV isolates appears to be one of the important additional prognostic criteria to early therapy. These assays in addition to parameters such as beta-2-microglobulin [9] and T cell reactivity [10] may prove to be clinically useful in the management of patients with HIV-1 infections.

References

1 Lange JMA, De Wolf F, Goudsmit J: Markers for progression in HIV infection. AIDS 1990;3(suppl 1):153–160.
2 Eyster ME, Gail MH, Ballard JO, Al-Mondiry H, Goedert JJ: Natural history of human immunodeficiency virus infections in hemophiliacs: Effects of T-cell subsets, platelet counts, and age. Ann Intern Med 1987;107:1–6.
3 Phillips AN, Lee CA, Elford J, Janossy G, Timms A, Bofill M, Kernoff PBA: Serial CD4 lymphocyte counts and development of AIDS. Lancet 1991;337:389–392.
4 Burcham J, Marmor M, Dubin N, Tindall B, Cooper DA, Berry G, Penny R: CD4% is the best predictor of development of AIDS in a cohort of HIV infected homosexual men. AIDS 1991;5:365–372.
5 Tersmette M, Lange JMA, De Goede REY, De Wolf F, Eeftink Schattenkerk JKM, Schellekens PTA, Coutinho RA, Huisman JG, Goudsmit J, Miedema F: Differences in risk for AIDS and AIDS mortality associated with biological properties of HIV variants. Lancet 1989;i:983–985.
6 Tersmette M, Miedema F: Interactions between HIV and the host immune system in the pathogenesis of AIDS. AIDS 1990;4(suppl):S57–S66.
7 Tersmette M, Gruters RA, De Wolf F, De Goede REY, Lange JMA, Schellekens PTA, Goudsmit J, Huisman JG, Miedema F: Evidence for a role of virulent HIV variants in the pathogenesis of AIDS obtained from studies on a panel of sequantial HIV isolates. J Virol 1989;63:2118–2125.
8 Tersmette M, De Goede REY, Al BJM, Winkel IN, Gruters RA, Cuypers HTM, Huisman HG, Miedema F: Differential syncytium-inducing capacity of human immunodeficiency virus isolates: frequent detection of syncytium-inducing isolates in patients with acquired immunodeficiency syndrome (AIDS) and AIDS-related complex. J Virol 1988;62:2026–2032.
9 Fahey JL, Taylor JM, Detels R: The prognostic value of cellular and serologic markers in infection with human immunodeficiency virus type 1. N Engl J Med 1990; 322:166–172.
10 Schellekens PTA, Roos MTL, De Wolf F, Lange JMA, Miedema F: Low T-cell responsiveness to activation via CD3/TCR is a prognostic marker for AIDS in HIV-1 infected men. J Clin Immunol 1990;10:121–127.

Peter Th.A. Schellekens, MD, Department of Clinical Immunovirology,
Central Laboratory of The Netherlands Red Cross Blood Transfusion Service,
PO Box 9190, NL-1006 AD Amsterdam (The Netherlands)

Janossy G, Autran B, Miedema F (eds): Immunodeficiency in HIV Infection and AIDS. EC/FERS/MRC Workshop on Immunodeficiency in HIV-1 Infections, Windsor, Surrey, 1991. Basel, Karger, 1992, pp 54–63

Opportunistic Diseases as Measure of Immunodeficiency in HIV Infection

Remko van Leeuwen[a], *Pieter J. van den Hurk*[a], *Jan Willem Mulder*[b], *Marijke Th.L. Roos*[c], *Peter Th.A. Schellekens*[c,d], *Peter Reiss*[a], *Joep M.A. Lange*[a,e]

[a]Department of Internal Medicine (AIDS Unit), Academic Medical Centre of the University of Amsterdam;
[b]Department of Infectious Diseases, Amsterdam Municipal Health Service;
[c]Department of Viral Immunology, Laboratory of Clinical and Experimental Immunology of the University of Amsterdam at the Central Laboratory of the Netherlands Red Cross Blood Transfusion Service;
[d]Department of Internal Medicine, Academic Medical Centre of the University of Amsterdam, and
[e]National AIDS Therapy Evaluation Centre (NATEC), Amsterdam, The Netherlands

Immunologic defects in HIV infection are accompanied by clinical manifestations which are initially relatively minor. As immune function deteriorates, severe opportunistic diseases may develop [1]. Clinicians often observe a particular sequence in the development of these opportunistic diseases. The various disease manifestations may therefore signify given stages of immunodeficiency. A staging system based on clinical observations could be of benefit for three reasons: Diagnostic and therapeutic approaches in AIDS patients are often complex; therefore, clinicians would benefit from further data indicating when their diagnostic procedures are most likely to be productive. Such data could also provide a basis for decisions concerning the precise timing of initiating prophylactic therapies. In addition, they may serve as a scoring system for clinical trials in order to provide more differentiated clinical endpoints.

One of the best laboratory parameters for assessing disease progression in HIV-1 infection is the number of circulating CD4+ lymphocytes in the blood of HIV-1 seropositive persons (CD4 count) [2–6]. Observations comparing clinical parameters, including opportunistic infections, with CD4 counts are, however, relatively scarce. We have therefore performed a study to define an order in the onset of opportunistic infections and malignancies, and relate these findings to CD4 counts. These observations are also relevant for the documentation of the clinical course of patients in relation to CD4 counts prior to the introduction of prophylaxis for *Pneumocystis carinii* pneumonia (PCP); they may provide 'historical' controls for patients who are now given PCP prophylaxis.

The Design of the Analysis

In this survey, we have studied 419 adult HIV-1-positive patients who were diagnosed to have AIDS at the Academic Medical Centre of the University of Amsterdam or at the Municipal Health Service (patients from the Amsterdam cohort of homosexual men [7]). This is a retrospective analysis which goes back in time as far as 1982, when the first AIDS case was recognized in the Netherlands. The cut-off for follow-up was the end of 1989; the study was therefore completed before the introduction of prophylaxis for PCP. The patients' medical records, systematically collected listings of cultures and biopsies, and reports to the Health Authorities were reviewed. The diseases included in this analysis were AIDS defining opportunistic infections and malignancies as proposed by the most recent Centers for Disease Control (CDC) criteria [8]. The diagnoses were made by confirming each condition according to the CDC recommendations [8]. All diseases are listed in table 1. Of these, PCP, Kaposi's sarcoma (KS), disseminated infection with *Mycobacterium avium* complex (MAC), progressive multifocal leukoencephalopathy (PML), cryptococcal meningitis (CYC), enteric cryptosporidiosis (CYS), high-grade malignant non-Hodgkin's lymphoma (LYM), extrapulmonary tuberculosis (E-TB), disseminated histoplasmosis (HIS), and enteric isosporidiosis (ISO) were diagnosed by microscopy, histology or culture. Candida oesophagitis (CAND) was diagnosed by endoscopic inspection. Cerebral toxoplasmosis (TOX) was diagnosed by the combination of the clinical picture, brain imaging and successful response to therapy. Cytomegalovirus chorioretinitis (CMV) was diagnosed by ophthalmoscopic examination. Although not listed in the CDC AIDS criteria, in our centre, enteric microsporidiosis (MIC) is also regarded as an opportunistic disease [9], and was therefore also included in this analysis. A diagnosis of MIC was based on microscopic examination of a small-intestinal biopsy [9]. The date of diagnosis was defined as the date on which a diagnostic confirmation of a disease was obtained, whether or not clinical suspicion had risen earlier.

Patients were excluded from the study (a) when the AIDS indicator diagnosis was not included in the study (because there were too few cases), or (b) when the AIDS indicator diagnosis and/or date of the AIDS indicator diagnosis was not known, or (c) if the documentation of the appropriate diagnostic method for the establishment of the AIDS indicator diagnosis was lacking. Only the first episode of each disease was included in the analysis. All studied events occurring within 30 days after the first diagnosis defining AIDS were regarded as AIDS indicator diagnoses. Autopsy findings were not included in the evaluation.

Table 1. The proportions of patients developing opportunistic diseases and the numbers of episodes seen since 1982: of the 419 adult AIDS patients, 388 were included

Disease	Abbre-viation	Number of episodes	Percent of patients	Percent AIDS indicator episodes[1]
Pneumocystis carinii pneumonia	PCP	213	55	75
Kaposi's sarcoma	KS	108	28	59
Candida oesophagitis	CAND	99	26	54
Mycobacterium avium complex	MAC	50	13	10
Cerebral toxoplasmosis	TOX	48	12	60
Cytomegalovirus chorioretinitis	CMV	44	11	11
Cryptococcal meningitis	CYC	26	7	54
Enteric cryptosporidiosis	CYS	23	6	43
High-grade malignant non-Hodgkin's lymphoma	LYM	21	5	71
Extrapulmonary tuberculosis	E-TB	16	4	56
Enteric microsporidiosis	MIC	13	3	23
Progressive multifocal leukoencephalopathy	PML	4	1	66
Disseminated histoplasmosis	HIS	3	0.7	66
Enteric isosporidiosis	ISO	1	0.2	0
Total		676		55

[1] This is the percentage within each episode to show how frequently the episode was the first to indicate AIDS.

Concomitant Therapy

Zidovudine was introduced in May 1987 for the treatment of symptomatic HIV-infected patients, and was usually started at a dosage of 1,000–1,200 mg/day. In case of clinical or haematologic intolerance this dose was generally reduced to 500–600 mg/day [10]. During zidovudine treatment, 104 episodes developed and 16 episodes were seen after zidovudine treatment was stopped. In patients who did not have previous antiretroviral therapy, 556 episodes developed. No primary prophylactic therapy against any of the studied diseases was used on a regular basis during the study period.

Analysis of T Cell Subsets

The lymphocyte subset count nearest to the date of diagnosis of a particular disease was used for evaluation provided that the analysis was performed within a period of 90 days before and 14 days after the date of the diagnosis. Blood was always sampled in the morning hours. Peripheral blood mononuclear cells were isolated from heparinized venous blood by density-gradient centrifugation on Ficoll-Paque, and the lymphocyte subsets were determined by flow cytometry at the Central Laboratory of the Netherlands Red Cross Blood Transfusion Service, Amsterdam. These subset counts were rounded off to the nearest multiple of $10/mm^3$, and in some cases before 1988 to the nearest multiple of $100/mm^3$.

Statistical Methods

Because the numbers of CD4-positive lymphocytes were not normally distributed, CD4 counts are expressed as a median. All significance levels correspond to those for two-tailed tests. Differences in CD4 cell numbers between multiple groups were assessed with a Kruskal-Wallis nonparametric one-way analysis of variance. When this test was significant at a $p \leq 0.05$ level, a Mann-Whitney U test was performed to compare the separate diseases, and significance was defined for this test at a $p \leq 0.01$ level.

Observed AIDS-Defining Conditions

Ten patients were excluded because their AIDS-defining diagnoses were unusual. These patients had AIDS dementia complex (3 patients), enteric CMV infections (3 patients), CMV pneumonitis (2 patients), and recurrent herpes simplex virus infections (2 patients). Another group of 21 patients was also excluded, because either the appropriate documentation for establishing the AIDS indicator diagnosis was lacking or the date of the episode was not known. The remaining 388 patients were predominantly homo- and bisexual men. The mean age of the studied patients was 38 years (range 22–66) and the number of episodes of the studied diseases was 676. Lymphocyte subset counts were available in 398 cases.

The number of episodes observed within each disease category and the percentage of patients who developed this particular disease are shown in table 1. PCP was the most commonly observed opportunistic disease, seen in 55% of patients, followed by KS (28%), CAND (26%), disseminated MAC (13%), TOX (12%) and CMV retinitis (11%). The remainder of diseases were each recognized in less than 10% of the patients.

Table 1 also shows the percentages of episodes where the appearance of symptoms signified AIDS development ('AIDS indicator episodes'). Fifty-five percent of all episodes were AIDS indicator diagnoses: PCP and LYM were the most frequently observed as a first manifestation of AIDS (75 and 71% of all episodes seen, respectively), while CMV retinitis and disseminated MAC infection were AIDS indicator diseases in only 10 and 11% of all episodes, respectively.

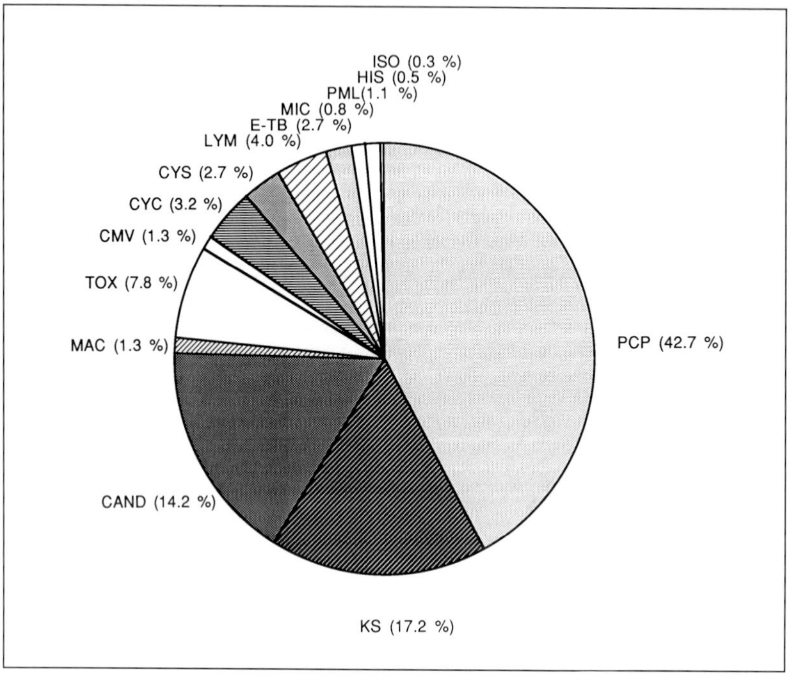

Fig. 1. Frequency distribution of the AIDS indicator diagnoses among 388 patients. Proportions of patients with the given diagnostic episodes are shown.

Figure 1 shows the frequency distribution of all AIDS indicator diagnoses among the 388 patients. PCP, KS, and CAND together make up 77% of the AIDS indicator diagnoses. TOX was an AIDS indicator diagnosis in 7%. Although CMV retinitis and disseminated MAC are frequently observed in AIDS, both each make up about 1% of all AIDS indicator diagnoses.

Correlation of AIDS-Defining Episodes with the CD4 Counts

The concomitant CD4 counts at the time of diagnosis of AIDS-defining episodes were available for more than 10 patients in five disease conditions: KS, LYM, CAND, TOX, and PCP (fig. 2). These observations have revealed a significant difference between the CD4 values seen at the

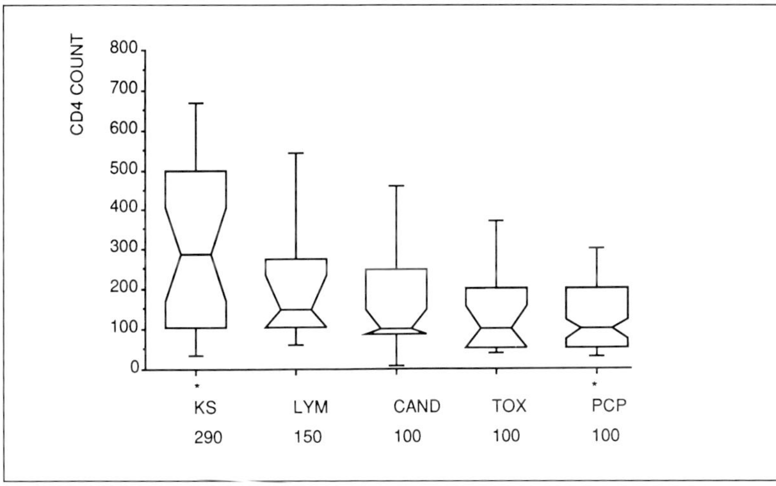

2

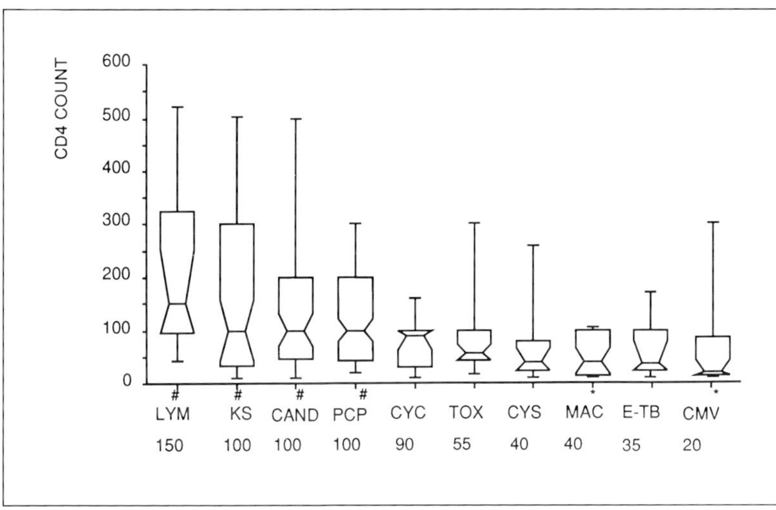

3

Fig. 2. The CD4 counts measured at the presentation of AIDS indicator episode expressed as boxplot diagram of the distribution of the 10th, 25th, 50th (median), 75th, and 90th percentiles. The median values are also shown below as numbers. Diseases marked * are significantly different from each other. Only those diseases are shown where the CD4 counts have been performed in groups of patients larger than 10, and when the CD4 counts were performed around the time of diagnosis.

Fig. 3. CD4 counts measured each time when opportunistic diseases were seen in patients with AIDS. In this diagram both the AIDS indicator and subsequent episodes are shown. For other details see figure 2. Diseases marked * and # are significantly different from each other.

presentation of KS and PCP, as expected from previous observations [11].

In the next analysis, all episodes have been included. In ten disease conditions more than 10 patients had their CD4 counts examined (fig. 3). The differences in CD4 counts between the diseases have become smaller as a consequence of the inclusion of later disease episodes. By using this method, however, diseases which were less frequently seen as AIDS indicator diagnosis could be characterized.

Discussion

Based upon the incidence of AIDS indicator diseases and the median CD4 count at the time of diagnosis, it appears that the AIDS-related opportunistic infections fall into three different categories: early, intermediate and late events (table 2). In table 2 only diseases occurring in more than 10% of the patients are included. The group of early opportunistic events includes KS, which is more frequently observed as an AIDS indicator disease at CD4 counts of greater than 100/mm^3. Intermediate opportunistic diseases include PCP, CAND, and TOX. This group of diseases was AIDS indicating in over 50% of the cases, and occurred at CD4 counts of approximately 100/mm^3. Late diseases are disseminated infection with MAC and CMV retinitis, occurring in less than 15% as an AIDS-defining event, at CD4 counts of less than 50/mm^3.

Our observations are in full agreement with the results of two other smaller studies. In a recent Australian study, opportunistic diseases were related to CD4 counts [12], and in a Danish study, opportunistic diseases were related to time [13]. The sequence of opportunistic diseases was similar to the pattern described in the present study and the episodes have occurred at the same CD4 levels as shown in table 2.

A further important observation in all these studies is that as many as 10–25% of all opportunistic diseases were associated with CD4 counts of 200/mm^3 or more at the time of diagnosis. The concept of a very low risk for PCP above a CD4 count of 200/mm^3 [14], which is generally used as the criterium for the start of primary PCP prophylaxis [15], is not fully supported by our study; 25% of the patients developing PCP as a first manifestation of AIDS had higher CD4 counts. Since similar findings have been described in recent other studies [12, 16], the stringent use of a CD4 count of 200/mm^3 as a cut-off value above which the risk of developing

Table 2. Categories of opportunistic diseases defining the clinical stage in progressing HIV-1 infection

	Category	Characteristics	Opportunistic diseases
1	Early	$> 50\%$ AIDS indicator CD4 count $> 100/mm^3$	KS
2	Intermediate	$> 50\%$ AIDS indicator CD4 count $\pm$ $100/mm^3$	PCP CAND, TOX
3	Late	$< 15\%$ AIDS indicator CD4 count $\leq 50/mm^3$	CMV MAC

PCP is low seems inadequate. It might be possible that functional assays such as the response to recall antigens or mitogens (see Clerici and Shearer, and Miedema et al., this volume) may contribute to the identification of patients who develop opportunistic infections at a greater than $200/mm^3$ level of CD4 counts.

We have the clinical impression that mucocutaneous manifestations of herpes simplex virus infections, which are frequently seen in HIV-infected patients, should also be regarded as an early opportunistic infection. Therapeutic intervention, however, prevents recurrent herpes simplex infections from serving as AIDS indicator diagnosis, because the CDC case definition for AIDS requires mucocutaneous ulcerations to persist for longer than 1 month [7].

The classification of diseases may be useful for three reasons: Firstly, it can help the clinician to decide which are the most important and useful diagnostic procedures at different stages. Secondly, it may help to find the appropriate moment to initiate primary prophylaxis against various infections. Finally, this classification may serve as a basis to develop a clinical scoring system, in order to obtain more differentiated clinical endpoints than just 'disease progression' in antiretroviral drug trials.

Acknowledgements

The authors would like to acknowledge the contributions of P. Portegies, J. de Gans, R.P. van Steenwijk, G.J. van den Horn, J. Bruins, M.J.H. van Oers, J.F.W.M. Bartelsman, J.K. Eeftinck Schattenkerk, J. Jöbsis, C.L. Kuiken and M.D. de Jong to this work.

References

1 Centers for Disease Control: Classification system for human T-lymphotropic virus type III lymphadenopathy-associated-virus infection. MMWR 1986;35:334–339.

2 Goedert JJ, Biggar RJ, Melbye M, Mann DL, Wilson S, Gail MH, Grossman RJ, DiGioia RA, Sanchez SH, Weiss SH, Blattner WA: Effect of T4 count and cofactors on the incidence of AIDS in homosexual men infected with the human immunodeficiency virus. JAMA 1987;257:331–334.

3 Moss AR, Bacchetti P, Osmond D, Krampf W, Chaisson RE, Stites D, Wilber J, Allain JP, Carlson J: Seropositivity for HIV and the development of AIDS or AIDS-related condition: three-year follow-up of the San Francisco General Hospital cohort. Br Med J 1988;296:745–750.

4 De Wolf F, Lange JMA, Houweling JTM, Coutinho RA, Schellekens PT, van der Noorda J, Goudsmit J: Numbers of CD4+ cells and the levels of core antigens of and antibodies to the human immunodeficiency virus as predictors of AIDS among seropositive homosexual men. J Infect Dis 1988;158:615–622.

5 Eyster ME, Gail MH, Ballard JO, Al-Mondhiry H, Goeddert JJ: Natural history of human immunodeficiency virus infections in hemophiliacs: effects of T-cell subsets, platelet counts, and age. Ann Intern Med 1987;107:1–6.

6 El-Sadr W, Marmor M, Zolla-Pazner S, Stahl RE, Lyden M, William D, D'Onofrio S, Weiss SH, Saxinger WC: Four-year prospective study of homosexual men: correlalation of immunologic abnormalities, clinical status, and serology to human immunodeficiency virus. J Infect Dis 1987;155:789–793.

7 De Wolf F, Goudsmit J, Paul DA, Lange JMA, Hooijkaas C, Schellekens P, Coutinho RA, van der Noordaa J: Risk of AIDS-related complex and AIDS in homosexual men with persistent HIV antigenaemia. Br Med J 1987;295:569–572.

8 Centers for Disease Control: Revision of the CDC surveillance case definition for acquired immunodeficiency syndrome. MMWR 1987;36(suppl S-1):3–15.

9 Eeftinck Schattenkerk JKM, van Gool T, van Ketel RJ, Bartelsman JFWM, Kuiken CL, Terpstra WJ, Reiss P: Clinical significance of small-intestinal microsporidiosis in HIV-1-infected individuals. Lancet 1991;337:895–898.

10 Van Leeuwen R, van den Hurk PJ, Jöbsis GJ, van der Wouw PA, Eeftinck Schattenkerk JKM, Danner SA, Lange JMA: Failure to maintain high-dose treatment regimens during long-term use of zidovudine in patients with symptomatic human immunodeficiency virus type 1 infection. Genitourin Med 1990;66:418–422.

11 Ammann AJ, Abrams D, Conant M, Chudwin D, Cowan H, Volberding P, Lewis B, Casavant C: Acquired immune dysfunction in homosexual men: immunologic profiles. Clin Immunol Immunopathol 1983;27:315–325.

12 Crowe SM, Carlin JB, Steward KI, Lucas CR, Hoy JF: Predictive value of CD4 lymphocyte numbers for the development of opportunistic infections and malignancies in HIV-infected patients. J AIDS 1991;4:770–776.

13 Pedersen C, Gerstoft J, Tauris P, Lundgren JD, Gøtzsche PC, Buhl M, Salim Y, Schmidt K: Opportunistic infections and malignancies in 231 Danish AIDS patients. AIDS 1990;4:233–238.

14 Masur H, Ognibene FP, Yarchoan R, Schelhamer JH, Baird BF, Travis W, Suffredini AF, Deyton L, Kovacs JA, Falloon J, Davey R, Polis M, Metcalf J, Baseler M, Westley R, Gill VJ, Fauci AS, Lane C: CD4 counts as predictors of opportunistic pneumonias in human immunodeficiency virus (HIV) infection. Ann Intern Med 1989;111:223–231.

15 Centers for Disease Control: Guidelines for prophylaxis against *Pneumocystis carinii* pneumonia for persons infected with human immunodeficiency virus. MMWR 1989;38(suppl S-5).1–9.

16 Phair J, Muñoz A, Detels R, Kaslow R, Rinaldo C, Saah A, and the Multicenter AIDS Cohort Study Group: The risk of *Pneumocystis carinii* pneumonia among men infected with human immunodeficiency virus type 1. N Engl J Med 1990;322:161–165.

Dr. J.M.A. Lange, National AIDS Therapy Evaluation Centre (NATEC),
Academic Medical Centre, Room F5-160, Meibergdreef 9,
NL–1105 AZ Amsterdam (The Netherlands)

Janossy G, Autran B, Miedema F (eds): Immunodeficiency in HIV Infection and AIDS. EC/FERS/MRC Workshop on Immunodeficiency in HIV-1 Infections, Windsor, Surrey, 1991. Basel, Karger, 1992, pp 64–75

The Use of in vitro T Cell Immune Function to Monitor the Course of HIV Infection

Mario Clerici, Gene M. Shearer

Experimental Immunology Branch, National Cancer Institute, National Institutes of Health, Bethesda, Md., USA

The acquired immunodeficiency syndrome (AIDS) is characterized by susceptibility to opportunistic infections, severe depletion of CD4+ cells, anergy to delayed skin reaction in vivo, and failure of peripheral blood leukocytes (PBL) to respond to antigens in vitro [1]. Progression to AIDS from the asymptomatic phase is complex and can take several years. Because an intact immune system is central to protection against infections and the development of a neoplastic condition, it would be of practical value to develop and utilize in vitro immunologic tests that would monitor: (a) progression toward AIDS; (b) the effect of therapeutic protocols on immune function in human immunodeficiency virus-infected (HIV+) patients on therapy; (c) the effect of candidate AIDS vaccines on the immune system, and (d) exposure to HIV-1 by the detection of virus-specific T cells. This report summarizes the results obtained using an in vitro T helper cell (Th) assay that, when appropriately used, addresses points a–d above.

Changes in Th function in AIDS Progression and Loss of CD4+ Cells

Although the immunologic hallmark of AIDS is the loss of CD4+ cells, several laboratories have reported that Th function declines before CD4+ cell numbers decrease to critical levels [2–8]. This decline in Th function is not uniform, however, and appears to involve the loss of Th function to recall antigens (Ag) with retention of responsiveness to HLA alloantigens (ALLO) and T cell mitogens. Our laboratory has extensively studied this phenomenon in which the loss of Th function is selective for the HLA self-restricted response to Ag [6]. During these studies, Th function was assessed for the ability of PBL from HIV+ individuals and from HIV– controls to proliferate and generate interleukin-2 (IL-2) responses by in vitro stimulation (fig. 1). Lymphocytes were activated by three types of stimuli: (a) Ag such as influenza A virus (e.g. the FLU A/Bangkok RX73 H3N2 strain) or tetanus toxoid; (b) ALLO stimulation provided by 5,000 rad irradiated allogeneic PBL from HIV– donors, and (c) the mitogen phytohemagglutinin (PHA) [6].

In these studies more than 400 asymptomatic HIV+ individuals were assessed as shown in figure 1. Among these people, four subgroups could be seen (table 1): (a) approximately 35% responded to Ag, ALLO, and PHA and are therefore classified as +/+/+; (b) 45% failed to respond to Ag but responded to both ALLO and PHA (classified as –/+/+); (c) 10% failed to respond to Ag and ALLO, but responded to PHA (classified –/–/+), and (d) 10% failed to respond to all three stimuli (classified as –/–/–). In a longitudinal study of this population of asymptomatic individuals we also found that there was a progression in loss of Th function. The response to Ag is lost first, followed by loss of Th function to ALLO, and finally by abrogation of response to PHA [6, 9] (table 2).

Furthermore, we have observed that the Th functional state is predictive for decline in CD4+ cell numbers [9]. Thus, a +/+/+ state was associated with a stable CD4 number when CD4+ cell numbers were determined 1 year later. In contrast, each of the other three states of Th dysfunction was associated with a decline in the number of CD4+ cells during the subsequent 12 months [9] that was proportionally more severe with the severity of Th dysfunction. In fact, the CD4+ cell number of –/–/– patients dropped below 400/μl ($<0.4 \times 10^9$/l) within the 1-year follow-up period. We also observed that patients who were considered to be in more advanced stages of AIDS progression by parameters such as low CD4 cell numbers and/or lymphadenopathy tended to exhibit more severe Th dysfunction [10, and Clerici and Shearer, unpubl. observations]. Although the loss of Th function per se was not associated with an increased incidence of opportunistic infections, we have observed a correlation between particular pattern of Th dysfunction and susceptibility to opportunistic and bacterial infections [11]. Based on the studies outlined, it appears that in vitro

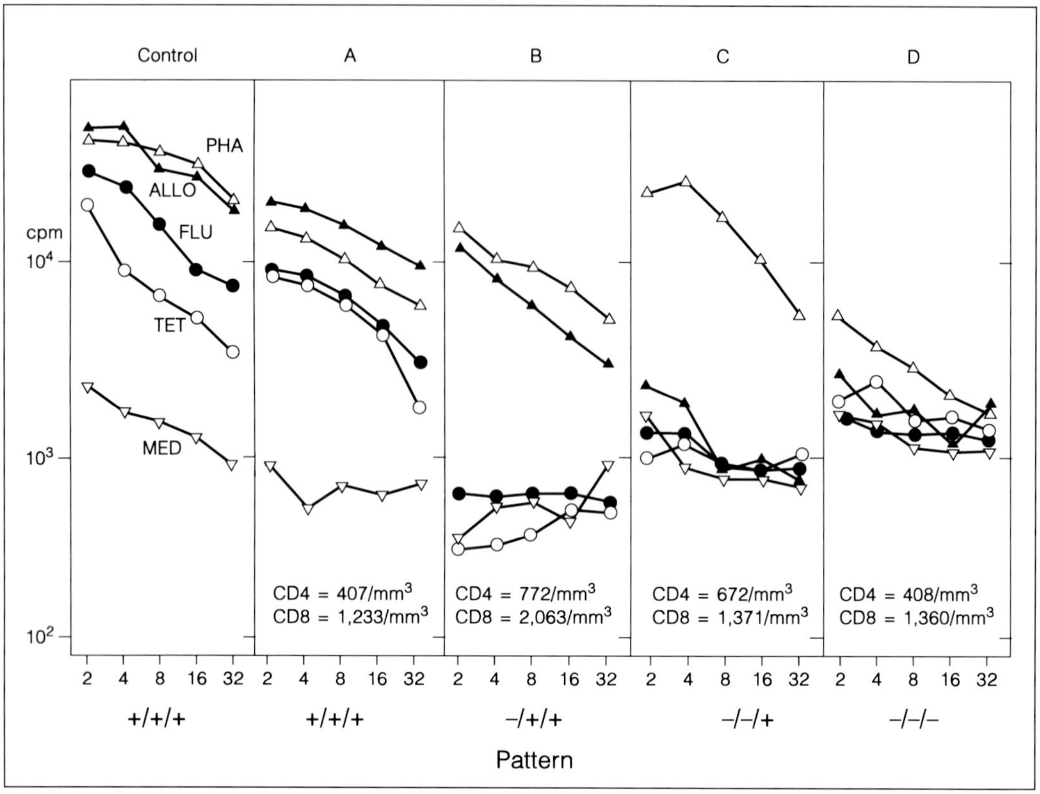

Fig. 1. Changes of lymphocyte activation during HIV-1 infection [from 6, with permission]. In cultures of peripheral blood mononuclear cells taken from normal HIV-1-seronegative donors as positive controls (Control) lymphocytes respond to polyclonal mitogens (PHA; △), alloantigens in mixed lymphocyte cultures (ALLO; ▲) and to recall antigens (Ag) such as influenza A (FLU; ●) and tetanus toxoid (TET; ○). Unstimulated controls (MED; ▽) are also shown. These responses can be analyzed by IL-2 secretion as depicted here or by thymidine incorporation shown in figure 2 below. In individuals with HIV-1 infections these responses are impaired to a different degree. Some individuals respond to all four stimuli (A). In others, responses to recall antigens are selectively impaired (B). A proportion of patients show responses to only mitogens such as PHA (C) or none at all (D). Please note that in patients with higher than 400/mm^3 CD4 counts, as in patients A–D, these functional defects show no direct correlation with the CD4 numbers detected. For example, patients A and D have similar CD4 and CD8 values and still show different performance in vitro. The Th functional state is, however, predictive for the decline in CD4+ numbers [9]. Below 200/mm^3 absolute CD4 counts the lymphocyte performance is generally poor – although again exceptions are seen.

Table 1. Pattern of Th function in asymptomatic HIV+ individuals [data from 6]

Percent of individuals	T cell stimulus		
	Ag[1]	ALLO	PHA
35	+	+	+
45	–	+	+
10	–	–	+
10	–	–	–

[1] Influenza A virus (e.g. Bangkok RX73 H3N2 strain).

Table 2. Progressive loss of Th function [data from 9]

Th function defect	T cell stimulus		
	Ag	ALLO	PHA
None	+	+	+
1st	–	+	+
2nd	–	–	+
3rd	–	–	–

tests of Th function to a panel of T cell stimuli can be used as an indicator of immunologic decline and progression towards AIDS, even within the asymptomatic period. Results similar to these have been reported using anti-CD3 monoclonal antibody as the T cell stimulus [4, 8] (see also Miedema et al., this volume).

Th Function in Therapeutic Trials

Several phase 1 protocols have been completed and phase 2 protocols are in progress to test the efficacy of antiretroviral drugs such as zidovudin (AZT) [12, 13], 2′,3′-dideoxyinosine (ddI) [14, 15] and 2′,3′-dideoxycytidine (ddC) [16] in slowing the progression to AIDS. A number of surrogate

Table 3. Surrogate markers used in AIDS

CD4:CD8 ratio
CD4 cell number
HIV-1 viral load
p24 antigen level
Neopterin
β_2-Microglobulin
Ig levels
Th function[1]

[1] Proposed immunological marker.

markers (table 3) have been used to monitor progression to AIDS, and these markers are now being utilized to assess the efficacy of therapeutic protocols. However, none of these markers correlate completely with variations in the clinical status of the patient, and several of the markers do not reflect changes in immune function. Among the parameters in table 3, CD4+ cell numbers are likely to best correlate with immune function but the increase of CD4 counts in patients treated with the antiretroviral drugs is frequently only minimal if observed at all. These changes often require months before they are detected. Therefore, a marker is needed which is sensitive enough to detect changes in immune function within weeks of initiating therapy. The ability of Ag, ALLO, or PHA to stimulate IL-2 production by PBL from HIV+ patients on AZT or ddI has been recently evaluated. We observed that more than 50% of children on ddI [Clerici et al., submitted], and 75% of adult patients on AZT [Clerici et al., submitted] exhibited improved Th function as evaluated by in vitro IL-2 production. The improvement is by a factor of 3- to 4-fold in some of these patients. Improved Th function could be detected in some patients within 8 weeks of starting therapy. In a proportion of individuals, improved Th function continued to be evident more than 2 years into therapy. Only 7% of control HIV+ patients not on therapy showed any improved Th function, and the majority of untreated patients continued to show a decline in Th function. Furthermore, in the pediatric study, improved Th function was also associated with a reduced incidence of bacterial and opportunistic infections [11]. Therefore, it is probable that an in vitro Th functional analysis of the type summarized here will be a valuable addition to the list of surrogate markers currently being used to monitor therapeutic trials.

Th Function in AIDS Vaccine Trials

A considerable amount of effort and resource are being devoted to the development of a safe and efficient vaccine for AIDS, and vaccine trials have begun in volunteers inoculated with candidate AIDS vaccines [17–19] for the purpose of determining their safety and ability to elicit immunity in the T and B cell compartments of the immune response. Our laboratory has utilized the in vitro proliferative and IL-2 production assays outlined above to assess Th function in a group of uninfected volunteers, who received multiple intramuscular inoculations of 40 or 80 μg of rgp160 (MicroGeneSys, Inc. West Haven, Conn.) in alum or alum alone [20]. We found that PBL from the vaccinees, but not from control donors, generated strong Th and CTL responses to the HIV-1 synthetic peptides up to 1 year after the last immunization [20, 21]. The four peptides used throughout these investigations were *env*T1 (T1; amino acid residues 428–443 of gp120) [22], *env*T2 (T2; aa 112–124 of gp120), *env*Th4 (Th4; aa 834–848 of gp160 [23]) and *env*P18 (P18; aa 315–319 [24]). Reimmunization with rgp160 elevated both helper and effector activity. It is interesting that the Th responses in the vaccinees were 10- to 100-fold greater than in peptide-responsive, asymptomatic HIV+ individuals [20]. It should be noted that, unlike other murine and human viral specific CTL responses, HIV peptide-specific CTL activity could be detected without in vitro restimulation. These findings demonstrate that in vitro Th and CTL response to HIV-1 antigenic determinants can be detected in HIV vaccinees, and indicate that such an approach could be used to test whether a candidate vaccine had resulted in immunization. It is also noteworthy that in this group of individuals the vaccine given contained relatively small amounts of rgp160, and antibody responses were mainly negative [20, 21]. Subsequent protocols that involved immunization with higher doses of rgp160 resulted in activation of both HIV-specific T and B cell responses [25]. Thus, it is possible that low doses of antigen will activate T cell responses without necessarily triggering antibody production [20]. It may therefore be important to test for Th function as well as for antibodies in any situation in which an individual has been exposed to HIV-1.

Early Detection of Exposure to HIV-1 Antigens

As noted above it has been possible in vaccinees immunized with low dose antigen to detect HIV-specific Th function in the absence of serum antibodies to the virus. This observation raises the possibility that there may be individuals who have been exposed to low doses of HIV-1 and are still antibody negative but who would exhibit Th reactivity to HIV antigenic determinants. In fact, we have detailed the findings in one homosexual male in order to document the sensitivity of these techniques [26]. In this study, frozen cells and serum samples have been available throughout a 19-month period prior to seroconvertion (fig. 2). In the first sample (month 1) only response to recall Ag (FLU) was detectable. This is an indication of intact Th cell function, and this response to FLU remained high throughout the study. The cultures at month 1 were Th unresponsive to HIV synthetic peptides T1, T2, Th4 and P18 both by proliferation and IL-2 production assays (fig. 2a, e). By month 5, however, cultures responded to three of the four peptides by proliferation (fig. 2b) and two of the peptides by IL-2 production (fig. 2f). The lack of full concordance here is probably due to an early immune response and/or few responding cells because our experience from >120 HIV-seropositive individuals [27] shows excellent concordance. Such a good agreement was also seen at months 13 and 19 for all peptides in both assays (fig. 2). The importance of these findings is highlighted by the parallel observations that this man became seropositive by HIV-1 ELISA, rgp160 ELISA and Western blot only at month 19. Furthermore, by polymerase chain reaction (PCR) the HIV positivity was established at the same time (month 19) while the PCR was negative in samples taken at months 0, 5, 10, 13 and 16. Thus, serology and PCR tests identified HIV-1 involvement >1 year after the positive results documented by Th responses.

We have recently extended the study to include seronegative homosexual men who have continued to practice high-risk sexual behavior and

---

Fig. 2. Proliferation (A–D) and IL-2 production (E–H) by peripheral blood mononuclear cells from a HIV-1-seronegative homosexual male during a 19-month interval [modified from 28]. From this individual, both frozen cells and serum samples were available and analyzed for response to recall antigens such as influenza A (FLU; full bars) and to four HIV synthetic peptides. These were *env*T1 (T1), *env*T2 (T2), *env*Th4 (Th4) and *env*P18 (P18). The empty bars (c) are negative controls with no stimulant added. Months of the blood samples are shown. T cell responses are first detected at month 5

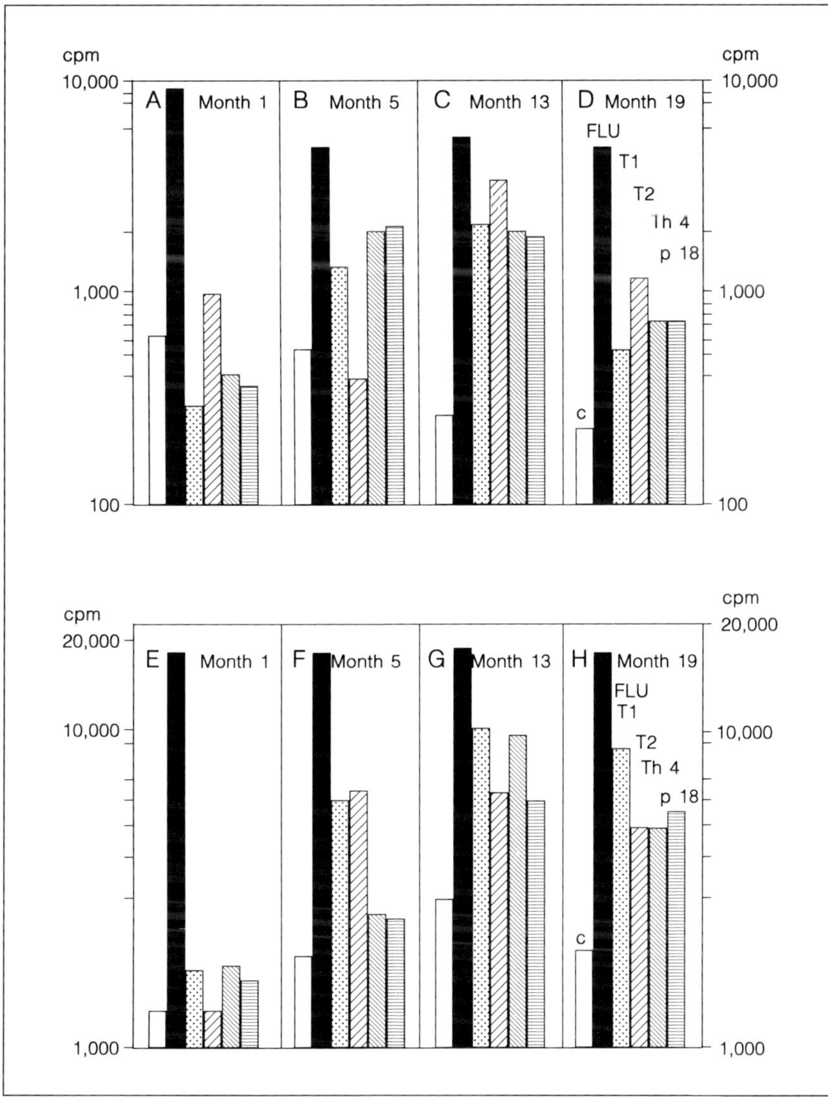

where a concordance in stimulation by the two types of tests is apparent only with the T1 peptide. At months 13 and 19, all cultures with HIV peptides are positive and the stimulation indices are high. The important finding of this study is that only the sample at month 19 shows HIV-1 positivity by HIV-1 ELISA, rgp160 ELISA, Western blot and PCR using *gag* primers SK38/39. Samples tested at months 0, 5, 10, 13 and 15 are negative by serology and PCR. The patient shows normal CD4 counts and good response to recall Ag throughout the study.

Table 4. HIV-specific Th function in high-risk seronegative individuals

High-risk group	Serum antibodies to HIV	PCR to HIV	Th cell response to HIV
Homosexual men[1]	0/6	0/6	6/6
Drug abusers[2]	0/7	n.t.	4/7

[1] Two of the six subsequently seroconverted and became positive by PCR.
[2] Results from Clerici et al. [28] and unpublished data obtained in collaboration with Dr. S.H. Weiss.

who are presumed to have been recently and repeatedly exposed to HIV. In collaboration with Dr. Carol O. Tacket (University of Maryland, Baltimore, Md.) [26] and Dr. Janis V. Giorgi (UCLA, Los Angeles, Calif.) [28], we have observed that in cultures of peripheral blood from 6/6 such individuals Th responses could be detected by IL-2 production when stimulated with four synthetic peptides corresponding to HIV *env* gp120 and gp160 [22–24]. The four synthetic peptides chosen have previously been documented to be immunogenic in the majority of asymptomatic HIV+ individuals as well as in HIV vaccinees [20, 27] but did not induce in vitro Th responses in more than 100 low-risk HIV-seronegative donors [6, and Clerici, unpubl. observations]. In collaboration with Dr. Stanley H. Weiss (Medical School of New Jersey, Newark, N.J.), we have also observed that blood lymphocytes from 4/7 seronegative high-risk intravenous drug abusers studied thus far generated Th responses to these peptides (table 4).

Encouraged by the sensitivity and potential clinical relevance of these tests, we have recently begun to investigate peripheral blood cultures from health care workers who have been accidentally exposed to HIV but remained seronegative after the accident. Some of these people do indeed show Th reactivity to the peptides indicating antigenic exposure to HIV despite apparent seronegativity. These observations, taken together, demonstrate the sensitivity and potential value of this in vitro functional assay in testing for exposure to HIV-1. The at-risk groups who may exhibit HIV-specific Th reactivity are recipients of HIV+ blood products, sexual partners of HIV+ individuals, intravenous drug users, professional people who had accidental exposure, infants of HIV+ mothers and vaccinated volunteers.

The significance of these observations are: Firstly, the analysis of Th cell responses may be more sensitive in detecting exposure to HIV as well as the early decline in specific immunity to recall antigens than other available tests. Secondly, these assays are likely to be sensitive measures of improving Th functions in patients who undergo clinical trials with newly developed drugs. Thirdly, when individuals are identified early during their course of HIV infection, not only antiviral therapy can start at the time of low viral load but vaccination can also be offered *before* the loss of Th function to recall antigens, especially to HIV antigens.

In summary: Th responses generated in vitro by PBL from individuals infected with HIV-1 and/or exposed to HIV antigenic determinants can be used to monitor immunological decline in AIDS progression and recovery of Th immune function as a result of therapy. These assays can also be used to assess HIV-specific immunization with candidate AIDS vaccines and exposure to HIV antigens in exposed individuals who may have come in contact with the virus but exhibit no evidence of HIV infection by other currently employed techniques such as serological tests and PCR.

References

1 Fauci AS: The human immunodeficiency virus: Infectivity and mechanisms of pathogenesis. Science 1988;239:617–622.

2 Smolen JS, Bettleheim P, Koller U, McDougal S, Graninger W, Luger TA, Knapp W, Lechner K: Deficiency of the autologous mixed lymphocyte reaction in patients with hemophilia treated with commercial factor VIII concentrate: correlation with T cell subset distribution, antibodies to lymphadenopathy-associated or human T lymphotropic virus, and analysis of the basis of the deficiency. J Clin Invest 1985;75: 1828–1834.

3 Shearer GM, Bernstein DC, Tung KS, Via CS, Redfield R, Salahiddin SZ, Gallo RC: A model for the selective loss of major histocompatibility complex self-restricted T cell immune response during the development of acquired immunodeficiency syndrome (AIDS). J Immunol 1986;137:2514–2521.

4 Miedema F, Chantal Petit AJ, Terpstra FG, Eeftinek Schattenkerk JKM, de Wolf F, Al BJM, Ross M, Lange JMA, Danner SA, Goudsmit J, Schellekens PTA: Immunological abnormalities in human immunodeficiency virus (HIV)-infected asymptomatic homosexual men: HIV affects the immune system before CD4+ T cell depletion occurs. J Clin Invest 1989;82:1908–1914.

5 Hofmann B, Jakobsen KD, Odum N, Dickmeiss E, Platz P, Ryder LP, Pedersen C, Mathiesen L, Bygbjerg I, Faber V, Sveffaard A: HIV-induced immunodeficiency. Relatively preserved phytohemagglutinin as opposed to decreased pokeweed mitogen responses may be due to possibly preserved responses via CD2/phytohemagglutinin pathway. J Immunol 1989;142:1874–1880.

6 Clerici M, Stocks NI, Zajac RA, Boswell RN, Lucey DR, Via CS, Shearer GM: Detection of three distinct patterns of T helper cell dysfunction in asymptomatic, human immunodeficiency virus-seropositive patients. J Clin Invest 1989;84:1892–1899.

7 Schnittman SM, Lane HC, Greenhouse J, Justment JS, Basler M, Fauci AS: Preferential infection of CD4+ memory T cells by human immunodeficiency virus type 1: Evidence for a role in the selective T-cell functional defects observed in infected individuals. Proc Natl Acad Sci USA 1990;87:6058–6062.

8 Schellekens PT, Roos MT, de Wolf F, Lange JM, Miedema F: Low T-cell responsiveness to activation via CD3/TCR is a prognostic marker for acquired immunodeficiency syndrome (AIDS) in human immunodeficiency virus-1 (HIV-1)-infected men. J Clin Immunol 1990;10:121–127.

9 Lucey DR, Melcher GP, Hendrix CW, Zajac RA, Goetz DW, Butzin CA, Clerici M, Warner RD, Abbadessa S, Hall K, Jaso R, Woolford B, Miller S, Stocks NI, Salinas CM, Wolfe WH, Shearer GM, Boswell RN: Human immunodeficiency virus infection in the US Air Force. Seroconversions, clinical staging and assessment of a T-helper cell functional assay to predict change in CD4+ T-cell counts. J Infect Dis 1991;164:631–637.

10 Clerici M, Landay AL, Kessler HA, Zajac RA, Boswell RN, Muluk SC, Shearer GM: Multiple patterns of alloantigen presenting/stimulating cell dysfunction in patients with AIDS. J Immunol 1991;146:2207–2213.

11 Roilides E, Clerici M, DePalma L, Rubin M, Pizzo PA, Shearer GM: Helper T-cell responses in children infected with human immunodeficiency virus type A. J Pediatr 1991;118:724–730.

12 Volberding P, Lagakos SW, Koch MA, Pettinelli C, Myers MW, Booth DK, Balfour HH, Reichaman RC, Bartlett JA, Hirsch MS, Murphy RL, Hardy WD, Soeiror Fische MA, Bartlett JG, Merigan TC, Hyslop NE, Richman DD, Valentine FT, Corey L: Zidovudine in asymptomatic human immunodeficiency virus infection. A controlled trial in persons with fewer than 500 CD4-positive cells per cubic millimeter. N Engl J Med 1990;322:941–949.

13 Fishl MA, Richman DD, Hansen N, Collier AC, Carey JT, Para MF, Hardy WD, Dolin R, Powderly WG, Allan JD, Wong B, Merigan TC, McAuliffe VJ, Hyslop NE, Rhame FS, Balfour HH, Spector SA, Volberding P, Pettinelli C, Anderson J: The safety and efficacy of zidovudine (AZT) in the treatment of subjects with mildly symptomatic human immunodeficiency virus type 1 (HIV) infection: A double-blind placebo-controlled trial. Ann Intern Med 1990;112:727–737.

14 Lambert JS, Seidlin M, Reichman RC, Plank CS, Laverty M, Morse GD, Knupp C, McLaren C, Pettinelli C, Valentine FT, Dolin R: 2′,3′-Dideoxyinosine (ddI) in patients with the acquired immunodeficiency syndrome or AIDS-related complex: A phase I trial. N Engl J Med 1990;322:1333–1340.

15 Cooley TP, Kunches LM, Saunders CA, Ritter JK, Perkins CJ, McLaren C, McCaffrey RP, Liebman HA: Once-daily administration of 2′,3′-dideoxyinosine (ddI) in patients with the acquired immunodeficiency syndrome or AIDS-related complex: Results of a phase I trial. N Engl J Med 1990;322:1340–1345.

16 Mêrigan TC, Skowron G, Bozzette SA, Richman D, Uttamchandani R, Fischl M, Schooley R, Hirsch M, Soo W, Pettinelli C, Schaumberg H, the ddC Study Group of the AIDS Clinical Trials Group: Circulating p24 antigen levels and responses to

dideoxycytidine in human immunodeficiency virus (HIV) infections: A phase I and II study. Ann Intern Med 1989;110:189–194.
17 Bolognesi DP: Progress in vaccines against AIDS. Science 1989;246:1233–1234.
18 Newmark P: AIDS complexities and strategies. Nature 1989;341:566–567.
19 Dolin R, Graham BS, Greenberg SB, Tacket CO, Belshe RB, Midthun K, Clements ML, Gorse GJ, Horgan BW, Atmar RL, Karzon DT, Bonnez W, Fernie BF, Montefiori DC, Stablein DM, Smith GE, Koff WC, NIAID AIDS Vaccine Clinical Trials Network: The safety and immunogenicity of a human immunodeficiency virus type 1 (HIV-1) recombinant gp160 candidate vaccine in humans. Ann Intern Med 1991; 114:119–127.
20 Clerici M, Tacket CO, Via CS, Lucey DR, Muluk SC, Zajac RA, Boswell RN, Berzofsky JA, Shearer GM: Immunization with subunit human immunodeficiency virus vaccine generates stronger T helper cell immunity than natural infection. Eur J Immunol 1991;21:1345–1349.
21 Tacket CO, Baqar S, Munoz C, Murphy JR Jr: Lymphoproliferative responses to mitogens and HIV-1 envelope glycoprotein among vaccinated with recombinant gp160 AIDS. Res Hum Retroviruses 1990;4:535–542.
22 Delisi C, Berzofsky JA: T cell antigenic sites tend amphipathic structures. Proc Natl Acad Sci USA 1985;82:7048–7052.
23 Cease KB, Margalit H, Cornette JL, Putney SD, Robey WG, Ouyang C, Streicher HZ, Fishchinger PJ, Gallo RC, Delisi C, Berzofsky JA: Helper T-cell antigenic site identification in the acquired immunodeficiency syndrome virus gp120 envelope protein and induction of immunity in mice to the native protein using a 16-residue synthetic peptide. Proc Natl Acad Sci USA 1987;84:4249–4253.
24 Takahashi H, Cohen J, Hosmalin A, Cease KB, Houghten R, Cornette JL, Delisi C, Moss B, Germain RN, Berzofski JA: An immunodominant epitope of the human immunodeficiency virus envelope glycoprotein gp160 recognized by class I major histocompatibility complex molecule-restricted murine cytotoxic T lymphocytes. Proc Natl Acad Sci USA 1988;85:3105–3109.
25 Keefer MC, Bonnez W, Roberts NJ Jr, Dolin R, Reichman RC: Human immunodeficiency virus (HIV-1) gp160 specific lymphocyte proliferative responses of mononuclear leukocytes from HIV-1 recombinant gp160 vaccine recipients. J Infect Dis 1991;163:448–453.
26 Clerici M, Berzofsky JA, Shearer GM, Tacket CO: Exposure to HIV-1 indicated by HIV-specific T helper cell responses before detection of infection by polymerase chain reaction and serum antibodies. J Infect Dis 1991;164:178–182.
27 Clerici M, Stocks NI, Zaja RA, Boswell RN, Bernstein DC, Mann DL, Shearer GM, Berzofsky JA: Interleukin-2 production used to detect antigenic peptide recognition by T-helper lymphocytes from asymptomatic HIV-seropositive individuals. Nature 1989;339:383–385.
28 Clerici M, Giorgi JV, Chou C-C, Gudeman VK, Zack JA, Gupta P, Ho H-N, Nishanian PG, Berzofski JA, Shearer GM: Specific anti-HIV-1 T cell immunity in seronegative homosexuals with recent sexual exposure to seronegative partners. J Infect Dis, in press.

Dr. Mario Clerici, Department of Health and Human Services,
Public Health Service, National Institutes of Health, Building 10, Room 4B-17,
Bethesda, MD 20892 (USA)

Janossy G, Autran B, Miedema F (eds): Immunodeficiency in HIV Infection and AIDS. EC/FERS/MRC Workshop on Immunodeficiency in HIV-1 Infections, Windsor, Surrey, 1991. Basel, Karger, 1992, pp 76–84

Decreased T Cell Proliferation to Pokeweed Mitogen Is a Prognostic Factor in Asymptomatic HIV-Positive Patients[1]

Claus Bohn Christiansen[a], *Ebbe Dickmeiss*[a], *Lars P. Ryder*[a], *Jan Gerstoft*[b], *Johannes Gaub*[b], *Per Platz*[a], *Bo Hofmann*[a], *Arne Svejgaard*[a]

[a] Department of Clinical Immunology, and [b] Department of Infectious Diseases, State University Hospital, Copenhagen, Denmark

A crucial element in the pathogenesis of the developing immunodeficiency is the progressive depletion of the CD4+ T cell compartment, mainly caused by cytopathic HIV infection of activated CD4+ T cells [1, 2]. It should be stressed, however, that the abnormality in the T cell compartment is not only manifested in the quantitative decrease in the number of this T cell population. Additional functional defects can also be demonstrated in vitro in isolated CD4+ as well as isolated CD8+ cells even in the early asymptomatic stage of the infection before CD4 cell depletion becomes obvious [3–5, and Christiansen, unpubl. data]. The clinical and pathologic relevance of these superimposed functional defects may become clear when a functional test provides prognostic information that is independent of predictive value of the CD4+ T cell counts.

In our previous short-term study on a cohort of 55 initially asymptomatic HIV-infected homosexuals, we have demonstrated that the in vitro proliferative T cell response to pokeweed mitogen (PWM) yielded significant prognostic assessment about the progress of HIV-1 disease. In this previous study the development of symptoms such as reaching the CDC stage IV was taken as the end point [6, 7]. In the present study we confirm

[1] This work has been supported by grants 12-7922, 12-6746 and 12-6747 from the Danish Medical Research Council.

and extent this observation in a larger cohort of 166 individuals with a substantially longer observation time which allow us to assess the full course of the disease.

Experimental Design

166 homosexual men with a positive anti-HIV-1 serology who regularly attended the AIDS Clinic at the State University Hospital in Copenhagen from 1984 to 1986 were selected and followed at intervals. The first 42 patients were included because their CD4/CD8 ratio had been below 1.0 in a prescreening blood sample 1 month before entry to the study. These patients were also tested retrospectively by analyzing stored serum samples by serology and found to be HIV-1 seropositive as soon as these assays have become available. The study has been closed at the end of 1990. There is partially overlap between this cohort and those reported previously [6, 7].

Cultures of 50,000 peripheral blood mononuclear cells were incubated in 150 μl RPMI 1640 medium in the presence of stimulants such as phytohemagglutinin (PHA) or PWM. Tritiated thymidine was added to all the cultures at 72 h and incorporated activity was measured in cells harvested 24 h later. The results are medians of the triplicate determinations expressed as percentages (PHA% and PWM%) of the corresponding responses detected in 2 healthy control samples studied simultaneously. The CD4+ and CD8+ T lymphocyte percentages were determined by FACS analyzer on gated lymphocytes. Total lymphocyte count was performed by routine methods and the absolute CD4 and CD8 counts were also calculated. A commercial solid-phase enzyme immunoassay (Abbott Laboratories) was used for detection of HIV p24 antigen.

To evaluate the immunological parameters as predictors for the development of AIDS and the demise of patients due to immunological deficiency caused by HIV-1, the proportional hazards model of Cox (BMDP Software, Los Angeles, Calif., USA) was used. The following parameters were incorporated as *continuous* variables: PWM%, PHA%, CD4/CD8 ratio, CD4%, CD8%, and total CD4 count. The HIV p24 antigen was scored as present or absent. AIDS-related demise of patients was applied as the sole end-point in this investigation.

Laboratory and Clinical Observations

The original observation by Hofmann et al. [6] about the bimodal distribution of PWM response in HIV-1-positive patients was reproduced in our data. On the other hand, PHA response was unimodally distributed (fig. 1, 2).

The mean observation time for the surviving individuals was 62 (38–84) months. During the observation time, 30 of the 166 patients had died, 29 with the diagnosis of AIDS, with a mean observation time to death of

54 (2–59) months. A total of 46 patients developed AIDS, with a mean observation time until diagnosis of 33 months.

Mean values of immunological parameters at the patient entry are given in table 1. In order to take into account the covariation among the parameters investigated, we used the multivariate proportional hazards model of Cox (table 2). The basic data of this model for survival showed a highly significant correlation between the entrance values of the CD4/CD8 ratio, PWM%, and CD4% ($\chi^2 > 20$), whereas the total CD4 count, CD8% and PHA% were less strongly correlated with subsequent survival ($\chi^2 < 15$). The covariation between the parameters is indicated by the first step of the Cox analysis (forward selection) where the CD4/CD8 ratio is selected as prognostic parameter (table 2). Only insignificant correlation is left for the other parameters, except for the PWM%. This indicates that the PWM% carries prognostic information which is independent of the covariation with the CD4/CD8 ratio. Both the CD4/CD8 ratio and the PWM% provide significant independent prognostic information in the final model, whereas the remaining parameters do not contribute with further prognostic information.

Table 1. Initial immunological data on 166 HIV-positive men and 40 HIV-negative healthy controls: values are medians (25–75 percentiles)

Immunological parameter	Seropositive (n = 166)	Seronegative (n = 40)
CD4%	25 (18–29)	46 (35–51)
CD8%	39 (30–47)	18 (16–23)
CD4/CD8 ratio	0.60 (0.40–0.90)	2.15 (1.70–2.92)
PWM% relative response	33 (20–49)	100 (50–200)
PHA% relative response	71 (52–94)	100 (50–200)

Fig. 1. Probability plot of the relative transformation response to PWM% in HIV-positive individuals. (x, y) values are: x = PWM%; y = the fraction of patients with a relative response value equal to or lower than the corresponding x value. The bimodal distribution of the PWM% is indicated by the intersecting lines which demonstrates that the cohort can be divided into two overlapping gaussian distributions with respect to PWM%, with PWM% = 25% being the border between the two populations.

Fig. 2. Relative transformation response to PHA% in HIV-positive individuals. The relative PHA response appears to be unimodally distributed.

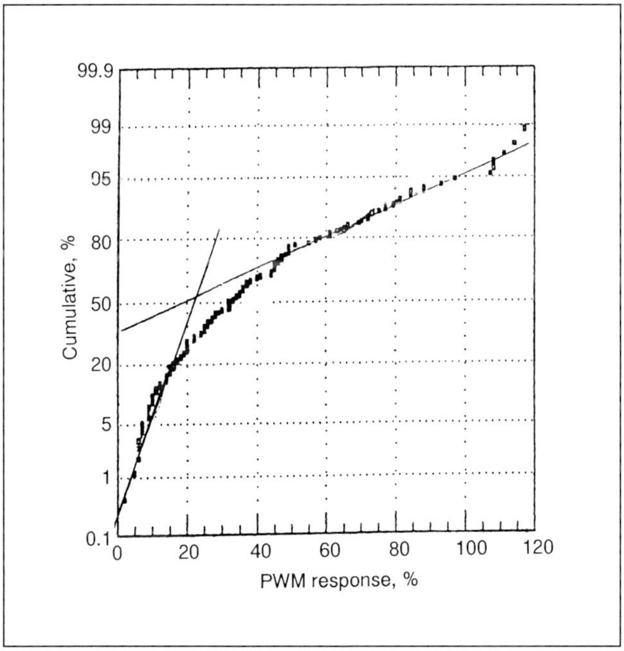
99.9
99
95
80
50
20
5
1
0.1
Cumulative, %
0
20
40
60
80
100
120
PWM response, %

1

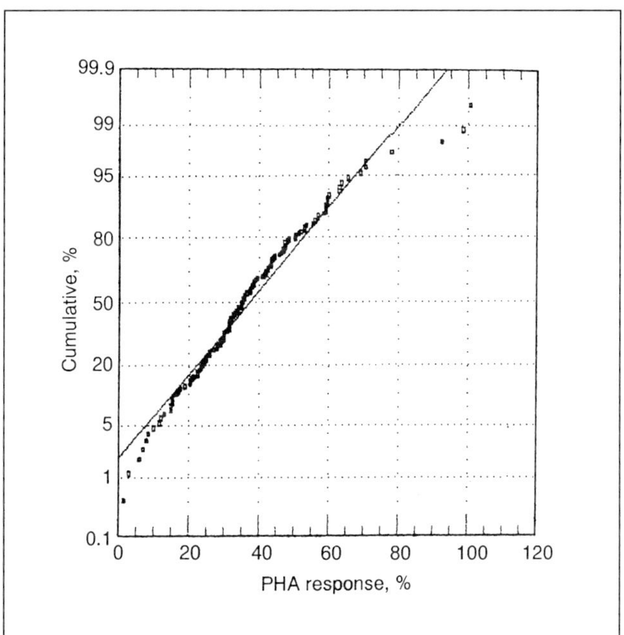
99.9
99
95
80
50
20
5
1
0.1
Cumulative, %
0
20
40
60
80
100
120
PHA response, %

2

Table 2. Analysis of data by the multivariate proportional hazards model of Cox

	Variable	χ^2	p value	β
Basic data[1]	CD4/CD8 ratio	23.9	<0.00001	
	PWM%	23.2	<0.00001	
	CD4%	21.8	<0.00001	
	Total CD4 count	14.3	0.0002	
	CD8%	9.6	0.002	
	PHA%	8.8	0.003	
CD4/CD8 ratio selected[2]	CD4/CD8 ratio	23.9	<0.00001	
	PWM%	8.9	0.003	
	Total CD4 count	2.8	0.1	
	PHA%	1.5	0.2	
	CD4%	1.0	0.3	
	CD8%	1.0	0.3	
Final model[3]	CD4/CD8 ratio	9.6	0.002	2.25
	PWM%	0.9	0.003	0.034
	Total CD4 count	2.0	0.2	
	CD8%	0.6	0.5	
	CD4%	0.2	0.6	
	PHA%	0.2	0.6	

[1] Basic data: The CD4/CD8 ratio, PWM% and CD4% showed the highest χ^2 values, indicating a strong correlation to subsequent survival. However, all of the parameters were significantly correlated with outcome.
[2] CD4/CD8 ratio selected: When CD4/CD8 ratio is selected as a prognostic parameter (forward selection), the Cox analysis shows that only PWM% provides further prognostic information (p value 0.003), whereas the information provided by the other parameters has been absorbed by the covariation with the ratio.
[3] Final model: When the CD4/CD8 ratio and the PWM% are selected in the model, it is shown that they both carry independent and significant prognostic information, whereas no significant information is left for the rest of the parameters. The β values of the final model can be used to estimate prognosis for individual patients in this cohort (see text).

Information on HIV p24 antigen status at entrance was available in 106 patients. There was no correlation between HIV-1 antigen status at entrance and outcome after 5 years (table 3), and thus in the present cohort the HIV p24 antigen status had no prognostic significance with respect to survival >5 years.

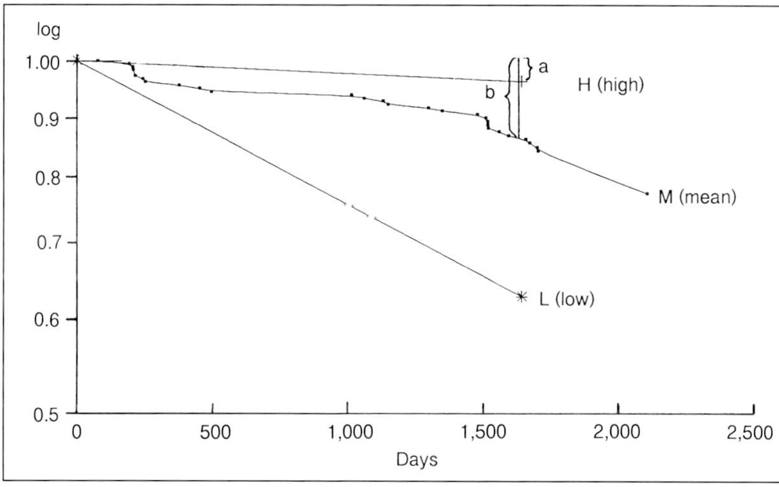

Fig. 3. The observed survival (M) and estimated survival at 4.5 years for patients with PWM% = 60%; CD4/CD8 ratio = 1.1 at entrance (H) and for patients with PWM% = 20%; CD4/CD8 ratio = 0.5 at entrance (L). X-axis: days after entrance; Y-axis: log survival ratio. a/b = calculated risk factor.

Table 3. Correlation between HIV p24 antigenemia at entrance and outcome after 5 years in 106 patients

Values at entrance	Status after 5 years	
	alive	dead
HIV p24 positive	14	6
HIV p24 negative	73	13

There was no significant correlation between the HIV antigenemia and subsequent outcome (Fischer's exact test, $p = 0.11$).

The β values of the final model (table 2) can be used to estimate prognosis for individual patients. As an example indicated in figure 3 in a semilogarithmic plot, the observed survival of the present cohort is given (M). The mean values of PWM and CD4/CD8 ratio in this cohort at entrance were 38% and 0.7, respectively. The survival expectancy for a

Table 4. Calculated risk factors for patients with different values of PWM% and CD4/CD8 ratios at entrance

CD4/CD8 ratio	Relative PWM response		
	60%	38%	20%
1.1	(1)[1]	2.1	3.9
0.72	2.4	5.0	9.2
0.5	3.9	8.2	15.0

[1] A risk factor of one is assigned to patients with fairly normal values at entrance. The other risk factors are calculated in relation to these patients, using the expression $e^{-[0.034(60-PWM\%)+2.25(1.1-ratio)]}$ (see text).

patient with certain values of PWM% and ratio can then be estimated on the basis of a risk factor calculated as:

$$e^{-[0.034(38-PWM\%)+2.25(0.72-CD4/8\ ratio)]}$$

The figure indicates that asymptomatic patients with a ratio of 1.1 and a PWM% of 60% at entrance have a 96% survival rate over 4.5 years (H), whereas patients with a ratio of 0.5 and a PWM% of 20% have a 63% survival rate. Table 4 gives risk factors for patients with different values for PWM% (columns) and for CD4/CD8 ratios (rows) at entrance, assigning a risk factor of one to patients with near normal values.

Discussion

The prognostically most informative quantitative CD4+ T cell parameters in the present study were the CD4/CD8 ratio and the CD4 percentage. The information given by the ratio was marginally, but not significantly, stronger than that given by the CD4 percentage, and therefore the ratio was selected by the Cox analysis in the present study. It is noteworthy that the absolute CD4 count was less predictive and it was completely included in the information carried by the ratio. The CD4 count was based on a lymphocyte count derived from a differential white blood count performed by several different hospital laboratories using either manual or automatic methods, whereas the CD4 and CD8 percentages were derived in our laboratory using standardized FACS analysis. The inaccuracy introduced by the lymphocyte count has presumably deprived the CD4 count of some of its predictive value. This has important practical implications and should caution against the use of fixed CD4 levels taken from the literature as a point of reference in the decision-making process for clinical interventions.

The PWM response in our hands has consistently been the functional parameter which has provided the strongest information. The reason for this is not known [8, 9]. PWM responses of a cohort of HIV-infected subjects are distinctly different when compared to responses elicited by the other mitogens that we have tested (PHA and ConA) because only the PWM response shows a bimodal distribution [6]. In this sense, PWM response resembles the stimulatory features of recall antigens including *Candida albicans,* PPD (tuberculosis) and cytomegalovirus which we have investigated in our laboratory [6]. Nevertheless, the standardization of antigen-specific responses as diagnostic tests is difficult since the magnitude of such responses heavily depends on the immunization history of the subject in question as well as on the strength and features of the particular antigen preparations used. These technical considerations render the PWM assay a more convenient routine test than the use of specific responses to recall antigens.

In conclusion, the present study indicates that the combined use of quantitative CD4 parameters and functional T cell assays such as the PWM response has a strong predictive value with respect to subsequent clinical outcome as indicated by the hazard estimate in table 4. As an example, it can be estimated that an asymptomatic HIV-1-infected person presenting with a relative PWM response of 20% and a CD4/CD8 ratio of 0.5 runs about 15 times greater hazard of perishing from AIDS over time than another individual with a relative PWM response of 60% and a CD4/CD8 ratio of 1.1.

References

1 Lane HC, Masur H, Edgar LC, Whalen G, Rook AH, Fauci AS: Abnormalities of B-cell activation and immunoregulation in patients with the acquired immunodeficiency syndrome. N Engl J Med 1983;309:453–458.

2 Lane HC, Depper JL, Greene WC, Whalen G, Waldmann TA, Fauci AS: Qualitative analysis of immune function in patients with the acquired immunodeficiency syndrome: evidence for a selective defect in soluble antigen recognition. N Engl J Med 1985;313:79–84.

3 Miedema F, Chantal Petit AJ, Terpestra F, Schattenkerk J, Wolf F, Al B, Roos M, Lange J, Danner S, Goudsmit J, Schellekens PTA: Immunological abnormalities in HIV-infected asymptomatic homosexual men. HIV affects the immune system before CD4+ T-helper cell depletion occurs. J Clin Invest 1988;82:1908–1914.
4 Petersen C, Dickmeiss E, Ryder LP, Platz P, Gaub J, Lindhardt BO, Lundgren J: T subset alterations, lymphocyte responsiveness and B cell alteration during primary infection with human immunodeficiency virus. AIDS, in press.
5 Hofmann B, Odum N, Platz P, Ryder LP, Svejgaard A, Nielsen JO: Immunological studies in AIDS, functional studies of lymphocyte subpopulations. Scand J Infect Dis 1985;17:235–243.
6 Hofmann B, Lindhardt BO, Gerstoft J, Petersen CS, Platz P, Ryder LP, Odum N, Dickmeiss E, Nielsen PB, Ullman S, Svejgaard A: Lymphocyte transformation response to pokeweed mitogen as a predictive marker for development of AIDS and AIDS-related symptoms in homosexual men with HIV antibodies. Br Med J 1987; 295:293–296.
7 Hofmann B, Bygbjerg I, Dickmeiss E, Faber V, Frederiksen B, Gaub J, Gerstoft J, Jakobsen BK, Jakobsen KD, Lindhardt BO, Petersen CS, Pedersen C, Platz P, Ryder LP, Odum N, Skinhoj P, Svejgaard A: Prognostic value of immunological abnormalities and HIV antigenemia in asymptomatic HIV-infected individuals: Proposal of immunologic staging. Scand J Infect Dis 1989;21:633–643.
8 Hofmann B, Jakobsen KD, Odum N, Dickmeiss E, Platz P, Ryder LP, Pedersen C, Mathiesen L, Bygbjerg I, Faver V, Svejgaard A: HIV-induced immunodeficiency: Relatively preserved phytohemagglutinin opposed to decreased pokeweed mitogen response may be due to possibly preserved responses via CD2/phytohemagglutinin pathway. J Immunol 1989;142:1874–1880.
9 Ceuppens JL, Meurs L, Baroja ML, van Wauwe JP: Effect of T3 modulation on pokeweed mitogen induced T cell activation: Evidence for an alternative pathway of T cell activation. J Immunol 1986;136:3346–3350.

Claus Bohn Christiansen, MD, Rigshospitalets Blodbank,
Klinisk Immunologisk afd., Blegdamsvej 9, DK–2100 Copenhagen (Denmark)

Janossy G, Autran B, Miedema F (eds): Immunodeficiency in HIV Infection and AIDS.
EC/FERS/MRC Workshop on Immunodeficiency in HIV-1 Infections, Windsor, Surrey, 1991.
Basel, Karger, 1992, pp 85–95

Low T-Cell Responsiveness to Activation via CD3/TCR: A Prognostic Marker for AIDS Useful for Monitoring Antiviral Therapy

Frank Miedema[a], *Marijke Th.L. Roos*[a], *Rob Gruters*[a],
Peter Th.A. Schellekens[a,b]

[a] Central Laboratory of The Netherlands Red Cross Blood Transfusion Service, incorporating the Laboratory of Experimental and Clinical Immunology, and
[b] Department of Internal Medicine, University of Amsterdam, The Netherlands

A large proportion of individuals infected with HIV-1 ultimately develops clinical symptoms and the acquired immunodeficiency syndrome (AIDS). However, the symptomless period varies considerably among HIV-1-infected individuals [1]. This variability in latency period may be explained by various virus-host interactions, i.e., virus virulence and the capacity of the immune system to mount an anti-HIV-1 response at different stages of infection [2]. Depletion of CD4+ T cells is the immunological hallmark of development of disease. Besides the decline in CD4+ T-cell numbers, which occurs relatively late after infection (see Schellekens et al., this volume), qualitative defects in the functional performance of peripheral mononuclear cells can also be detected. These occur relatively early after infection [3–7]. T-cell reactivity in vitro to influenza virus-infected cells is selectively lost in a substantial number of symptomless HIV-1-infected persons and in almost all AIDS patients [4] (see also Gruters et al., this volume). It has also been demonstrated that pokeweed mitogen (PWM) responses are low in asymptomatic individuals who are at risk for progression toward clinical symptoms [5] (see Christiansen et al., this volume). In a transsectional study we have demonstrated that in long-term asymptomatic HIV-1-infected men, T cells are relatively unresponsive to activation signals induced by anti-CD3 monoclonal antibodies (Mab) via the CD3/T-cell receptor (CD3/TCR) complex [3, 8, 9]. These assays are likely to have a prognostic value for the development of AIDS. The prospective study to establish this clinical aim in a large cohort of HIV-1-infected men is reviewed below.

Study Population

Between October 1984 and July 1986, 975 asymptomatic homosexual men were enrolled in a prospective study of risk factors for AIDS as previously described [10]. Every 3 months, clinical and epidemiological data were collected and serum and peripheral blood mononuclear cells (PBMC) were sampled. Because investigations concerning lymphocyte stimulation by anti-CD3 Mab started after July 1987, at a time when the cohort study was already in progress for about 30 months, we were restricted for observation of the early changes in anti-CD3 Mab reactivity to the seroconversions that occurred since that time.

Determination of T-Cell Reactivity by Whole Blood Culture

The whole lymphocyte culture technique has recently been described in detail for utilizing CD3 Mab as a mitogen [11]. The advantage of this method is its simplicity and that it assesses functional performance per unit blood rather than within a lymphocyte-enriched population. The cells are either stimulated with phytohemagglutinin-P or CD3 Mab CLB T3/4E (subclass IgE; final dilution 1:10^4) [12]. The proliferative responses are measured after 4 days of culture by incorporation of [^{3}H]thymidine of moderate specific activity, added 24 h before harvesting. A linear correlation between counts per minute and T-cell numbers [13] was demonstrated in T-cell titration experiments and by documenting in healthy volunteers that after a single oral dose of prednisolone (10 mg) CD3-induced T-cell activation declines proportionally with the prednisolone-induced lymphocytopenia. By the virtue of this near linearity, counts per minute values can be interpreted as mean reactivity per CD3+ (T) lymphocyte. Reactivity was expressed as the percentage of mean responses detected in concurrently running control cultures of healthy volunteers.

T-cell proliferation induced by anti-CD3 Mab of the IgE isotype is to some extent dependent on accessory-cell function delivered by monocytes [12]. Therefore, in whole-blood cultures from HIV-1-infected men, severe functional monocyte defects may also contribute to the observed decrease in T-cell proliferative responses [3]. Nevertheless, we have previously dem-

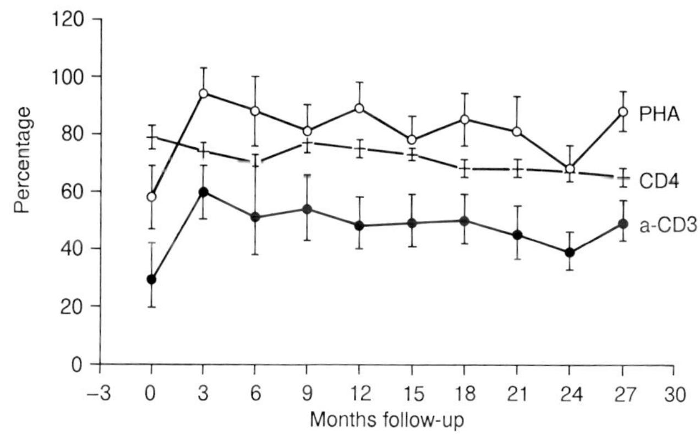

Fig. 1. Proliferative responses in whole blood induced by PHA (○) and anti-CD3 Mab (●) in asymptomatic HIV-1-infected men that seroconverted during the study (n = 15–30). Zero denotes the first immunoblot-positive anti-HIV-1 antibody reaction. Results represent the mean reactivity per T lymphocyte ± SEM related to the mean of the responses of healthy controls. 10th–90th percentiles in healthy controls: PHA (50–160); anti-CD3 (30–190).

onstrated that, when studied in the CD3 assay, in asymptomatic HIV-1-infected men accessory-cell function of monocytes was only marginally decreased when compared to the severity of the impairment in T-cell functions.

Lymphocyte Reactivity in HIV-1-Infected Asymptomatic Men

Lymphocyte responses to anti-CD3 Mab and PHA in HIV-1-infected men who seroconverted are shown in figure 1. Around the time of seroconversion (t = 0) responses to both mitogens were depressed. PHA-induced proliferation after having recovered to 60–80% of the normal response within 6 months after seroconversion, declined slowly till reaching 50–60% of normal values at 45 months of follow-up. The responses to anti-CD3 Mab have, however, not recovered but gradually decreased to 30–35% during the follow-up period.

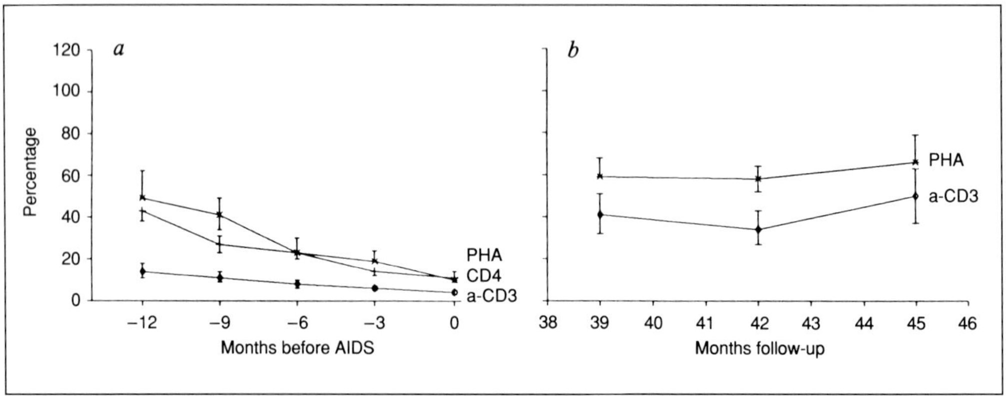

Fig. 2. Proliferative responses in whole blood induced by PHA (x) and anti-CD3 Mab (●) in patients with AIDS-OI (n = 13–22) (*a*) and in asymptomatic HIV-1-infected men (n = 17–25) with CD4+ cell numbers > 600/mm³ (*b*). The results represent the mean reactivity per T lymphocyte ± SE related to the mean of the responses of healthy controls.

Loss of Lymphocyte Reactivity at Progression to AIDS-OI

During these investigations two main issues were addressed: (1) the lymphocyte reactivity to PHA and anti-CD3 Mab in progressors versus nonprogressors in a retrospective study, and (2) the prognostic significance of anti-CD3 Mab reactivity compared to CD4 counts in blood. In the first study, progressors, individuals who were asymptomatic at the start of the study and developed AIDS-OI (CDC group IVC1), were compared with nonprogressors, persons who remained asymptomatic during the follow-up of 45 months. In figure 2a, the time of AIDS diagnosis is t = 0 and T-lymphocyte reactivity of progressors is shown before manifestation of clinical symptoms over the 12-month period. It can be seen that 12 months before AIDS diagnosis, T-cell reactivity to PHA in the progressors is still comparable to the reactivity seen in nonprogressors. However, from that time onwards it starts to decline steadily and is almost absent at the time of diagnosis. In contrast, T-cell reactivity to anti-CD3 Mab is severely decreased to about 10% of the normal response already 12 months before AIDS diagnosis. At that time, CD4+ T-cell numbers are only slightly decreased, to about 42% of the normal value, and only later, 6 months before diagnosis, show a sharp decline to 20% of the normal values.

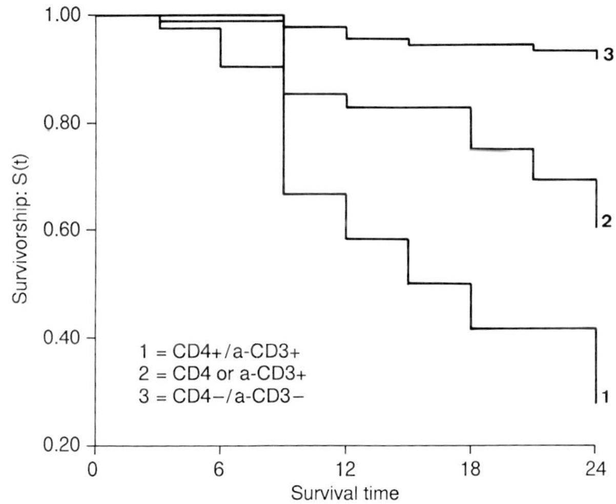

Fig. 3. Disease-free survival curves in HIV-1-infected men with regard to diagnosis of AIDS-OI according to the parameters. Top curve: normal CD4 cell numbers and normal anti-CD3-induced T cell reactivity; middle curve: low CD4+ cell numbers or low anti-CD3 Mab reactivity; bottom curve: both CD4 cell numbers and T-cell reactivity decreased.

In the group of nonprogressors the results are shown during the last 6 months of follow-up (fig. 2b). Individuals in this group remained asymptomatic for > 45 months, and the CD4 counts were > 600/mm^3, i.e. higher than in the progressors above. At the end of the follow-up period, 39–45 months after seroconversion, reactivity to anti-CD3 Mab and PHA of nonprogressors is significantly higher (fig. 2b) compared with the reactivity of the progressors (fig. 2a) and does not differ from the reactivity of the whole asymptomatic group (fig. 1). In individuals who progressed to AIDS responses to anti-CD3 Mab were virtually absent at least 12 months before diagnosis, whereas CD4+ cell numbers and reactivity to PHA were still comparable to values observed for nonprogressors and the asymptomatic population as a whole. Hence a specific loss of T-cell responsiveness to activation via the CD3/TCR complex appears to precede, by at least 1 year, the development of AIDS.

We regard anti-CD3 Mab responses severely depressed when they are decreased to $<15\%$ of the healthy control values. Next the predictive value of this low reactivity for progression to AIDS was analyzed and compared to that of the low ($<300/mm^3$) CD4+ lymphocyte numbers (fig. 3). In this figure the start of the observation period ($t = 0$ on the abscissa) is 30 months after the initiation of the whole study depicting a 18- to 21-month period. Both low CD4+ lymphocyte numbers and low anti-CD3 Mab reactivity portened progression to clinical disease: in about 45 and 35% of the subjects, respectively, within 21 months. In individuals who do not show these poor prognostic markers, 10% progress to AIDS (fig. 3a, b). Cox regression analysis reveals that both low CD4+ lymphocyte numbers and anti-CD3 Mab reactivity are poor prognostic factors for development of AIDS with relative risks of 3.8 and 2.9, respectively. CD4+ numbers and anti-CD3 Mab reactivity are independent parameters. Patients who show both poor prognostic features progress even faster to AIDS with a relative risk of 11.0 when compared to individuals who have neither of those features (fig. 3).

On the basis of these and other observations [14], it can be suggested that an early diminished T cell immunity may contribute to a poor control of viral infections allowing the development and replication of high virulent HIV-1 variants associated with rapid CD4 decline and development of AIDS [14]. Taken together, these data implicate that decreased cellular immunity in HIV-1 infection is a progression marker that may be of direct pathogenic relevance.

T-Cell Reactivity for Monitoring Zidovudine Treatment

Once we had demonstrated that loss of T-cell responsiveness to activation by anti-CD3 Mab is an early independent indicator of progression and predicts the development of clinical symptoms [15], we evaluated the effect of zidovudine treatment on T-cell function in asymptomatic men over a 2-year follow-up period [16]. Two related questions were investigated: These were whether the differences in clinical course among 4 groups related to differences in CD4+ T-cell numbers or T-cell functional capacities at the time of intake and whether there was a different response to zidovudine in the treated groups. Initially in this study all the treated men were selected for being at high risk for AIDS because of p24 antigenemia [17, 18] and the functional analysis was retrospectively evaluated on the basis of tests routinely carried out on all patients during the observation period. Four groups were selected retrospectively: (1) a group comprising men treated with zidovudine but still developing AIDS during the 2-year follow-up (progressors, n = 5); (2) a group of zidovudine-treated men who remained asymptomatic (nonprogressors, n = 4), matched for the treatment protocol during the first 6 weeks [18]; groups 3 and 4 were nontreated groups, 5 progressors and 5 nonprogressors, matched for lymphocyte numbers and CD4+ and CD3+ T-cell numbers at the start of the study.

There were no significant differences between the 4 groups in CD3+ T-cell and lymphocyte numbers at intake, 1 year before start of treatment, or at the start of zidovudine treatment. In treated progressors, CD4+ T-cell numbers 1 year before and at the start of zidovudine treatment were not significantly different from CD4+ T-cell numbers in nonprogressors.

Our previous results summarized above have already indicated that individuals with a progressive course of HIV infection have very low T-cell responses to stimulation with anti-CD3 Mab 15 months before AIDS [15]. This observation has been confirmed and extended in the present study. Already at the beginning of the study in the nontreated control group T cells from progressors showed significantly lower anti-CD3 response (fig. 4a, c) ($p < 0.02$). In the zidovudine-treated groups a similar difference in T-cell reactivity was observed between progressors and nonprogressors at the start of the study. Progressors had a decreased T-cell responsiveness (fig. 4b, d) which was lower than the values seen in nonprogressors and this difference reached statistical significance ($p < 0.01$) despite the fact that only few patients were available for this comparative study.

Next the efficacy of zidovudine therapy was analyzed for restoring CD4+ numbers and T-cell responses. After the start of zidovudine treatment in nonprogressors an increase of CD4+ counts was seen and the counts remained stable thereafter. In individuals who progressed to AIDS, however, the increase in CD4+ T-cell numbers induced by zidovudine treatment was less pronounced and transient after which CD4+ T-cell counts further declined until AIDS was diagnosed. The proliferative responses to anti-CD3 Mab remained stable in nonprogressors over the follow-up period, with the exception of patient 472 (fig. 4a). In nonprogressors the effect of zidovudine treatment on T-cell proliferation was limited, 3 subjects showed a temporary increase whereas in 1 person a

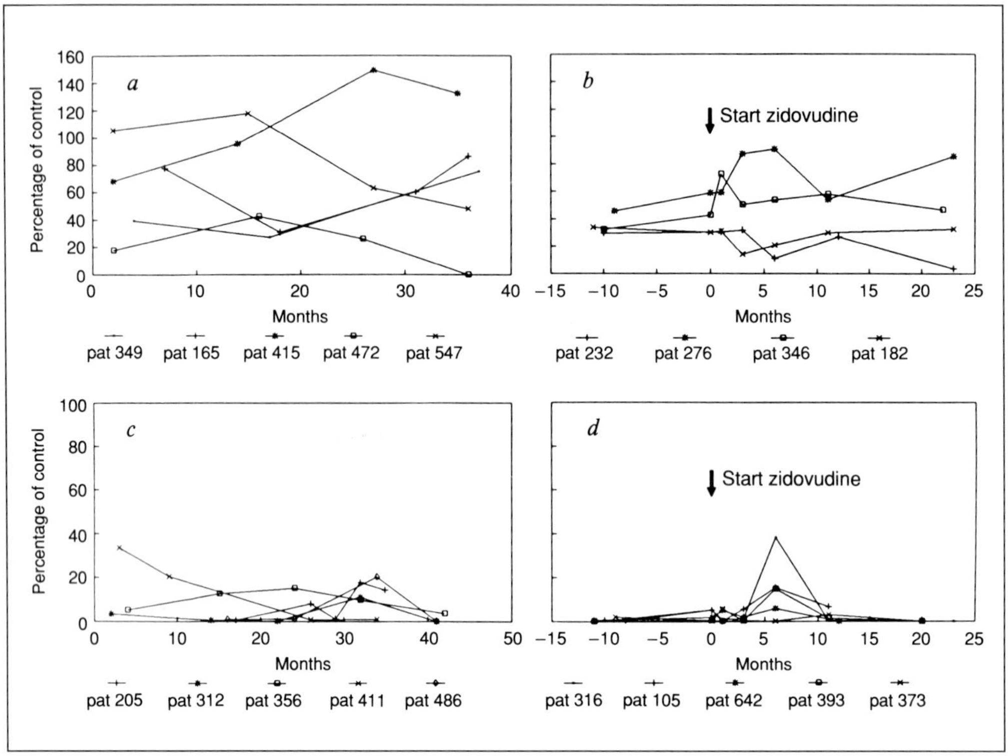

Fig. 4. T-cell proliferation in sequential cultures from nonprogressors (*a, b*) and progressors (*c, d*) who remain untreated (*a, c*) or given zidovudine treatment (*b, d*). T-cell proliferative response to immobilized anti-CD3 Mab (5 μg/ml) was determined by ^{3}H-thymidine incorporation and compared to values seen in cultures of HIV-1 seronegative controls running on the same day. For time scale and designations of the groups see the legend to figure 1. There is a significant difference in T-cell response between groups *a* and *c* ($p < 0.01$) at all time points, and between groups *b* and *d* ($p < 0.01$) at all time points except 6 months after the start of zidovudine treatment.

decrease was observed (patient 232; fig. 4b). The effect of zidovudine on the T-cell response of progressors was, however, clear. In all subjects 6 months after the start of treatment, T-cell proliferation was significantly improved. Nevertheless, this effect was transient and it leveled off to the initial low level 12 months after the start of zidovudine treatment (fig. 4d). In nontreated progressors this transient rise was not observed and T-cell proliferation remained low throughout the observation period until AIDS was diagnosed (fig. 4c).

It is possible that zidovudine suppresses HIV replication resulting in immune improvement but that after prolonged treatment zidovudine-resistant HIV variants emerge [19]. Also, differences in sensitivity to zidovudine of HIV biological variants may contribute.

These findings show that differences in clinical response to zidovudine treatment in a group of asymptomatic HIV seropositive men appeared to be due to the heterogeneity of this group with respect to their immunological status as determined by the response to anti-CD3 Mab. The treated men were all p24 antigenemic, by definition at the entry of the study. They did not significantly differ for CD4+ T cell numbers, thus anti-CD3 induced T-cell proliferation is an early marker for progression of HIV infection that is independent of CD4+ counts. This conclusion is in agreement with results from a prospective cohort study [15]. It is also apparent that zidovudine treatment is effective for only a period of time but after this period it does not dramatically change the natural history of HIV-1 infection. These findings further emphasize that in addition to CD4+ T-cell numbers, T-cell functional capacity should be considered as an important laboratory test which contributes to both the stratification and evaluation of anti-HIV therapy during the early phases of HIV infection.

Conclusions

In HIV-1-infected homosexual men, studied longitudinally, T-cell proliferative responses show a prognostic value for development of AIDS. In individuals who progress to AIDS, lymphocyte responses to anti-CD3 Mab are virtually absent already 12 months before diagnosis whereas at that time CD4+ cell numbers and reactivity to PHA are still comparable to values observed for nonprogressors and the asymptomatic population as a whole. In a disease-free survival analysis, low anti-CD3 Mab reactivity predicts progression to AIDS. Using this assay to monitor zidovudine treatment we also show that preventive therapy during the asymptomatic phase of HIV infection results only in a transient improvement of T-cell function. The differences in the clinical course among zidovudine-treated asymptomatics is associated with the heterogeneity of this group with respect to T-cell functional capacity that is already manifest at the start of treatment.

References

1 Taylor JGM, Schwartz K, Detels R: The time from infection with human immunodeficiency virus (HIV) to the onset of AIDS. J Infect Dis 1986;154:694–697.

2 Miedema F, Tersmette M, Van Lier RAW: AIDS pathogenesis: a dynamic interaction between HIV and the immune system. Immunol Today 1990;11:293–297.

3 Miedema F, Petit AJC, Terpstra FG, Schattenkerk JKME, De Wolf F, Al BJM, Roos M, Lange JMA, Danner SA, Goudsmit J, Schellekens PTA: Immunological abnormalities in human immunodeficiency virus (HIV)-infected asymptomatic homosexual men. HIV affects the immune system before CD4+ T helper cell depletion occurs. J Clin Invest 1988;82:1908–1914.

4 Shearer GM, Bernstein DC, Tung KSK, Via CS, Redfield R, Salahuddin SZ, Gallo RC: A model for the selective loss of major histocompatibility complex self-restricted T cell immune responses during the development of acquired immune deficiency syndrome (AIDS). J Immunol 1986;137:2514–2521.

5 Hofmann B, Orskov Lindhardt B, Gerstoft J, Sand Petersen C, Platz P, Ryder LP, Odum N, Dickmeiss E, Nielsen PB, Ullman S, Svejgaard A: Lymphocyte transformation response to pokeweed mitogen as a marker for the development of AIDS and AIDS-related symptoms in homosexual men with HIV antibodies. Br Med J 1987;295:293–296.

6 Giorgi JV, Fahey JL, Smith DC, Hultin LE, Cheng HL, Mitsuyasu RT, Detels R: Early effects of HIV on CD4 lymphocytes in vivo. J Immunol 1987;138:3725–3730.

7 Terpstra FG, Al BJM, Roos MThL, De Wolf F, Goudsmit J, Schellekens PT, Miedema F: Longitudinal study of leukocyte functions in homosexual men seroconverted for HIV: rapid and persistent loss of B-cell function after HIV infection. Eur J Immunol 1989;19:667–673.

8 Van Noesel CJM, Gruters RA, Terpstra FG, Schellekens PTA, Van Lier RAW, Miedema F: Functional and phenotypic evidence for a selective loss of memory T cells in asymptomatic HIV-infected men. J Clin Invest 1990;86:293–299.

9 Gruters RA, Terpstra FG, De Jong R, Van Noesel CJM, Van Lier RAW, Miedema F: Selective loss of T-cell functions in different stages of HIV infection. Eur J Immunol 1990;20:1039–1044.

10 De Wolf F, Lange JMA, Houweling JTM, Coutinho RA, Schellekens PT, Van der Noordaa J, Goudsmit J: Numbers of CD4+ cells and the levels of core antigens of and antibodies to the human immunodeficiency virus as predictors of AIDS among seropositive homosexual men. J Infect Dis 1988;158:615–622.

11 Bloemena E, Ross MThL, van Heijst JLAM, Vossen JMJJ, Schellekens PThA: Whole-blood lymphocyte cultures. J Immunol Methods 1989;122:161–167.

12 Van Lier RAW, Boot JHA, Verhoeven AJ, De Groot ER, Brouwer MA, Aarden LA: Functional studies with anti-CD3 heavy chain isotype switch variant monoclonal antibodies. J Immunol 1987;139:2873–2879.

13 Bloemena E, Van Oers RHJ, Weinreich S, Stilma-Meinesz AP, Schellekens PTA, Van Lier RAW: The influence of cyclosporin A on the alternative pathways of human T cell activation in vitro. Eur J Immunol 1989;19:943–946.

14 Tersmette M, Lange JMA, De Goede REY, De Wolf F, Eeftink JKM, Schattenkerk JKME, Schellekens PTA, Coutinho RA, Huisman JG, Goudsmit J, Miedema F: Differences in risk for AIDS and AIDS mortality associated with biological properties of HIV variants. Lancet 1989;i:983–985.
15 Schellekens PTA, Roos MTL, De Wolf F, Lange JMA, Miedema F: Low T-cell responsiveness to activation via CD3/TCR is a prognostic marker for AIDS in HIV-1-infected men. J Clin Immunol 1990;10:121–127.
16 Gruters RA, Terpstra FG, Lange JMA, Roos MTL, Harkema T, Mulder JW, De Wolf F, Schellekens PTA, Miedema F: Differences in clinical course in zidovudine-treated asymptomatic HIV-infected men associated with T-cell function at intake. AIDS 1991;5:43–47.
17 Lange JMA, Paul DA, Huisman HG, et al: Persistent HIV antigenaemia and decline of HIV core antibodies associated with transition to AIDS. Br Med J 1986;293: 1459–1462.
18 De Wolf F, Lange JMA, Goudsmit J, Cload P, de Gans J, Schellekens PTA, Coutinho RA, Fiddian AP, Van der Noordaa J: Effect of zidovudine on serum human immunodeficiency virus antigen levels in symptom-free subjects. Lancet 1988;i: 373–376.
19 Larder BA, Darby G, Richman DD: HIV with reduced sensitivity to zidovudine (AZT) isolated during prolonged therapy. Science 1989;243:1731–1734.

Frank Miedema, Central Laboratory of the Netherlands Red Cross Blood Transfusion Service, PO Box 9190, NL-1006 AD Amsterdam (The Netherlands)

Janossy G, Autran B, Miedema F (eds): Immunodeficiency in HIV Infection and AIDS. EC/FERS/MRC Workshop on Immunodeficiency in HIV-1 Infections, Windsor, Surrey, 1991. Basel, Karger, 1992, pp 96–114

Proliferative Defects of CD4 and CD8 Lymphocytes in HIV-1 Infection[1]

George Janossy, Nicola Borthwick, Ruth Lomnitzer, Margarita Bofill

Department of Clinical Immunology, Royal Free Hospital School of Medicine, London, UK

The simplest arrangement for lymphocyte analysis is to study freshly obtained cells in short-term cultures in the presence of nutrients and serum without any extra stimulus added [1, and see Gougeon et al., this volume]. These circumstances closely reflect the functional state of recirculating cells in vivo. In HIV-1-seropositive patients minimally enhanced lymphocyte cycling activity is seen: the cells in proliferative cycle include a proportion of CD4+ lymphocytes [2–4]. At the same time the lymphocyte survival is poor, mainly among CD8+ cells [5]. The other concept is to analyse functional changes by stimulating lymphocytes in vitro. These effects probably mimic the function of lymphocytes upon activation with foreign antigens in vivo, a relevant proposition because lymphocyte functional assays in vitro correlate well with skin tests for delayed type hypersensitivity and vaccine responses [6–8]. The in vitro tests can be standardized in a number of ways; often the prime aim is to set up methods for optimal discrimination between the normal and HIV-1 infected or immunosuppressed – most readily achieved when relatively weak polyclonal mitogens are utilized. This pragmatic approach was used to chose pokeweed [8, 9, and see Christianssen et al., this volume]. Others utilize Con-

[1] This work was supported by the MRC AIDS Directed Programme of Great Britain (to G.J. & M.B.). We are grateful to Drs. C. Lee, S.B. Squire and Margaret A. Johnson for allowing us to investigate patients under their care and to Mr. Anthony Timms and Ms. Louisa Baptista for technical help.

canavalin-A (Con-A), a much weaker mitogen for human than for murine T cells [e.g. 10]. Alternatively, phytohaemagglutinin (PHA), added in suboptimal concentrations [11, 12], Streptolysin O [13] or immobilized anti-CD3 antibodies [14] are applied. These mitogens do not activate T cells maximally, e.g. the mitogenicity of immobilized anti-CD3 is further increased by co-stimulating CD28 antibodies [14]. Others use antigen-specific assays such as allogeneic cells, tetanus toxoid, PPD, mannoproteins from *Candida albicans* [15], or viral antigens [see Clerici and Shearer, this volume].

Despite the heterogeneity of stimuli used, the functional assessment of blood lymphocytes from HIV-1 donors uniformly reveal that many asymptomatic patients exhibit reduced proliferative responses early in the disease before the onset of clinical symptoms and prior to a severe loss CD4+ cells [5–16]. Nevertheless, in these studies only limited efforts have been made to quantitate the numbers of responding lymphocytes individually: isotope techniques such as thymidine uptake do not give direct read-outs in this respect, and the counts are not linearly related to the stimulated cell numbers.

Our main interest has been to quantitate the numbers of cells capable of responding to strong activation signals under optimized culture conditions in vitro; cytokines and conditioned medium were also added to avoid IL-2 deficiency. Mostly short-term cultures of 3–5 days' duration were carried out to avoid the overcrowding by the stimulated cells. The methods shown below are single cell assays which discriminate individual resting lymphocytes from lymphoblasts, and also document the absolute numbers of these populations as they survive and proliferate in the culture [17]. One further advantage of these single cell studies, as opposed to bulk assays of thymidine incorporation, is that lymphocyte activation phenomena can be assessed within the different cell populations, i.e. among CD4+ and CD8+ subsets, while these cells grow or die in the company of other cell types. It has recently been shown that the higher molecular weight (MW) isoforms of CD45 antigens, CD45RA, are expressed on 'unprimed' or naive T cells while the lower MW isoform, CD45RO, is expressed on 'primed' or memory T cells [18–20]. Methods similar to those used in this study demonstrated the phenotypic shifts from 'naive' CD45RA+ to 'memory' CD45RO+ T cells during priming in vitro (fig. 1) [17, 19]. Consequently, in the analysis summarized below the performance of lymphocytes is related to both CD4/CD8 populations and to these cells' CD45R phenotype during the various stages of HIV-1 infection.

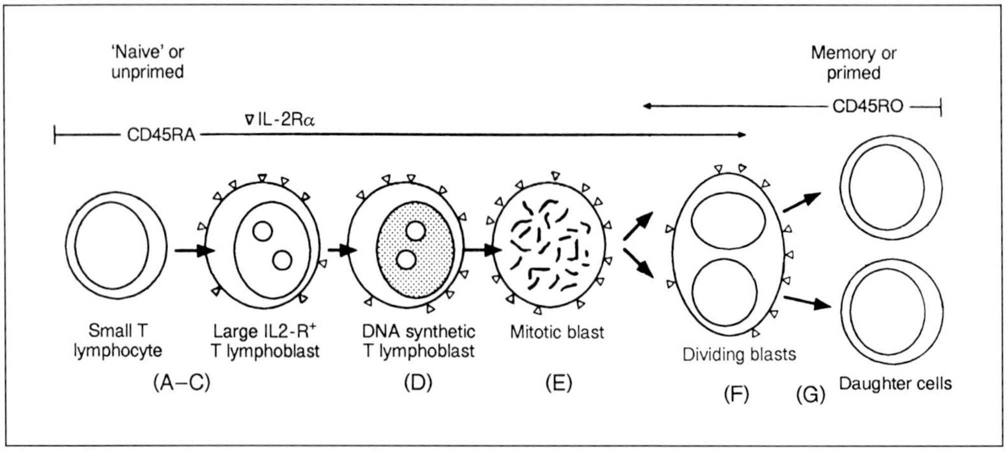

Fig. 1. The events of lymphocyte stimulation in cultures in the presence of IL-2 and conditioned medium using powerful mitogens such as PHA or anti-CD3. Large IL-2R blasts develop in 48 h (Stage C), start DNA synthesis and mitotic activity (Stages D–F) which can be readily quantitated in the presence of Cytochalasin-B by the development of multinucleated blast cells [17].

Selection of Methods

Complementary techniques provide a full assessment of the events during the first 3–4 days of lymphocyte activation by mitogens (fig. 1). These steps of activation are investigated by five methods as follows:

(1) With flow cytometers events of the proliferative cell cycle are documented by measuring DNA content with propidium iodide (PI) [21]. This method is rendered very precise by using chicken and trout erythrocytes as DNA standards and measuring events on FACScan with the CellFit program. The labelling can be performed as double staining with fluoresceinated CD4 (CD4-FITC), CD8 (CD8-FITC), or Ki67, another marker of proliferating cells [22]. An example of this analysis is shown in figure 2a, where mononuclear cells from a normal HIV-seronegative individual were stimulated with anti-CD3 antibody and the harvested cells were labelled with CD4-FITC 5 days later. In order to count the progeny of activated T blasts, the 'clones' are kept together as multinucleated blasts by adding Cytochalasin-B (see below). In this experiment the measurements of the DNA content with PI were performed on gated CD4+ cells: 50.4% of the

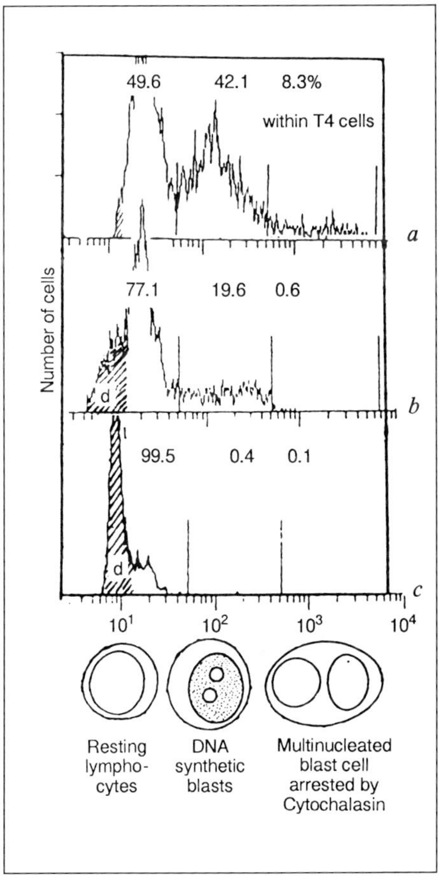

Fig. 2. Investigation of the proliferative cell cycle in cultures stimulated with anti-CD3 antibody for 5 days in the presence of Cytochalasin-B added at 72 h. The membranes of live cells were stained for CD4 in suspension with CD4-FITC, fixed with methanol and restained intracellularly for PI. The gates were set to include exclusively CD4+ cells and these are shown in cultures of HIV-1-seronegative donors cells *(a),* an HIV-1-seropositive asymptomatic person *(b),* and a patient hospitalized with AIDS *(c).* Cells in the proliferative cycle are shown. The shaded area at the left ('d') correspond to dying cells with DNA fragmentation.

normal CD4+ cells entered the cell cycle (stage D in fig. 1), and few cells are detected at the left part of the curve ('d') where dying cells are seen (but see below).

(2) Other methods also utilize flow cytometry. After activation with PHA or anti-CD3 multiparameter analysis is set up for cell scatter and double or triple immunofluorescence (IF) labelling. In the experiment shown in figure 3 the PHA-stimulated cells were double labelled for phycoerythrin conjugated CD4 (CD4-PE) and fluoresceinated CD25 (CD25-FITC) detecting IL-2 receptor α-chain (IL-2Rα; Tac antigen). CD4+ populations are gated on the PE channel (not shown in fig. 3), and the cells are

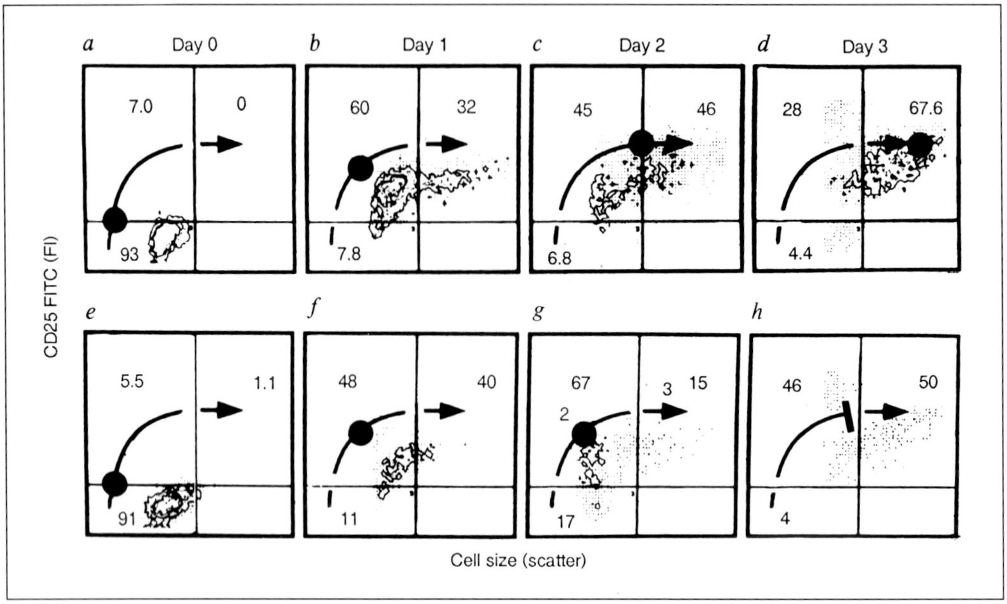

Fig. 3. Multiparameter analysis of PHA-stimulated cultures taken from HIV-1– normal control *(a–d)* and from an asymptomatic HIV-1+ person *(e–h)* and incubated for various periods of time in the presence of conditioned medium. The cells were stained for CD4 (phycoerythrin; PE) and CD25 (fluorescein; FITC). The gating included only CD4+ cells and the forward cell scatter (x-axis) versus CD25 positivity (y-axis) is shown. The numbers inside the quadrants refer to the releative percentages of cells and not their absolute numbers (but see fig. 4).

analysed for scatter (x-axis) and IL-2R expression (y-axis; fig. 3a–c). Within 24 h CD4+ cells start expressing IL-2Rα (CD25+) but are still small to intermediate sized (fig. 3b). By day 3, however, most of the CD4+ cells are CD25+ lymphoblasts (67.6% and only few CD4+, CD25– small lymphocytes remain (4.4% in fig. 3d).

(3) One important use of the flow cytometric method is to render the method quantitative by admixing standard numbers of unlabelled microspheres or beads into the culture and then investigate the numbers of lymphoblasts developing in the culture in relation to a fixed number of beads (fig. 4) [23]. In normal unstimulated cultures lymphoid cells tightly clustered; monocytes are also away from the lymphoblast gate (fig. 4a). In the

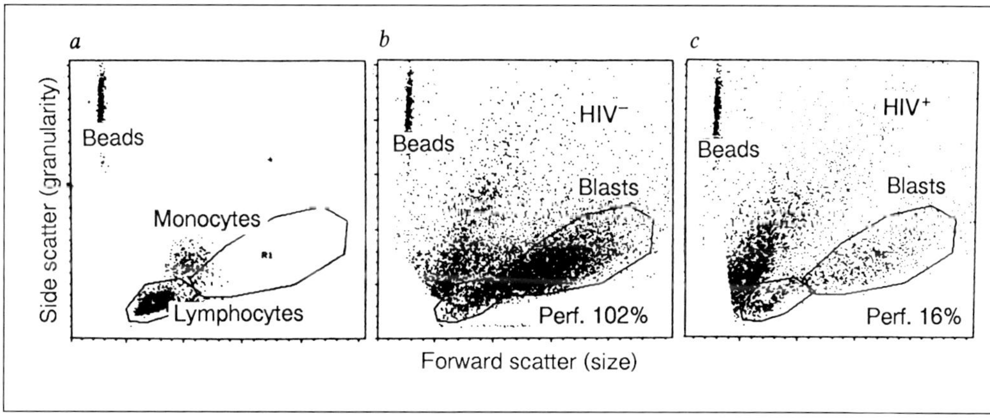

Fig. 4. Setting of gates for lymphocytes and lymphoblasts (R1) on freshly isolated mononuclear blood cells *(a)* and the analysis of the absolute numbers of activated blasts after 3 days' stimulation with PHA *(b, c)* in samples taken from a normal donor *(a, b)* and from an HIV-1+ asymptomatic patient *(c)*. To these lymphocyte cultures (10^6/ml) half the numbers of inert beads (5×10^5/ml 'beads') were added as described elsewhere [23]. The numbers of cells falling into the 'blasts' gate are recorded after 3-day activation by PHA – and related to a standard bead count read in the 'beads' gate, which corresponds to the original input of lymphocytes. In a representative normal stimulated culture shown in *b*, 102% blasts/input is recorded while in HIV-1+-stimulated culture depicted in *c*, 16% blasts/input is seen. These percentage values are referred to as 'lymphocyte performance', and the patients are grouped as normal (> 70%), poor (30–70%) and very poor (< 30%) performers (see fig. 6).

PHA and anti-CD3-activated normal cultures many blasts develop which are precisely counted (fig. 4b).

(4) Microscopic techniques include incubation of cells with bromodeoxyuridine (BrdU) in order to analyse the DNA synthetic BrdU+ cells in cytocentrifuged preparations in combination with CD4 and CD8 expression [24] and to analyse p24 in the BrdU+ blast cell population. One relevant observation is that in normal HIV-1– cultures BrdU+ blast cells preserve a healthy lymphoblastic morphology but among PHA-stimulated BrdU+ blasts from HV-1-seropositive donors morphological signs of disintegration and nuclear fragmentation can be seen. Thus, blasts may start DNA synthesis and die later in these strongly activated cultures.

(5) Finally, mitotic and dividing blasts, shown as Stages E and F in figure 1, can be better observed by adding colcemid and Cytochalasin-B to

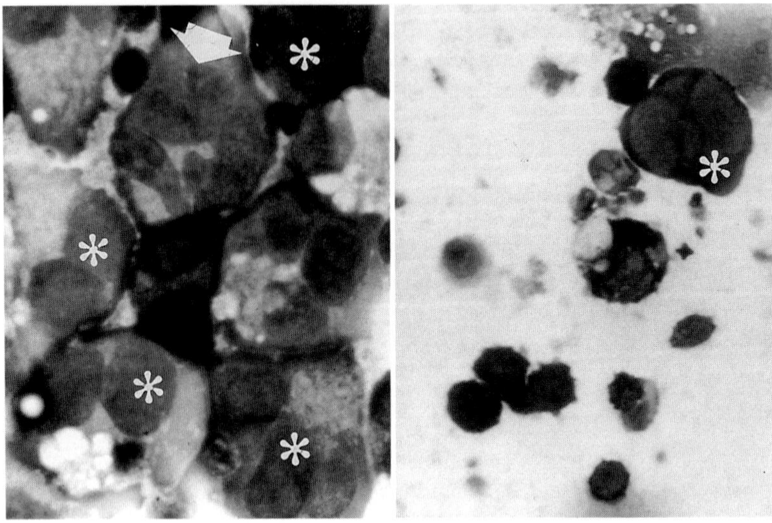

a b

Fig. 5. To cultures stimulated with PHA for 72 h, Cytochalasin-B was added and the cells were harvested at 90 h. Cytochalasin disrupts the cytoskeleton and dividing cells form multinucleated cells. In the HIV-1– cultures *(a)* some cells divide twice with four nuclei (large arrow) and other cells divide once (*). In the cell culture taken from an HIV-1-seropositive patient *(b)* only a single double nucleated cell is seen (*) amidst pyknotic nuclei. When cultured for longer periods of time such cells can 'take off' and divide repeatedly, probably corresponding to residual T-colony-forming cells reported to be present in HIV-1+ individuals [25].

the cultures. After 72–84 h of activation some T-blasts start to proliferate rapidly, corresponding to the population frequently referred to as T-colony-forming cells [25]. Cytochalasin-B permits this proliferation but the cytoskeleton is disrupted and dividing cells remain together as multinucleated forms. In this way it can be assessed how many times the cells have divided (fig. 5).

Quantitation of Transformation Defects in HIV-1 Infection

In the early phases of stimulation between 3 and 5 days among the cells taken from HIV-1 donors only few (<1%) expressed p24 proteins even when the analysis was focussed on blast cells – in agreement with published observations [26, 27]. In cultures incubated for longer periods of time, p24+ cells appeared in variable numbers and occasionally CD8+ blasts with p24 expression could also be detected [28]. The early defects of cell proliferation demonstrated below are not attributable to rampant acute HIV-1 production as was the case in previous seminal studies when activated lymphoblasts were exposed to large doses of live HIV-1 virus in vitro [29, 30].

During our studies both stimulation with anti-CD3 andibodies and PHA has been performed with similar results. PHA activates both 'naive' and 'memory' T cells [18, 19], while anti-CD3 preferentially activates CD45RO+ 'memory' cells [31]. As PHA activates larger proportions of T cells, in the review below these results are shown and observations with anti-CD3 are published elsewhere [32].

The results below are expressed in relation to CD4 lymphocyte counts (fig. 6) and according to the CDC staging (fig. 7). A total of 474 HIV-1+ individuals were studied including 74 haemophilic, 350 homosexual and 50 other patients. The 77 controls included 40 laboratory controls and 37 HIV-1– homosexuals. If repeatedly investigated, a single representative value is shown for each person. The quantitative assessment of blast transformation carried out with the method shown in figure 4 reveals a significantly poorer performance in HIV-1+ patients with normal (>600/mm^3) or near normal levels of CD4+ cells (400–600/mm^3) than observed in controls with similar normal CD4 cell counts. While in 82% of HIV-1– cases the PHA performance is >70%, in the HIV-1+ group with >600/mm^3 CD4 counts only 31% have good performance and 21% exhibit very poor (<30%) performance. There is a significant deterioration of performance values when HIV-1+ patients with >600, 200–600 and <200/mm^3 CD4 counts are compared. Surprisingly, the values between the 400–600 and 200–400/mm^3 groups of patients are not significantly different. We interpret this finding to show that the patients in the former group already demonstrate a low functional status with approximately half of people revealing <30% performance (fig. 6).

The early damage to lymphoid stimulation can also be confirmed when patients in the asymptomatic CDC2 category and in the group with persistent generalized lymphadenopathy (PGL; CDC3 category) are studied. In these groups the mean lymphocyte performance is 45 and 51%, respectively – with a further decrease in the CDC4 group to 26% level (fig. 7). The corresponding mean normal value is in the range of 95%.

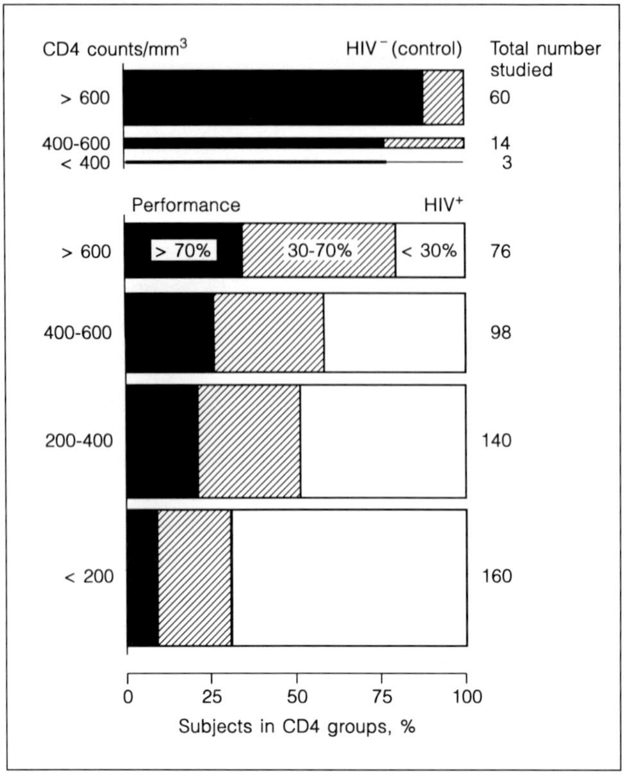

Fig. 6. The performance of lymphocytes taken from 77 HIV-1-seronegative donors including 39 homosexual males (HIV– control) and from 474 HIV-1-seropositive individuals (HIV+). The cultures were stimulated with PHA for 4 days in the presence of IL-2 and conditioned medium. The percentages of blast cells were quantitated as shown above in figure 4 and depicted as 'performance' values: > 70% (■), 30–70% (▨) and < 30% (□). The subjects are grouped by their CD4 counts (> 600, 400–600, 200–400 and < 200/mm³). The width of the bars in each group is proportional to the total number of cases studied in each group (see on the right). It can be seen that the lymphocytes from HIV-1+ cases perform relatively poorly irrespective of the CD4 counts.

Both CD4 and CD8 Lymphoid Cells Perform Poorly

The activation of CD4 and CD8 populations by PHA was separately measured by cell cycle analysis (fig. 2), blast transformation and IL-2Rα expression (fig. 3) and by morphology utilizing peroxidase membrane

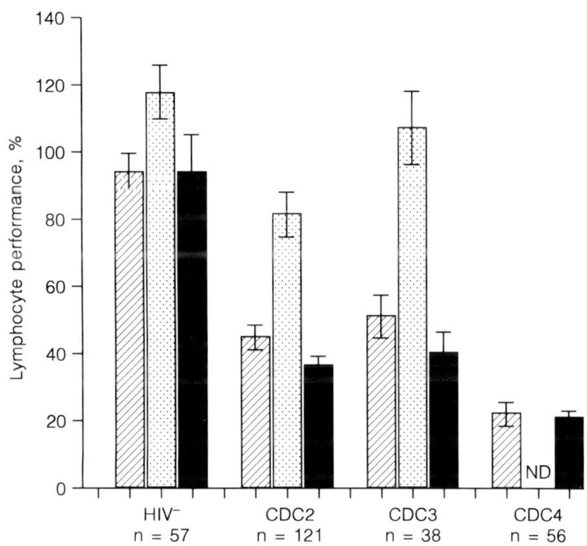

Fig. 7. In cultures of lymphocytes taken from 215 HIV-1+ individuals (▨) not only the CD4+ (▩) but also the CD8+ cells (■) fail to perform. The assay is shown in figure 4 and the performance of CD4+ and CD8+ cells are separately assessed in the CDC2, CDC3 and CDC4 groups as depicted. In the CDC4 category there were too few CD4+ cells for meaningful analysis. ND = Too few cells to read, value is not determined. The bars are means ± SEM.

labelling for CD8 antigens on Cytochalasin-B arrested lymphoblasts (fig. 5). In addition, during the performance test (fig. 4) CD4+ or CD8+ lymphoblasts were counted by selective gating (fig. 7). In all of these assays both the CD4+ and CD8+ lineages of HIV-1+ cases were affected despite the fact that rIL-2 and conditioned medium have been added.

There was a trend that CD8+ cells were more seriously affected than the CD4+ population (fig. 7). In the absence of inhibitors the numbers of CD4+ or CD8+ lymphoblasts seen were related to the original input of the same cell types, i.e. corrected for the CD4 lymphopenia seen in HIV-1+

cases. As CD4+ cells respond slightly faster to PHA than the CD8+ cells, by day 4 the normal values for CD4+ blasts have been 118%. The HIV-1+ values in the CDC2 group were significantly lower (78%) while CD4+ cells in the PGL group (CDC3) showed near normal values (103%). At the same time the performance of the CD8+ populations dropped from the 95% control values to 37 and 40% performance values in the CDC2 and the CDC3 groups, respectively. Obviously, as the numbers of CD4+ cells dwindle with time during HIV-1 infection, the lymphocyte performance in CDC4 cultures, around 18–20%, primarily reflects the features of residual CD8+ cells.

Poor Lymphocyte Performance in the Presence of Normal CD4+ Cells and Cytokines

As the proliferative stimulus given by PHA is stronger and the response is faster than seen in mixed lymphocyte responses, PHA response effectively masks allogeneic effects. A series of experiments on 6 normal donors and 12 HIV-1-seropositive patients have been performed by mixing equal numbers of cells from different donors and growing them in PHA-stimulated cultures for varying lengths of time. The pairs were selected by the virtue of their different HLA-A2 antigen expression. In Figure 8 the HLA-A2+ cells were from a HIV-1-seronegative person and the HLA-A2– cells were from a HIV-1-seropositive asymptomatic homosexual. The cells from the respective donors were identified in cultures by staining with an anti-HLA-A2 antibody used in double labelling combinations with CD8 or CD4. In mixtures of cells taken from HIV-1– pairs the contributions of the 2 donors (i.e. HLA-A2+ or HLA-A2–) remained around 50% throughout the 5-day cultures period as expected (not shown). Nevertheless, in the mixed 50% plus 50% cultures of HIV-1– and HIV-1+ donors harvested at days 3–5 the contribution of the seropositive partner declined to 27 and 18% values, respectively (fig. 8a). The rapid decline of lymphoblasts from the seropositive donor at day 3 cannot be explained by a HIV-1– proliferative burst because at this early period only a few lymphoblasts started to divide and the numbers of the HIV-1– cells had not yet increased substantially. Therefore, these results indicate that 50–80% of lymphocytes from HIV-1+ donors die 1–3 days after activation by PHA.

There was no evidence of suppressive activity from HIV-1+ donors' cells upon proliferating normal HIV-1– CD4+ or CD8+ populations [32].

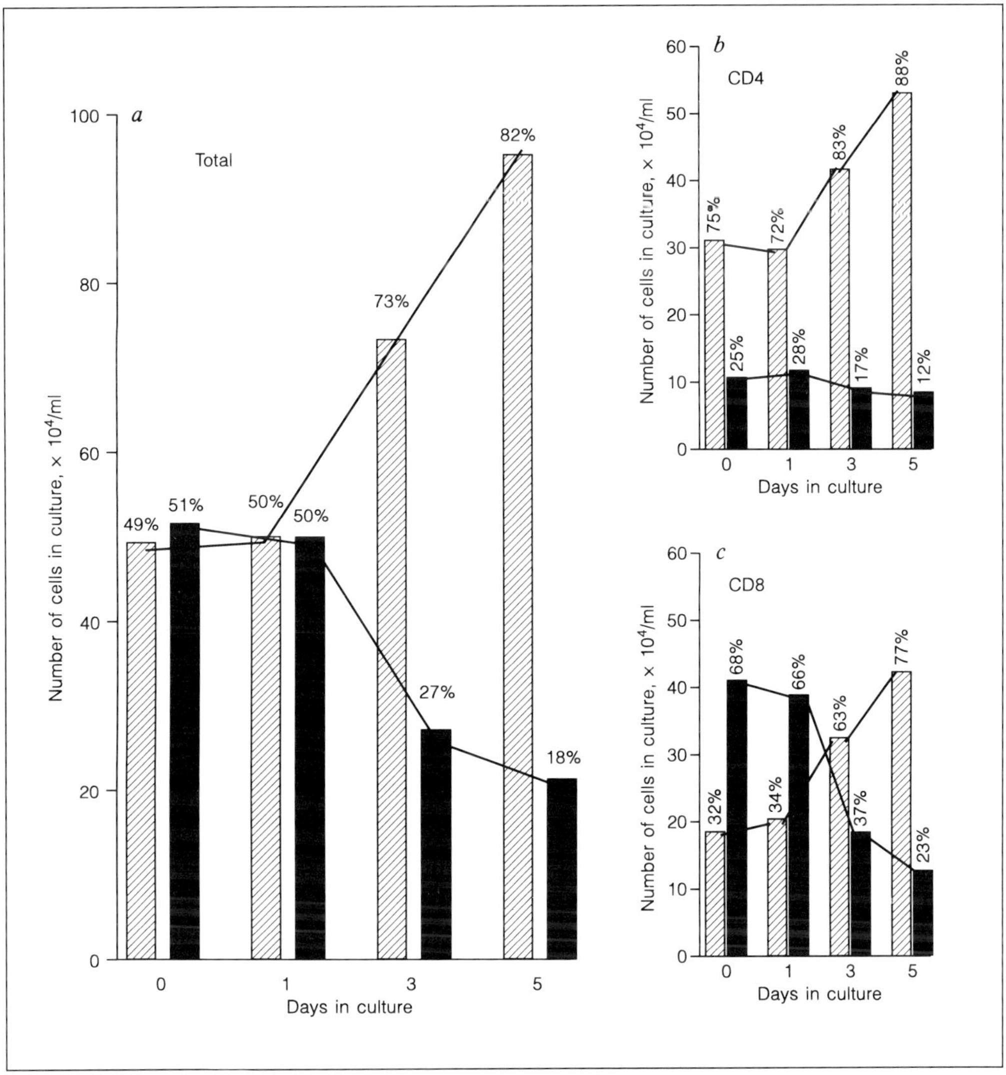

Fig. 8. a–c Both CD4 and CD8 lymphocytes taken from HIV-1+ individual show a proliferative handicap when compared to the cells from a HIV-1– donor which are grown in the same culture in the presence of IL-2. Lymphocytes from a HLA-A2+ HIV-1– donor (▨) were admixed to equal number of lymphocytes from an HIV-1+ asymptomatic person of a different HLA-A2– tissue type (■). The proportions of HLA-A2+ and HLA-A2– cells were then counted on the different days as shown – both within the CD4+ and CD8+ subsets. It can be seen that while the HLA-A2+ cells proliferate, the HLA-A2– cells decrease in number.

Simple calculations have correctly predicted the numbers of blasts seen in mixed cultures from the values already seen in cultures grown separately at different cell concentrations.

These mixing experiments also confirm the preferential demise of CD8+ populations taken from HIV-1-seropositive donors (fig. 8c). Due to their inverted CD4/CD8 ratios the initial input of CD8+ cells from seropositive patients is higher (68%) than that of the HIV-1– donors (32%). Despite the higher HIV-1+ CD8 starting values at day 0, during the 5-day culture period CD8+ cells decline to 23% while the proportions of HIV-1–, CD8+ cells increase to 77%. These values reflect cell death in the former and an active proliferation in the latter population within the same microenvironment in vitro. The addition of cytokines, such as IL-2, and IL-4, or antibodies to tumour necrosis factor (TNFα and β) and interferon-γ or the incubation of cells with enriched adherent macrophages do not influence the relative survival of the HIV-1– and HIV-1+ populations of lymphocytes, excluding the possibility that the bulk of these effects are cytokine mediated [32].

Cells of 'Memory' Type Are More Vulnerable

A great heterogeneity in the response pattern of CD8+ populations from HIV-1-seropositive donors has been revealed. During HIV-1 infection CD8+ cells with CD38 antigen expression accumulate [reviewed by Autran and Giorgi, this volume] and many of these also express CD45RO and HLA-DR antigens indicating 'memory' features [Janossy , Jiang et al., this volume]. Prince and Jensen [5] have demonstrated that cell death in unstimulated cultures of HIV-1+ donors occurred preferentially in subsets with phenotypes indicative of activation in vivo. Some of these cells with HLA-DR+, CD57+, CD38+ phenotypes could be rescued by adding IL-2, but others could not [7]. Pantaleo et al. [reviewed in this volume] have extended these studies and reported that freshly isolated CD8+ cells expressing HLA-DR gave poor proliferative responses to anti-CD3 mitogens even in the presence of irradiated allogeneic macrophages, IL2 and IL-4; the CD8+, HLA-DR– populations proliferated better. In our study we have analysed the time course of the survival of the CD45RA+ ('naive') and CD45RO+ ('memory' type) populations in PHA- and anti-CD3-activated cultures. The cultures at start contained either purified CD45RA+ or CD45RO+ lymphocytes; >80% of these were CD8+. At days 3–4 the

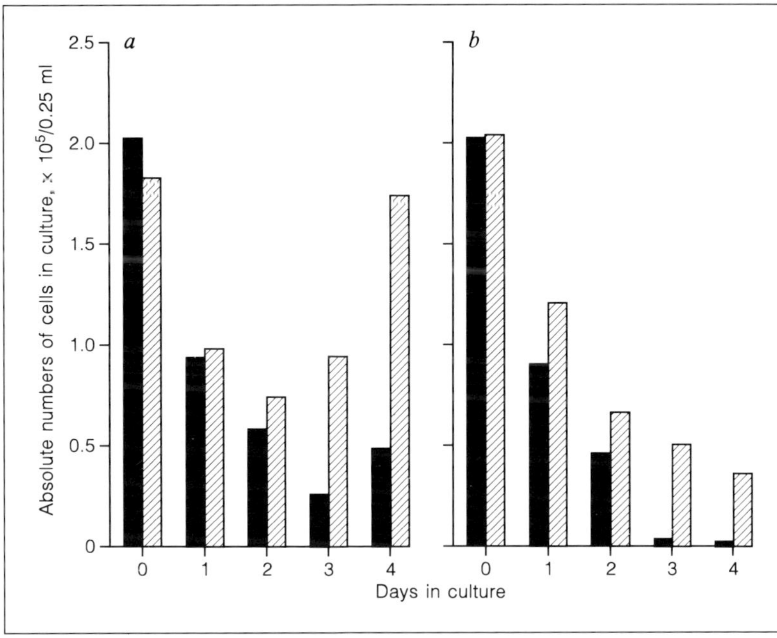

Fig. 9. The analysis of the performance within the CD45RA+ (▨) and CD45RO+ (■) populations. Two representative HIV-1-seropositive donors *(a, b)* with intermediate to low but still detectable performance values (50 and 30% at day 3) were selected and re-investigated for the survival of their negatively selected CD45RA+ and CD45RO+ cells following PHA stimulation in the presence of rIL-2. In both subjects the isolated CD45RA+ cells survive better than the CD45RO+ populations.

CD8+ cells were mostly blasts. As documented in representative experiments on 2 HIV-1+ donors in PHA cultures (fig. 9), CD45RA+ cells showed a preferential survival when compared to the CD45RO+ cells. In some patients the CD45RA+ population started to proliferate. This was shown by increasing cell counts at day 4 (fig. 9a), the presence of multinucleated CD8+ blasts, similar to the cell shown in figure 5b, and also by the recovery of thymidine uptake in cultures harvested at days 4–5 and later. These proliferating CD8+ cells mostly derive from a subset of resting CD45RA+, CD8+ cells which are still capable, under conditions in vitro, of switching to 'memory' cells along the lines of changes documented in figure 1 [33].

The Abortive Activation of Both CD4 and CD8 Populations

The final question here is to determine how rapidly the deteriorating changes manifest themselves in the activation cycle of lymphocytes from HIV-1+ donors. In cultures incubated with mitogens such as PHA and anti-CD3 the appearance of CD25 (IL-2Rα chain)-positive cells was analysed on CD4+ and CD8+ cells together with these cells' size (fig. 3). By day 1 of activation strong CD25 expression is induced in both the CD4+ (fig. 3b, f) and the CD8+ cells taken from HIV-1– and HIV-1+ donors. During days 2–3 in the HIV-1– cultures these cells progress to the CD25+ blast cells stage (fig. 3c, d) while in the samples from HIV-1+ patients a variable proportion, and frequently the majority, of CD25+ cells die (fig. 3g, h).

These observations are similar to the findings seen in a murine model of graft-versus-host disease (GvHR) elicited by the injection of class I/II MHC disparate parental cells into unirradiated F_1 recipient mice [34]. In this study, single cell methods, similar to ours, rather than bulk assays were utilized to document the development of immune dysfunction. Three weeks after GvHR initiation splenic T cells from GvHR mice displayed normal CD3 and IL-2R expression but elevated class II antigens. Despite culture with normal F_1 accessory cells, both CD4+ and CD8+ GvHR T cells exhibited low levels of proliferation to both Con-A, a strong mitogen, and anti-CD3, a moderately powerful mitogen. Following exposure for 12 h to these stimuli, GvHR T cells initially expressed normal levels of IL-2R, but between 24 and 48 h IL-2R expression decreased. At no time-point was secretion of IL-2 detectable by GvHR T cells, an observation which is paralleled in human HIV-1 infection where IL-2 secretion in subjects with ARC and AIDS is regularly low [35, 36]. Finally, in this animal model no suppressive activity in the purified GvHR T cell subsets could be detected in mixing experiments and the proliferative activity could not be restored by cytokines of Con-A supernatants or rIL-7, a finding similar to observations seen above.

Conclusions

These observations extend previous studies by optimally activating T lymphocytes with powerful mitogens in the presence of cytokines and quantitating the proliferative handicap of both CD4+ and CD8+ populations which already develop in asymptomatic patients in vivo following HIV-1 infection. The observations, which document a severe disorder in PHA and anti-CD3 stimulated cultures providing that the assays are standardized to assess the performance of individual cells, do not contradict previous observations which point out that under other stimulating conditions the disorders might be even more pronounced. Other factors further contribute to immunodeficiency measured in more complex and sensitive assays, and these include cytokine deficiency [5, 14, 27, 35, 36], defects of antigen presentation by monocytic cells [10, and reviewed by Knight et al., this volume] and suppressor factors [37, and reviewed by Sadat-Sofwi et al., this volume]. Nevertheless, we believe these are additional mechanisms which further aggravate an already damaged immune status.

The cells might already be committed in the patients to a suicidical pathway of behaviour, which do not seem to be altered by the other cells also present in the culture. Such an interpretation is in line with the demonstration of poor viability in unstimulated cultures [reviewed by Gougeon et al., this volume]. Nevertheless, this cell death is not an early event but appears to be predominant only after the cells initiate their CD25 expression in their membrane, similarly to animal models of certain types of GvHR [34]. It will be particularly interesting to investigate the events in these clinical and experimental conditions in vivo but the answer(s) have to be found not only for CD4 defects but also for the damage seen in the CD8 cell populations. The quantitative tests described above will be useful for these clinical aims.

References

1 Prince HE, Czaplicki CD: Preferential loss of Leu-8–, CD45R–, HLA-Dr+ CD8 cell subsets during in vitro culture of mononuclear cells from HIV-1-seropositive former blood donors. J Clin Immunol 1989;9:421–428.

2 Lane HC, Fauci AS: Immunologic abnormalities in the acquired immunodeficiency syndrome. Annu Rev Immunol 1985;3:477–500.

3 deShazo RD, Morgan JE, Daul CB, Andes A, Hyslop N, Palmer C: Studies of spontaneous proliferation of mononuclear cells in human immunodeficiency virus (HIV-1)-infected individuals. J Allergy Clin Immunol 1989;84:34–39.

4 Hornicek FJ, Malinin GI, Thornthwaite JT, Whiteside ME, MacLeod CL, Malinin TI: Effect of mitogens on the cell cycle progression and the quantification of T-lymphocyte surface markers in acquired immune deficiency syndrome. J Leukoc Biol 1987;42:122–127.

5 Prince HE, Jensen ER: HIV-related alterations in CD8 cell subsets defined by in vitro survival characteristics. Cell Immunol 1991;134:276–286.

6 Redfield RR, Tramont EC: Toward a better classification system for HIV infection. N Engl J Med 1989;320:1414–1416.
7 Birx DL, Rhoads JL, Wright JC, Burke DS, Redfield RR: Immunologic parameters in early-stage HIV-seropositive subjects associated with vaccine responsiveness. J Acquir Immune Defic Syndr 1991;4:188–196.
8 Opravil M, Fierz W, Matter L, Blaser J, Luthy R: Poor antibody response after tetanus and pneumococcal vaccination in immunocompromised, HIV-infected patients. Clin Exp Immunol 1991;84:185–189.
9 Klimas NG, Caralis P, LaPerriere A, Antoni MH, Ironson G, Simoneau J, Schneiderman N, Fletcher MA: Immunologic function in a cohort of human immunodeficiency virus type 1-seropositive and -negative healthy homosexual men. J Clin Microbiol 1991;29:1413–1421.
10 Melendez Guerrero LM, Nicholson JK, McDougal JS: Infection of humanmonocytes with HIV-1Ba-L. Effect on accessory cell function for T-cell proliferation in vitro. AIDS Res Hum Retroviruses 1991;7:465–474.
11 Prince HE, Czaplicki CD: In vitro activation of T lymphocytes from HIV-seropositive blood donors. II. Decreased mitogen-induced expression of interleukin-2 receptor by both CD4 and CD8 cell subsets. Clin Immunol Immunopathol 1988;48:132–139.
12 Odum N, Hofmann B, Jacobsen N, Langhoff E, Moller J, Platz P, Ryder LP, Svejgaard A: The immunodeficiency of bone marrow-transplanted patients. II. CD8-related suppression by patient lymphocytes of the response of donor lymphocytes to mitogens, antigens, and allogeneic cells. Scand J Immunol 1987;26:247–253.
13 Creemers PC, O'Shaughnessy M, Boyko WJ: Analysis of absolute T helper cell number and cellular immune defects in HIV antibody positive and negative homosexual men, AIDS Res Hum Retroviruses 1988;4:269–278.
14 Gruters RA, Terpstra FG, De Jong R, Van Noesel CJ, Van Lier RA, Miedema F: Selective loss of T cell functions in different stages of HIV infection. Early loss of anti-CD3-induced T cell proliferation followed by decreased anti-CD3-induced cytotoxic T lymphocyte generation in AIDS-related complex and AIDS. Eur J Immunol 1990;20:1039–1044.
15 Quinti I, Palma C, Guerra EC, Gomez MJ, Mezzaroma I, Aiuti F, Cassone A: Proliferative and cytotoxic responses to mannoproteins of *Candida albicans* by peripheral blood lymphocytes of HIV-infected subjects. Clin Exp Immunol 1991;85:485–492.
16 Schellekens PT, Roos MT, De Wolf F, Lange JM, Miedema F: Low T-cell responsiveness to activation via CD3/TCR is a prognostic marker for acquired immunodeficiency syndrome (AIDS) in human immunodeficiency virus-1 (HIV-1)-infected men. J Clin Immunol 1990;10:121–127.
17 Akbar AN, Timms A, Janossy GJ: Cellular events during memory T cell activation in vitro: the UCHL1 (180,000 MW) determinant is newly synthesized after mitosis. Immunology 1989;66:213–218.
18 Sanders ME, Makgoba MW, Sharrow SO, Stephany D, Springer TA, Young HA, Shaw S: Human memory T lymphocytes express increased levels of three cell adhesion molecules (LFA-3, CD2 and LFA-1) and three other molecules (UCHL1, CDw29 and Pgp-1) and have enhanced IFN-gamma production. J Immunol 1988;140:1401–1407.

19 Akbar AN, Terry L, Timms A, Beverley PCL, Janossy G: Loss of CD45R and gain of UCHL1 reactivity is a feature of primed T cells. J Immunol 1988;140:2171–2178.
20 Beverley PCL: Is T cell memory maintained by cross-reactive stimulation? Immunol Today 1990;11:203–205.
21 Vindelov LL, Christensen IJ, Nissen NI: Standardization of high-resolution flow cytometric DNA analysis by the simultaneous use of chicken and trout red blood cells as internal reference standards. Cytometry 1983;3:328–331.
22 Baisch H, Gerdes J: Simultaneous staining of exponentially growing versus plateau phase cells with the proliferation-associated antibody Ki-67 and propidium iodide: analysis by flow cytometry. Cell Tissue Kinet 1987;20:387–391.
23 Bofill M, Janossy G: Quantitation of proliferative activity in unstimulated and stimulated cultures of lymphocytes from HIV-1-infected persons. J Immunol Methods, submitted.
24 Campana D, Coustan-Smith E, Janossy G: Double and triple staining methods for studying the proliferative activity of human B and T lymphoid cells. J Immunol Methods 1988;107:79–88.
25 Winkelstein A, Kingsley LA, Weaver LD, Machen LL: Defective T cell colony formation and IL-2 receptor expression in HIV-infected homosexuals: relationship between functional abnormalities and CD4 cell numbers. J Acquire Immune Defic Syndr 1989;2:353–358.
26 Schnittman SM, Psallidopoulos MC, Lane HC, Thompson L, Baseler M. Massari F, Fox CH, Salzman NP, Fauci AS: The reservoir for HIV-1 in human peripheral blood is a T-cell that maintains expression of CD4. Science 1989;245:305–308.
27 Bentin J, Tsoukas CD, McCutchan JA, Spector SA, Richman DD, Vaughan JH: Impairment in T-lymphocyte responses during early infection with HIV. J Clin Immunol 1989;9:159–168.
28 Tsubota H, Ringler DJ, Kannagi M, King NW, Solomon KR, Mackey JJ, Walsh DG, Letvin NL: CD8+CD4– lymphocyte lines can harbor the AIDS virus in vitro. J Immunol 1989;143:858–863.
29 Zagury D, Bernard J, Leonard R, Cheynier R, Feldman M, Sarin PS, Gallo RC: Long-term cultures of HTLV-III– infected T cells: a model of cytopathology of T-cell depletion in AIDS. Science 1986;231:850–853.
30 Laurence J, Friedman SM, Chartash EK, Crow MK, Posnett DN: Human immunodeficiency virus infection of helper T cell clones. Early proliferative defects despite intact antigen-specific recognition and interleukin-4 secretion. J Clin Invest 1989; 83:1843–1848.
31 Akbar AN, Salmon M, Janossy G: The synergy between naive and memory T cells during activation. Immunol Today 1991;12:184–188.
32 Janossy G, Bofill M, Lumnitzer R, Bortwick N, Squire B, Johnson M, Lee C: Early inhibition of $CD8^+$ cell proliferation in asymptomatic HIV-1-positive patients – a quantitative study (in press).
33 Pinto L, Covas MJ, Victorino RM: Loss of CD45RA and gain of CD45RO after in vitro activation of lymphocytes from HIV-infected patients. Immunology 1991;73: 147–150.
34 Levy RB, Jones M, Cray C: Isolated peripheral T cells from GvHR recipients exhibit defective IL-2R expression, IL-2 production, and proliferation in response to activation stimuli. J Immunol 1990;145:3998–4005.

35 Murray HW, Welte K, Jacobs JL, Rubin BY, Mertelsmann R, Roberts RB: Production of and in vitro response to interleukin-2 in the acquired immunodeficiency syndrome. J Clin Invest 1985;76:1959–1964.

36 Kirkpatrick CH, Davis KC, Horsburgh CR Jr, Cohn DL, Penley K, Judson FN: Interleukin-2 production by persons with the generalized lymphadenopathy syndrome or the acquired immune deficiency syndrome. J Clin Immunol 1985;5:31–37.

37 Ammar A, Cibert C, Bertoli AM, Tsilivakos V, Jasmin C, Georgoulias V: Biological and biochemical characterization of a factor produced spontaneously by adherent cells of human immunodeficiency virus-infected patients inhibitin IL-Rα chain (Tac) expression on normal T cells. J Clin Invest 1991;87:2048–2055.

Prof. G. Janossy, Department of Clinical Immunology, The Royal Free Hospital, School of Medicine, Pond Street, GB–London NW3 2QG (UK)

Janossy G, Autran B, Miedema F (eds): Immunodeficiency in HIV Infection and AIDS.
EC/FERS/MRC Workshop on Immunodeficiency in HIV-1 Infections, Windsor, Surrey, 1991.
Basel, Karger, 1992, pp 115–126

Apoptosis as a Mechanism of Cell Death in Peripheral Lymphocytes from HIV-1-Infected Individuals

Marie-Lise Gougeon, Sylvie Garcia, Denise Guétard, René Olivier, Charles Dauguet, Luc Montagnier

Unité d'Oncologie Virale et URA CNRS 1153, Département SIDA et Retrovirus, Institut Pasteur, Paris, France

Infection with HIV causes a wide variety of cellular and humoral immunologic abnormalities. One of the major obstacles in understanding the pathology of HIV infection is to explain how the presence of the HIV genome in a minority of cells leads to the progressive destruction of the immune system. The loss of CD4+ T cells appears to result primarily from their destruction following infection, whether by viral cytopathic effects, host-mediated cytotoxicity or both [1]. However, during the asymptomatic phase of HIV infection, the magnitude of immunologic defects observed in infected individuals is disproportionately high in comparison to the level of infectious virus in these patients, suggesting that factors other than direct virus-induced cytopathogenicity may be involved.

Cell death in a variety of development systems involves chromatin condensation, membrane blebbing and fragmentation of DNA after the activation of a calcium-dependent endogenous endonuclease. These events have been collectively referred to as apoptosis [2] and are believed to represent a mechanism of programmed cell death (PCD) which normally occurs during certain stages of development. Apoptosis has been shown to play an essential role during embryogenesis and adult life in tissue turnover including control of hemopoietic stem cell growth [3] and negative intrathymic selection of the T cell repertoire [4].

In preliminary experiments we had observed that peripheral blood lymphocytes (PBLs) from AIDS patients or from HIV-infected asymptomatic individuals displayed a rapid loss of viability upon culture in medium deprived of growth factors [5]. We report here that this cellular death occurs following endonucleosomal DNA fragmentation, characteristic of apoptosis. Furthermore, this death can be prevented by the in vitro stimulation of patients' lymphocytes with cytokines. These results suggest that in vivo this death process migth contribute, independently of any HIV-mediated cytopathic effect, for the progressive depletion of CD4+ T cells that leads to AIDS.

Patients Studied

The experiments studying the in vitro loss of viability of PBLs from patients were done on a large number of patients at various stages as detailed in figure 1. Patients were staged according to CDC surveillance definition [6]. The study of PCD was realized with PBLs from 16 HIV-seronegative blood donors and 34 HIV-1-seropositive patients. These patients corresponded to primo-infected subjects (stage I, n = 3), asymptomatic carriers (stage II, n = 10), lymphadenopathy-associated syndrome (stage III, n = 3), aids-related complex (stage IVA, n = 5), AIDS (stage IVC, IVD, n = 13). Most ARC and AIDS patients (90%) were receiving AZT. Some of the patients were included more than once in the PCD study.

Poor Survival of Stimulated Lymphocytes from HIV-1-Positive Donors

We had observed that PBLs from HIV-infected individuals died prematurely when cultured in a medium devoid of growth factors: after a 3-day culture the loss of viability (as evidenced by the trypan blue exclusion method) of lymphocytes from HIV-infected patients could reach up to 60% while it was less than 10% in lymphocytes from HIV-negative blood donors, under the same conditions (fig. 1). Furthermore, this in vitro induced premature death was observed in patients' lymphocytes even at the asymptomatic stage and was maintained in the ARC and AIDS stages (fig. 1).

To analyze the phenomenon further, we developed a viability test consisting of double staining with two nucleic acid dyes analyzed by flow cytometry (FACS). Ethidium bromide stains mostly DNA of dead cells,

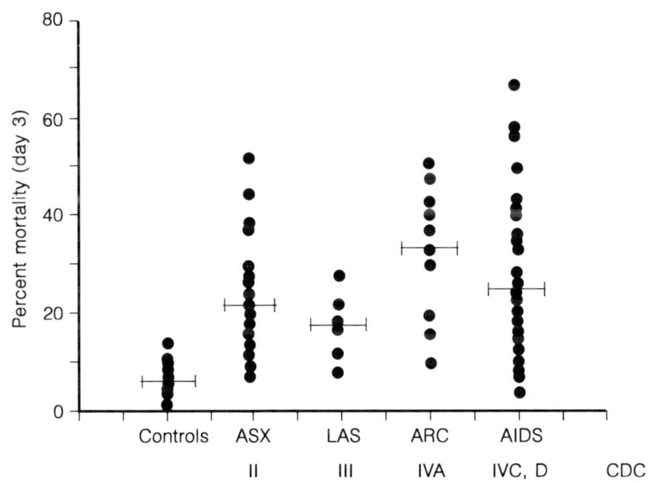

Fig. 1. In vitro loss of viability of lymphocytes from HIV-infected patients as compared to seronegative donors. PBLs from HIV-infected individuals and from controls were cultured during 72 h at 10^6 cells/ml in culture medium composed of RPMI-1640, 1 m*M* glutamine, 10 m*M* Hepes, 1% penicillin/streptomycin, 10% FCS. Cell death was assessed by trypan blue permeability and each circle represents a single donor lymphocyte culture. Patients were staged according to CDC surveillance definition: stage II, n = 22; stage III, n = 6; stage IVA, n = 11; stage IVC/IVD, n = 30; controls, n = 12.

while acridine orange stains nucleic acids of living cells. Immunofluorescence analysis of a 3-day culture of patient PBLs double stained with the above mentioned dyes, revealed the appearance of a living population with a poor staining by acridine orange (fig. 2). This population was either absent in cultures of PBLs from seronegative donors or was found in a much lower proportion than in cultures of PBLs from seropositive donors. The forward scatter analysis of the low stained population indicated the smaller size of these cells as compared with cells normally stained by the nucleic acid dye (not shown).

In order to determine if the appearance of the low stained population could be correlated with the evolution of the disease, we defined the ratio

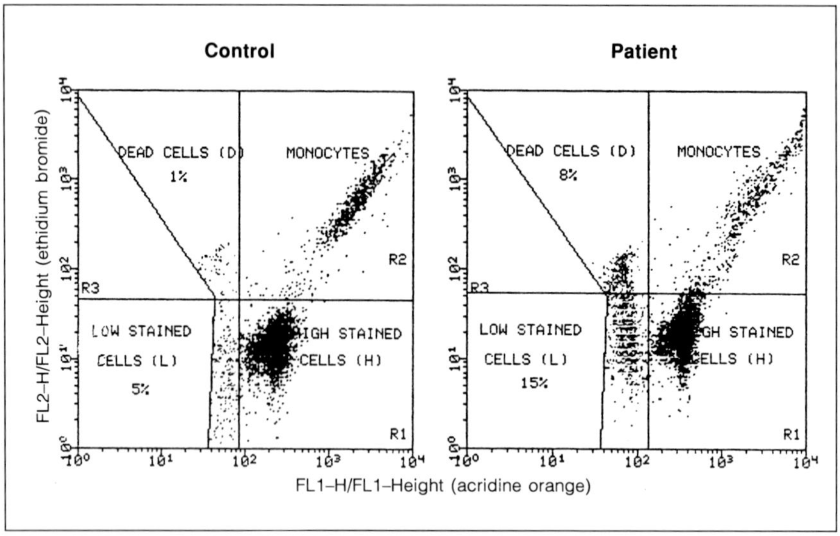

Fig. 2. Flow cytometry analysis of double stained lymphocytes. PBLs were cultured as described in figure 1 and stained with ethidium bromide (10 μg/ml) which enters the dead cells, and acridine orange (3 μg/ml) which stains nucleic acids of living cells. Stained cells were then analyzed on a FACScan (Becton-Dickinson). The cell population in the upper right angle of each plot is an aspecific labeled monocyte population. Dead cells are double stained by both nucleic acid dyes. In a 3-day culture of lymphocytes from a seronegative donor, the majority of cells is living and highly stained by acridine orange. In the culture of patient lymphocytes, more dead cells are found and a second living cell population, with a weak staining by orange acridine, appears in greater proportion than in control cultures.

D+L/H (% Dead cells + % Low stained cells/% High stained cells) which reflects a process leading to premature cell death. As shown in figure 3a, significant correlation between the magnitude of the ratio D+L/H and the ratio of CD4/CD8 cells could be demonstrated.

The significance of the emergence, during the culture of patient PBLs, of an additional living population weakly stained by acridine orange, could be explained by several hypotheses and among them we favored the idea that this population would represent cells committed to PCD. The reduction in acridine orange incorporation would then be due to the condensation of chromatin in the nuclei of these cells, a first step towards DNA fragmentation and the cell death [2].

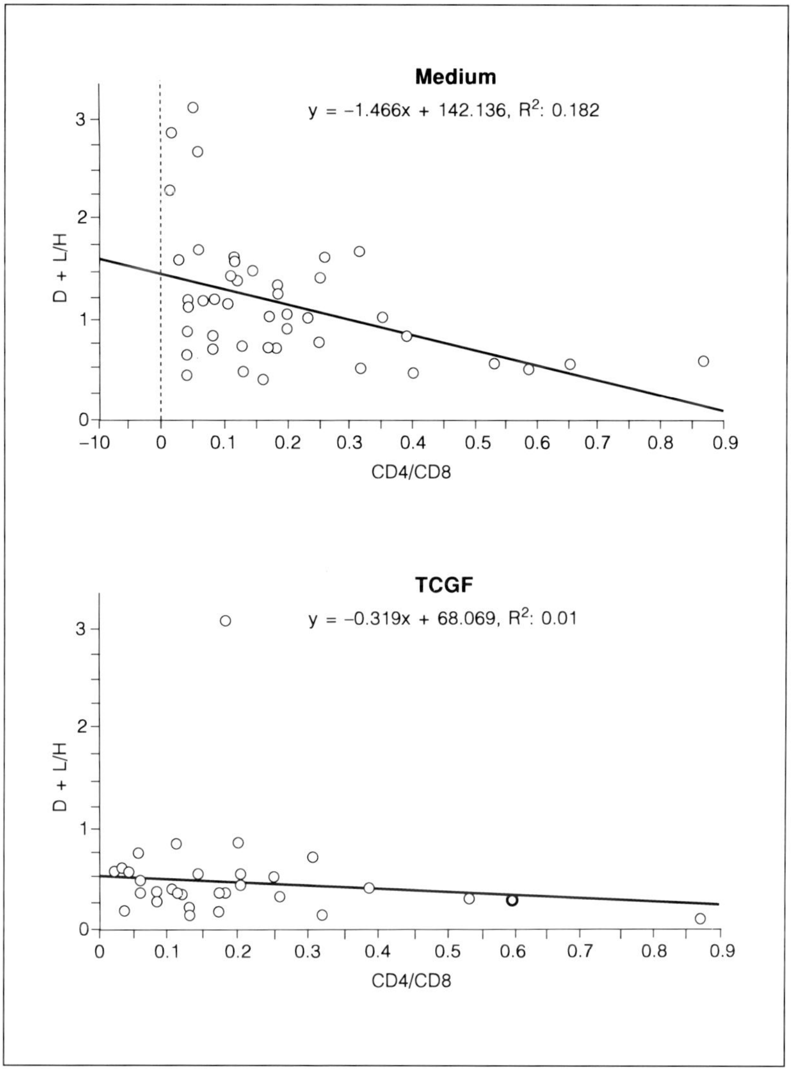

Fig. 3. Correlation between the ratio D+L/H and the ratio CD4/CD8 in HIV-infected patients. Three-day cultures were performed as described in figure 1 with PBLs from HIV-infected patients at various stages of the disease. The cultured PBLs were then stained as described in figure 2 and analyzed by flow cytometry. The ratio D+L/H was then determined where D corresponds to the % dead cells, L to the % of low stained cells and H to the % of high stained cells with acridine orange. The upper part of the figure corresponds to cultures performed in medium and the lower part to cultures performed in the presence of TCGF (Biotest, lymphocult, 10 U/ml).

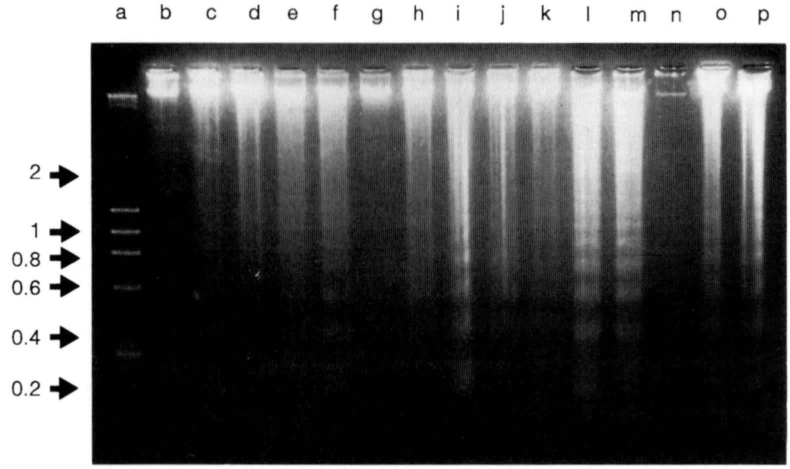

Fig. 4. Apoptosis of PBLs from HIV-infected individuals. Lymphocytes from seronegative donors (lanes b–d, k) or from seropositive patients (lanes e–j, l–p) were cultured during 3 days in the absence of activating agents (lanes k–p) or during 24 h in the presence (lanes c, f, i) or in the absence (lanes b, e, h) of ionomycin (Calbiochem, 1 μg/ml) or in the presence of ionomycin and TCGF (lanes d, g, j). At the end of the culture, cells were lyzed in a buffer containing SDS (10%) and proteinase K (10 mg/ml) and DNA was extracted by phenol-chloroform. After overnight incubation at −80 °C, DNA samples were incubated with RNAse A (0.5 mg/ml) and loaded on a 2% agarose gel. Electrophoresis was carried out during 6 h under a 50 V voltage. Molecular sizes in kilobases are indicated on the left of the gel in lane a. Oligonucleosomal fragments appear as ladders of degraded DNA bands whose molecular sizes are approximately multiples of 200 bp.

PCD among Lymphocytes from HIV-1-Positive Donors

The most characteristic feature of the activation of a PCD process is the fragmentation of cellular DNA at internucleosomal sites into regular multiples of an oligonucleosomal lenght unit of 200 bp [2, 7–9]. To determine whether in vitro death of patients' lymphocytes involves the process of apoptosis, DNA samples from patients' BPLs cultured for 3 days in the absence of exogenous stimuli were isolated and analyzed by gel electrophoresis: a characteristic 200-bp nucleosomal ladder was detectable in the DNA prepared from cultured patient's lymphocytes and not in the DNA from seronegative donors' cultures (fig. 4, lanes k–p). The phenomenon of apoptosis in patients' lymphocytes could be accelerated by increasing the

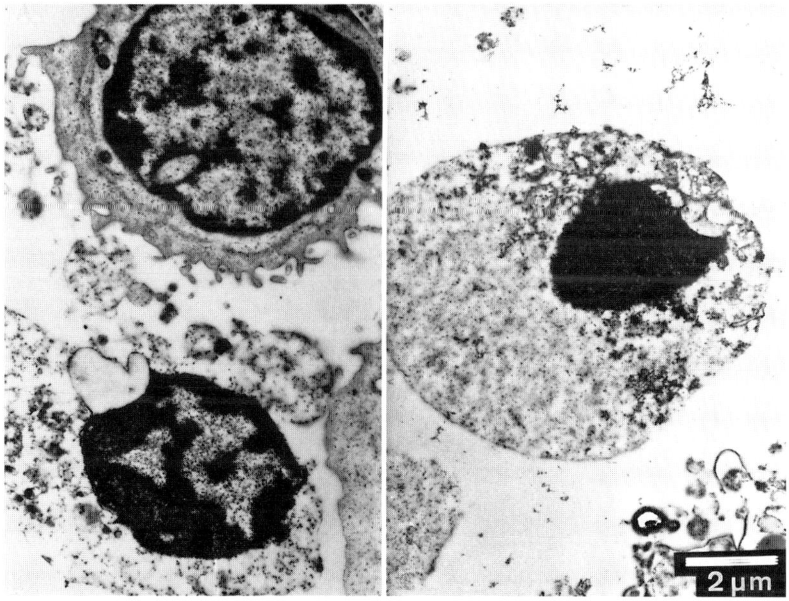

a *b*

Fig. 5. Electron micrograph of ultrathin sections (uranyl acetate staining) from HIV+ patient's lymphocytes cultured during 24 h in the presence of ionomycin. *a*, upper part: normal lymphocyte; lower part: lymphocyte with condensed chromatin. *b* Apoptotic lymphocyte.

intracellular Ca^{2+} mobilization by stimulating lymphocytes with ionomycin during 24 h, this Ca^{2+} ionophore being known to activate the endogenous endonuclease and to induce apoptosis in normal immature thymocytes [7, 10] (fig.4, lanes c, f, i). The implication of apoptosis in cell death of lymphocytes from HIV-infected patients was confirmed by the electron microscopy of lymphocytes stimulated during 24 h by ionomycin (fig. 5) [7]. Characteristic pictures of chromatin hypercondensation were observed.

To further identify the cell population undergoing PCD, experiments were done in which ionomycin-stimulated patients' lymphocytes were depleted either of CD4+ cells or of CD8+ cells prior to DNA extraction and gel electrophoresis. Preliminary results indicated that PCD was observed in both CD4+ and CD8+ populations (not shown).

In order to understand the implication of the PCD process in the evolution of the disease, we performed analysis of PCD on lymphocytes

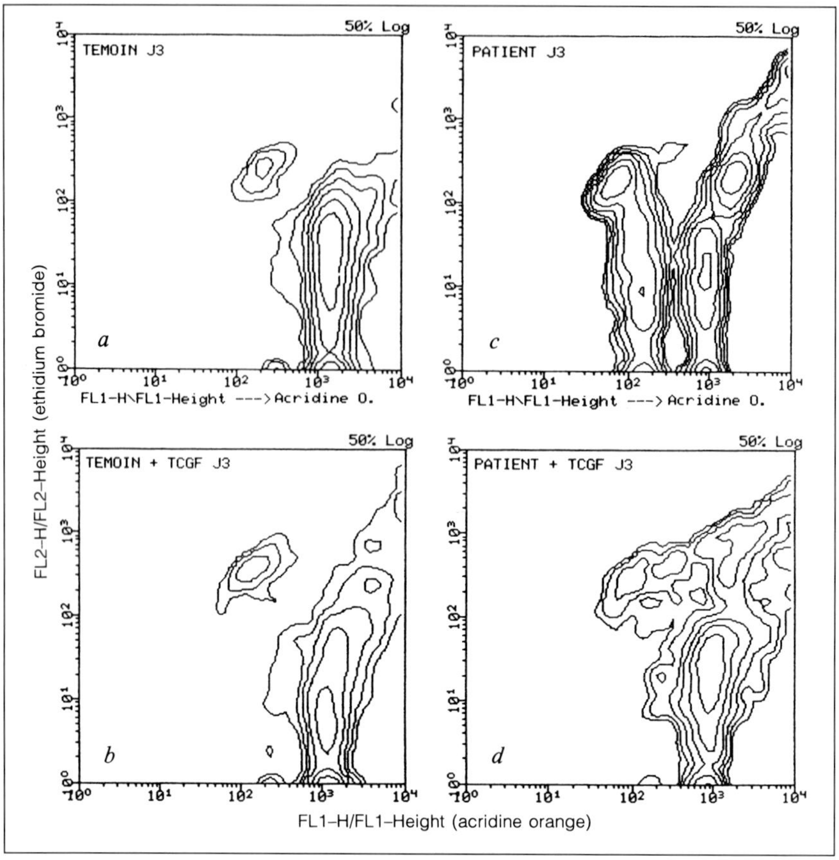

Fig. 6. Influence of TCGF on the apparition of the population weakly stained by acridine orange. PBLs from an HIV-infected patient or from a seronegative donor were cultured during 3 days in the presence *(b, d)* or in the absence *(a, c)* of TCGF (10 U/ml) and stained for analysis by flow cytometry as described in figure 2. Contour curves correspond, from the center to the periphery, to cellular density levels. Stimulation of patient's lymphocytes with TCGF *(d)* induce the disappearance of the population weakly stained with acridine orange observed in unstimulated cultures *(c)*.

from HIV-infected patients at different stages of the disease. PCD was observed very early after primo-infection. A longitudinal study of a patient recently infected revealed that PCD is already observed in vitro a few weeks after HIV infection and then follows a period where lymphocytes behave as seronegative controls. It appears that this death programme occurs discontinuously all along the disease. We also noticed that PCD is observed at all stages of the disease and whether the patient received AZT or not. Among 31 HIV-seropositive patients classified from stage II to stage IVC/IVD, only 3 of them had PBLs which behaved in vitro as the seronegative controls. Longitudinal studies of patients at various stages of the disease would allow to identify clinical events associated in vivo with this programme leading to apoptosis.

Signals Influencing PCD

A Ca^{2+} increase appears to serve as a trigger for PCD in patients' lymphocytes. Sustained Ca^{2+} increments, such as those observed in response to antigens or mitogenic lectins, also mediate lymphocyte proliferation [11]. However, these signals are only mitogenic when they are accompanied by other independent signals delivered by accessory cells and include PKC activation [11]. Agents that stimulate PKC, including IL-1, block endonuclease activation and PCD in mouse thymocytes [12]. We thus studied the influence of a crude preparation of cytokines secreted by PHA-activated T lymphocytes (TCGF) on the in vitro viability of patients' lymphocytes. Addition of TCGF at the initiation of the culture prevented the apparition of the population weakly stained by acridine orange (fig. 3, 6) and in parallel prevented the ionomycin-induced apoptosis (fig. 4, lanes g, j). The same protection was obtained in the absence of ionomycin in a 3-day culture (not shown). The identification of the cytokine(s) responsible for this protection is under investigation.

This study suggests that, in HIV-infected individuals, a relatively large fraction of lymphocytes are programmed for death by apoptosis, a process which implies that these lymphocytes are primed in vivo by mechanisms probably related directly or indirectly to HIV infection.

In immature thymocytes, PCD is the consequence of incomplete signal transduction related to the nature of antigen-presenting cells and to the absence of certain cosignals [13]. In HIV-infected patients the process of apoptosis may be related to abortive cell activation. It has been proposed that interaction of HIV gp120 envelope protein released in the serum with CD4 leads to defective signal transduction in several cell populations including Th cells [14]. This defective activation has recently been observed in lymphocytes from HIV-infected patients at various stages of the disease [15]. Soluble gp120 alone has been reported to suppress T cell activation [16] and the binding of gp120 to CD4 inhibits CD3-TCRαβ signal transduction pathway [17]. Recent experiments in the murine model have suggested that an altered signal in mature T cells leads to the re-emergence of the cell deletion program: Newell et al. [18] have shown that when preceded by ligation of CD4, signalling through TCRαβ results in T cell unresponsiveness due to the induction of activation-dependent cell death by apoptosis. In HIV-infected patients, the priming of T cells for a PCD response following activation could be induced by the virus itself [19] or by any CD4-triggering molecule such as gp120 or anti-CD4 antibodies [20].

Our results suggest that both CD4+ and CD8+ T cells are primed for PCD. PCD in CD8 T cells points to a CD4-independent mechanism, perhaps mediated by other viral proteins released from infected cells. These viral proteins could bind to uninfected cells and induce changes in signal transduction pathways leading to cell activation [14]. Indeed, activation of CD8+ T cell is seen in HIV infection [21] and could not only represent a CD8+ T cell response to HIV but also correspond to a defective activation leading to PCD. The identification of signals involved in vivo in the priming of T cells for PCD in HIV-infected individuals and the role of apoptosis in the progressive immunodeficiency which follows HIV infection remain important areas for future investigations.

Conclusions

PBLs from AIDS patients or from infected asymptomatic individuals display a rapid loss of viability upon culture in survival medium. This cell death occurs following endonucleosomal DNA fragmentation into multiples of a 200-bp unit, preceded by chromatin condensation and a lower staining with acridine orange. These features are characteristics of a programmed mechanism of cell death (PCD) referred to as apoptosis. The possible induction of apoptosis by Ca^{2+} ionophore in patients' lymphocytes is consistent with the active process of PCD. Furthermore, apoptosis could be prevented by the in vitro stimulation of lymphocytes with cytokines. Altogether, these results suggest that in vivo a significant fraction of lymphocytes from HIV-infected patients are committed for a suicide process by apoptosis, resulting from an abnormal activation due to defective or incomplete signal transduction.

Acknowledgments

This work was supported by grants from the ANRS, the CNRS and the Pasteur Institute. The authors wish to express their thanks to C. Axler, S. Chamaret, M.C. Cumont, H. Lecoeur and V. Rame for technical assistance; Drs. M. Adams, J.M. Bechet and O. Lopez for their contribution in the statistical analysis of the data; Drs. B. Dupont, G. Pialoux, G. Gonzales, R. Roue, M. Kierstetter and A.M. Berthier for their collaboration in the clinical aspects of this work (Hôpital de l'Institut Pasteur, Hôpital Begin, Centre médical Rey-Leroux).

References

1 Fauci AS: The human immunodeficiency virus: infectivity and mechanisms of pathogenesis. Science (Wash) 1988;239:617–622.
2 Wyllie AH, Kerr JFR, Currie AR: Cell death: the significance of apoptosis. Int Rev Cytol 1980;68:251–305.
3 Williams GT, Smith CA, Spooncer E, Dexter M, Taylor DR: Haemopoietic colony-stimulating factors promote cell survival by suppressing apoptosis. Nature 1990; 343:76–79.
4 Smith CA, Williams GT, Kingston R, Jenkinson EJ, Owen JJT: Antibodies to CD3/T cell receptor complex induce death by apoptosis in immature T cell in thymic cultures. Nature 1989;337:181–184.
5 Montagnier L, Guetard D, Rame V, Olivier R, Adams M: Virological and immunological factors of AIDS pathogenesis. Quatrième colloque des cents gardes, 1989, pp 11–17.
6 Center for Disease Control: Classification system for human T-lymphotropic virus type III/lymphadenopathy associated virus infections. Ann Intern Med 1986;105: 289–291.
7 Wyllie AH, Morris RG, Smith AL, Dunlop D: Chromatin cleavage in apoptosis: associated with condensed chromatin morphology and dependence on macromolecular synthesis. J Pathol 1984;142:67–77.
8 Duvall E, Wyllie AH: Death and the cell. Immunol Today 1986;7:115–119.
9 Arends MJ, Morris RG, Wyllie AH: Apoptosis. The role of the endonuclease. Am J Pathol 1990;136:593–607.
10 McConkey DJ, Hartzell P, Nicotera P, Orrenius S: Calcium activated DNA fragmentation kills immature thymocytes. FASEB J 1989;3:1843–1849.
11 Crabtree GR: Contingent genetic regulatory events in T lymphocyte activation. Science 1989;243:355–361.
12 McConkey DJ, Hartzell P, Amador-Perez JF, Orrenius S, Jondal M: Calcium-dependent killing of immature thymocytes by stimulation via the CD3/T cell receptor complex. J Immunol 1989;143:1801–1806.
13 McConkey DJ, Orrenius S, Jondal M: Cellular signalling in programmed cell death (apoptosis). Immunol Today 1990;11:120–121.
14 Pinching AJ, Nye K: Defective signal transduction – a common pathway for cellular dysfunction in HIV infection? Immunol Today 1990;11:256–259.

15 Nye KE, Knox KA, Pinching AJ: Lymphocytes from HIV-infected individuals show aberrant inositol polyphosphate metabolism which reverses after zidovudine therapy. AIDS 1991;5:413–417.
16 Diamond DC, Sleckman BP, Gregory T, Lasky LA, Greestein JL, Burakoff SJ: Inhibition of CD4+ T cell function by the HIV envelope protein gp 120. J Immunol 1988;141:3715–3717.
17 Cefai D, Debre P, Kaczorek M, Idziorek T, Autran B, Bismuth G: Human immunodeficiency virus-1 glycoproteins gp120 and gp160 specifically inhibit the CD3/T cell-antigen receptor phosphoinositide transduction pathway. J Clin Invest 1990;86: 2117–2124.
18 Newell MK, Haughn LJ, Maroun CR, Julius MH: Death of mature T cells by separate ligation of CD4 and the T cell receptor for antigen. Nature 1990;347:286–289.
19 Terai C, Kornbluth RS, Pauza CD, Richman DD, Carson DA: Apoptosis as a mechanism of cell death in cultured T lymphoblasts acutely infected with HIV-1. J Clin Invest 1991;87:1710–1715.
20 Ameisen JC, Capron A: Cell dysfunction and depletion in AIDS: the programmed cell death hypothesis. Immunol Today 1991;12:102–105.
21 Giorgi J, Detels R: T cell subset alteration in HIV-infected homosexual men: NIAD multicenter AIDS cohort study. Clin Immunol Immunopathol 1989;52:10–18.

Dr. Marie-Lise Gougeon, Unité d'Oncologie Virale et URA CNRS No 1153, Département SIDA et Retrovirus, Institut Pasteur, 28, rue du Docteur-Roux, F–75724 Paris Cedex 15 (France)

Janossy G, Autran B, Miedema F (eds): Immunodeficiency in HIV Infection and AIDS. EC/FERS/MRC Workshop on Immunodeficiency in HIV-1 Infections, Windsor, Surrey, 1991. Basel, Karger, 1992, pp 127–134

Programmed Death of Both CD4+ and CD8+ Cells in Asymptomatic HIV-1 Infection[1]

Linde Meyaard[a], Sigrid A. Otto[a], M. Janneke Mijnster[b], René P.M. Keet[c], Frank Miedema[a]

[a] Department of Clinical Viro-Immunology, Central Laboratory of The Netherlands Red Cross Blood Transfusion Service and Laboratory for Experimental and Clinical Immunology of the University of Amsterdam;
[b] Netherlands Institute for Brain Research and
[c] Department of Infectious Diseases, Municipal Health Service, Amsterdam, The Netherlands

Programmed cell death (PCD), also known as apoptosis or activation-induced cell death, is a physiological mechanism of cell deletion that differs morphologically and biochemically from necrosis [1]. PCD has been described to play a role in a wide variety of immunological regulatory processes [2]. The process is characterized by a typical cellular morphology and degradation of the chromatin into discrete fragments which are multiples of about 190 base pairs of DNA [3]. This pattern of cleavage is due to the vulnerability of the linker DNA running between nucleosomes. Already early in HIV infection, in clinically stable asymptomatic individuals, immunological abnormalities can be demonstrated. HIV infection affects both CD4+ and CD8+ T-cell function, such as interleukin-2 production and proliferation after stimulation with recall antigens and CD3 antibodies before CD4+ T-cell numbers are decreased [4–8]. Early in infection, this T-cell nonresponsiveness is caused by a selective deletion of T-memory cells [6] which cannot be explained by direct HIV infection, since only 1 in 10,000 to 1 in 1,000 T cells are infected [9]. In addition to deletion of T-memory cells, in long-term infected persons an intrinsic nonresponsiveness in both CD4+ and CD8+ cells is observed [6, 7, 10]. We have pre-

[1] This study was supported by a Grant from the Netherlands Ministry of Public Health.

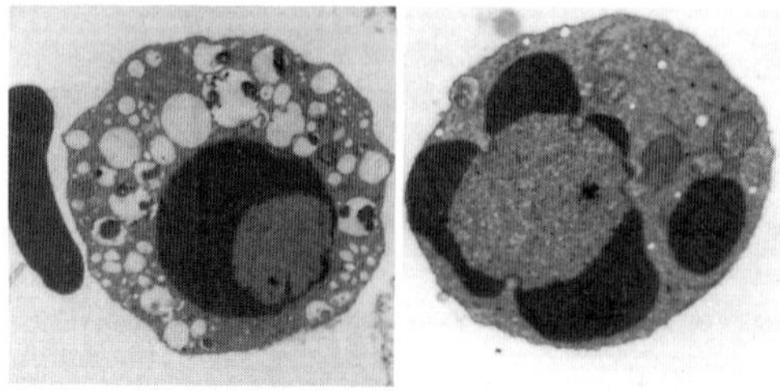

Fig. 1. Transmission electron microscopy analysis of the typical morphological changes in cells undergoing PCD. PBMC from a HIV-infected individual were cultured overnight in the presence of CD3 antibodies. Details showing extensive peripheral chromatin condensation, dilation of the endoplasmatic reticulum under the cell membrane but preservation of mitochondrial structures as features of PCD.

viously demonstrated that this defect is not at the level of induction of early activation events such as increased intracellular calcium concentration and activation of protein kinase C [6]. We have therefore investigated whether PCD contributes to T-cell nonresponsiveness in HIV infection.

Patients Studied

Sixteen HIV-1 asymptomatic seropositive (CDC class II or III) homosexual men were selected from a prospective cohort study in Amsterdam [11] for having normal numbers of circulating CD4+ cells (mean 540/mm^3, range 320–880) and being seropositive for more than 3.5 years. Seronegative controls were healthy male laboratory workers. The extent of DNA fragmentation was determined by a modification of the method of Sellins and Cohen [12]. High and low molecular weight DNA fractions from lysed cells were separated and the low molecular weight fraction, containing fragmented DNA, was analyzed in gel electrophoresis.

PCD of T-Lymphocytes from HIV-1-Positive Individuals Is Enhanced by Activation

First, the hypothesis whether T-cell nonresponsiveness in HIV infection is caused by PCD was tested. We studied peripheral blood mononuclear cells (PBMC) from 16 HIV-1-infected persons selected for having normal CD4+ T-cell counts. Proliferation to anti-CD3 antibodies ranged from <1 to 80% of that of the controls (data not shown) as previously reported [5–7, 10].

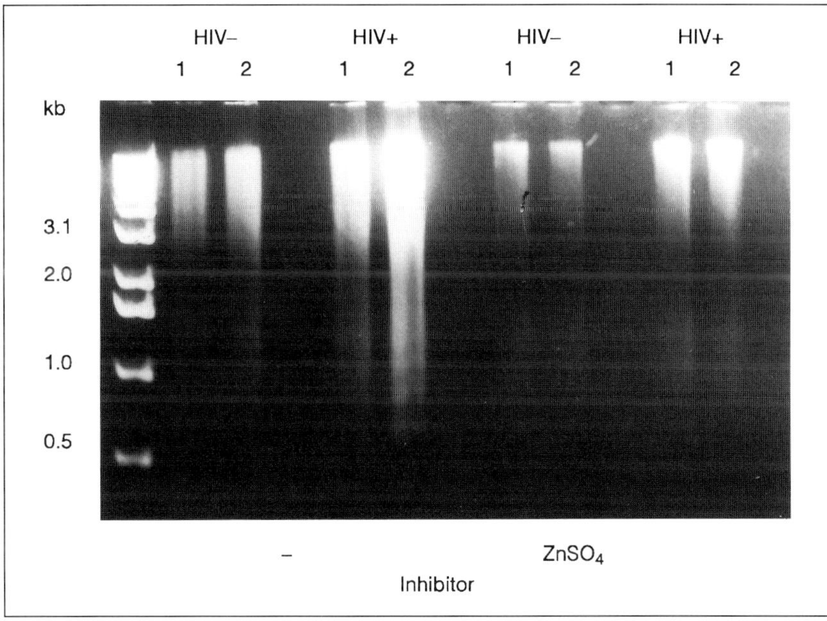

Fig. 2 Prevention of PCD in PBMC of HIV-infected individuals by $ZnSO_4$. PBMC, cultured in the absence (1) or presence (2) of CD3 antibodies, treated or not treated with 2 m*M* $ZnSO_4$.

In PBMC from HIV-infected persons, cultured overnight in the presence of anti-CD3 antibodies, the typical morphology characteristic for PCD was recognized by electron microscopy (fig. 1). Cells showed extensive peripheral chromatin condensation, dilation of the endoplasmatic reticulum under the cell membrane but preservation of mitochondrial structures as features of PCD [13].

When low molecular weight DNA fractions were isolated from lysed cells and put on gel electrophoresis, the DNA cleavage pattern specific for apoptosis was observed, which could be prevented by Zn^{2+}, known to inhibit endonuclease activity and thereby DNA fragmentation in PCD [14] (fig. 2). DNA fragments corresponding to 1–7 nucleosomes were identified by gel electrophoresis, but longer fragments were also detected, forming a smear in the gel near the origin of migration.

In unstimulated cultures DNA fragmentation was observed in 9 out of 16 of the HIV-infected men but in none of the seronegative controls. After

Table 1. Increases in percentage of dead cells after culture of PBMC from HIV-infected men and seronegative controls

Subjects	n	Day 0	Day 1 no mitogen	Day 1 anti-CD3
HIV–	10	2.1 ± 0.7	6.0 ± 2.1	11.5 ± 4.1
HIV+	11	1.9 ± 0.8	8.9 ± 4.1*	23.8 ± 6.4**

Values represent mean ± SD. Percentages cell death after overnight culture of PBMC from HIV-infected men were significantly higher than in seronegative controls (*$p < 0.05$, **$p < 0.005$ in the Student t test). Samples of 0.3×10^6 PBMC were taken directly after isolation (day 0) or after overnight culture (day 1). Dead cells were visualized by adding the DNA intercalating dye propidium iodide. Red fluorescence was analyzed on a FACScan flow cytometer. The percentage of propidium iodide-stained cells represents the percentage of dead cells in the samples.

stimulation with anti-CD3 antibodies, PBMC from all HIV-infected persons showed DNA fragmentation, compared with 5 out of 16 seronegative controls. The observed PCD in nonstimulated cultures could not be correlated to a mitogenic factor in the cultures, since varying the culture media or culturing in human serum did not alter the observations (data not shown).

Significant differences in cell death in PBMC from HIV-infected compared to seronegative persons could clearly be demonstrated by propidium iodide staining (table 1). The number of propidium iodide-stained dead cells increased after stimulation with CD3 antibodies.

PCD Is Observed in Both CD4+ and CD8+ Cells

In the next series of experiments, CD4+ and CD8+ cells were separated by negative selection. The cell types were removed using CD19 (B-specific), CD16 (NK-reactive) and CD8 or CD4 antibodies and anti-mouse IgG-coated magnetic beads as described [15]. DNA fragmentation was observed in cell fractions highly enriched for either CD4+ or CD8+ cells (fig. 3). PCD therefore does not seem to be restricted to CD4+ cells, in agreement with reported nonresponsiveness in both CD4+ and CD8+ T cells [5–7].

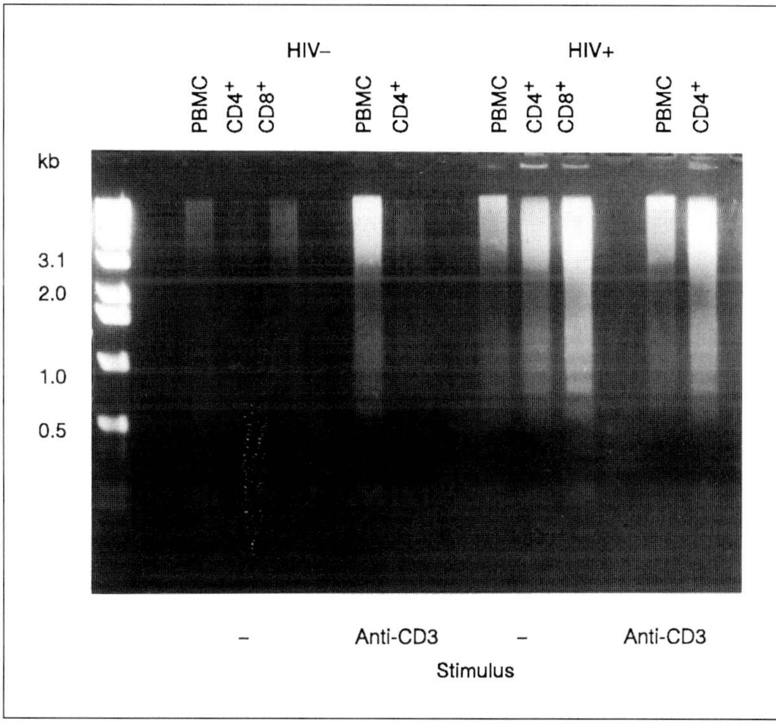

Fig. 3. PCD of cells of HIV-1-infected individuals occurs in both the CD4+ and the CD8+ subset after overnight culture in the absence or presence of anti-CD3 antibodies.

Discussion

With respect to the underlying mechanism by which T cells in HIV-infected men are programmed for death, several alternatives may be envisaged. Since only a few cells are infected and the mode of death due to lytic viral infection is necrosis, a systemic effect of HIV proteins or disturbances of cytokine regulatory networks rather than direct viral infection are likely to be involved. Recently, it has been proposed that in HIV infection interaction of soluble gp120 with CD4, previously shown to lead to impaired lymphocyte function [16], would prime CD4+ T cells for PCD [17]. This hypothesis is supported by results obtained with mature murine lymphocytes which die of PCD after stimulation via TCR/CD3 when CD4 was previously cross-linked by anti-CD4 antibodies [18]. CD4-gp120 ligation however, is not likely to be the only mechanism to account for PCD in HIV infection, because it does not explain PCD observed in CD8+ cells. The CD8+ T-cell subset in HIV-infected individuals contains increased percentages of cells with phenotypes associated with activation, that are severely defective in their proliferative response to polyclonal activators [19, 20] and are reported to die in culture [21].

In HIV infection, encountering antigen may cause a continuous PCD-mediated depletion of T cells in vivo. Compared to loss of memory cells due to direct HIV infection, which only affects small percentages of cells, PCD may be the principal mechanism for deletion of reactive T cells. In the early phase of infection rapid memory T-cell turnover initially is compensated for by increased T-cell renewal reflected by phenotypic changes but relatively stable CD4+ numbers [6, 7]. Gradual deletion of immune regulatory T cells by PCD may contribute to the attenuation of the immune system leading to the emergence and high-rate replication of virulent HIV variants associated with rapid CD4+ cell decline and progression to AIDS [8, 22].

Conclusions

After overnight culture both CD4+ and CD8+ cells from asymptomatic HIV-infected individuals showed condensed chromatin morphology and DNA fragmentation characteristic for PCD. Cell death was enhanced by stimulation with CD3 antibodies. Our data imply that PCD contributes to the observed in vitro T-cell dysfunction in asymptomatic individuals. PCD-mediated deletion of T cells after antigenic stimulation, in vivo, may result in a gradual loss of reactive T cells and thus disturb the balance between virus and host defense.

References

1 Duvall E, Wyllie AH: Death and the cell. Immunol Today 1986;7:115–119.

2 Smith CA, Williams GT, Kingston R, Jenkinson EJ, Owen JJT: Antibodies to CD3/T-cell receptor complex induce death by apoptosis in immature T cells in thymic cultures. Nature 1989;337:181–184.

3 Wyllie AH, Morris RG, Smith AL, Dunlop D: Chromatin cleavage in apoptosis: association with condensed chromatin morphology and dependence on macromolecular synthesis. J Pathol 1984;142:67–77.

4 Clerici M, Stocks N, Zajac RA, Boswell RN, Lucey DR, Via CS, Shearer GM: Detection of three different patterns of T-helper cell dysfunction in asymptomatic, human immunodeficiency virus-seropositive patients. J Clin Invest 1989;84:1892–1899.
5 Miedema F, Petit AJC, Terpstra FG, Schattenkerk JKME, De Wolf F, Al BJM, Roos M, Lange JMA, Danner SA, Goudsmit J, Schellekens PTA: Immunological abnormalities in human immunodeficiency virus (HIV)-infected asymptomatic homosexual men. HIV affects the immune system before $CD4^+$ T-helper cell depletion occurs. J Clin Invest 1988;82:1908–1914.
6 Van Noesel CJM, Gruters RA, Terpstra FG, Schellekens PTA, Van Lier RAW, Miedema F: Functional and phenotypic evidence for a selective loss of memory T cells in asymptomatic HIV-infected men. J Clin Invest 1990;86:293–299.
7 Gruters RA, Terpstra FG, De Jong R, Van Noesel CJM, Van Lier RAW, Miedema F: Selective loss of T-cell functions in different stages of HIV infection. Eur J Immunol 1990;20:1039–1044.
8 Miedema F, Tersmette M, Van Lier RAW: AIDS pathogenesis: a dynamic interaction between HIV and the immune system. Immunol Today 1990;11:293–297.
9 Schnittman SM, Psallidopoulos MC, Lane HC, Thompson L, Baseler M, Massari F, Fox CH, Salzman NP, Fauci AS: The reservoir for HIV-1 in human peripheral blood is a T cell that maintains expression of CD4. Science 1989;245:305–308.
10 Gruters RA, Terpstra FG, De Goede REY, Mulder JW, De Wolf F, Schellekens PTA, Van Lier RAW, Tersmette M, Miedema F: Immunological and virological markers in individuals progressing from seroconversion to AIDS. AIDS 1991;5:837–844.
11 De Wolf F, Goudsmit J, Paul DA, Lange JMA, Hooijkaas C, Schellekens PTA, Coutinho RA, Van der Noordaa J: Persistent HIV antigenemia is associated with rapid clinical and immunological deterioration. Br Med J 1987;295:569–572.
12 Sellins KS, Cohen JJ: Gene induction by γ-irradiation leads to DNA fragmentation in lymphocytes. J Immunol 1987;139:3199–3206.
13 Wyllie AH, Kerr JFK, Currie AR: Cell death: The significance of apoptosis. Int Rev Cytol 1980;68:251–306.
14 Duke RC, Chervenak R, Cohen JJ: Endogenous endonuclease-induced DNA fragmentation: An early event in cell-mediated cytolysis. Proc Natl Acad Sci USA 1983;80:6361–6365.
15 De Jong R, Brouwer M, Miedema F, Van Lier RAW: Human $CD8^+$ T lymphocytes can be divided into $CD45RA^+$ and $CD45RO^+$ cells with different requirements for activation and differentiation. J Immunol 1991;146:2088–2094.
16 Cefai D, Debré P, Kaczorek M, Idziorek T, Autran B, Bismuth G: Human immunodeficiency virus-1 glycoproteins gp120 and gp160 specifically inhibit the CD3/T cell-antigen receptor phosphoinositide transduction pathway. J Clin Invest 1990;86:2117–2124.
17 Ameisen JC, Capron A: Cell dysfunction and depletion in AIDS: the programmed cell death hypothesis. Immunol Today 1991;12:102–105.
18 Newell MK, Haughn LJ, Maroun CR, Julius MH: Death of mature T cells by separate ligation of CD4 and the T-cell receptor for antigen. Nature 1990;347:286–288.

19 Stites DP, Moss AR, Bacchetti P, Osmond D, McHugh TM, Wang YJ, Hebert S, Colfer B: Lymphocyte subset analysis to predict progression to AIDS in a cohort of homosexual men in San Francisco. Clin Immunol Immunopathol 1989;52:96–103.
20 Pantaleo G, Koenig S, Baseler M, Lane HC, Fauci AS: Defective clonogenic potential of $CD8^+$ T lymphocytes in patients with AIDS: Expansion in vivo of a nonclonogenic $CD3^+$ $CD8^+$ DR^+ $CD25^-$ T cell population. J Immunol 1990;144:1696–1704.
21 Prince HE, Jensen ER: HIV-related alterations in CD8 cell subsets defined by in vitro survival characteristics. Cell Immunol 1991;134:276–286.
22 Tersmette M, Miedema F: Interactions between HIV and the host immune system in the pathogenesis of AIDS. AIDS 1990;5:S57–S66.

Dr. Linde Meyaard, Department of Clinical Viro-Immunology,
Central Laboratory of The Netherlands Red Cross Blood Transfusion Service,
PO Box 9190, NL–1006 AD Amsterdam (The Netherlands)

Janossy G, Autran B, Miedema F (eds): Immunodeficiency in HIV Infection and AIDS.
EC/FERS/MRC Workshop on Immunodeficiency in HIV-1 Infections, Windsor, Surrey, 1991.
Basel, Karger, 1992, pp 135–144

T Lymphocyte Functional Defects in HIV Infection[1]

Rob A. Gruters, Frank Miedema

Department of Clinical Immunology and Virology, Central Laboratory of the Netherlands Red Cross Blood Transfusion Service and Laboratory for Clinical and Experimental Immunology of the University of Amsterdam, The Netherlands

Acquired immunodeficiency syndrome (AIDS) is the collective name for various disease manifestations in the final stages of an infection with the human immunodeficiency virus (HIV-1) [1, and reviewed by van Leeuwen et al., this volume]. Development of disease is associated with a severe depletion of CD4+ T cells, one of the main targets for HIV [2]. The central role of T-helper cells with CD4+ phenotype in regulation of several immune responses may explain the severe immune dysfunction in AIDS. However, it is unlikely that the mere depletion of CD4+ T cells can explain the wide range of humoral and cellular abnormalities, including T-helper function, specific and nonspecific cytotoxic activities, B-cell and monocyte function [2]. Furthermore, immunological abnormalities have been observed in asymptomatic HIV-infected people with normal CD4+ T-cell numbers and the possible pathomechanisms of these disorders are discussed below.

T-Cell Function in Early HIV Infection

T-lymphocyte proliferation can be induced using monoclonal antibodies (MAb) directed against the monomorphic CD3 part of the CD3/T-cell receptor (TCR) complex. This stimulating signal bypasses the poly-

[1] This work was supported by grants from the Netherlands Foundation for Preventive Medicine and the Ministry of Health. Frank Miedema is a senior fellow of the Royal Netherlands Academy of Arts and Sciences.

morphic TCR that recognizes its specific antigen in the context of self MHC [3]. Although this mitogenic signal is partially dependent upon monocytic cells [4], anti-CD3 induced proliferation reflects the capacity of the T-cell compartment for activation irrespective of specific antigen. Using soluble anti-CD3 Mab in asymptomatic HIV-1-infected individuals, we and others have shown that T-cell proliferation is affected before any CD4+ T-cell depletion can be observed [5–7]. The decreased response, due to intrinsic T-cell defects, has also been confirmed in an assay using immobilized anti-CD3 Mab [8, 9]. Under this condition, cross-linking of CD3 receptors is facilitated by the immobilized anti-CD3 and the T-cell proliferation is even more monocyte-independent. Furthermore, stimulation with a mitogenic combination of anti-CD2 MAb also results in decreased T-cell proliferation in asymptomatic HIV-infected men [8]. In contrast to T-cell proliferation, T-cell differentiation to cytotoxic T lymphocytes (CTL), induced by immobilized anti-CD3 MAb, is not severely affected in early HIV infection [9].

Hofmann et al. [20] observed that T-cell proliferation induced by the lectin pokeweed mitogen (PWM) is also decreased early in HIV-1 infection. Loss of T-cell responsiveness is also observed for T-helper activity in PWM-driven B-cell differentiation assays in vitro [5, 6]. In contrast, in asymptomatic individuals with stable CD4+ cell numbers the response to the mitogen phytohaemagglutinin (PHA) is not as strongly affected as the responses to anti-CD3 and PWM [11, and reviewed by Janossy, Borthwick et al., this volume].

Shearer et al. [12–14] demonstrated in asymptomatic HIV-infected men, in contrast to normal responsiveness to alloantigens (ALLO), a selective loss of T-cell reactivity with respect to both proliferation and generation of CTL in response to nominal antigen (Ag) presented by self MHC molecules. This nonresponsiveness to nominal Ag can be restored by rIL-2 or by costimulation by nominal Ag plus ALLO antigen [13]. This indicates that the antigen-specific T cells present cannot be properly activated by nominal Ag alone. Shearer et al. [13, 15] have interpreted these findings as a selective CD4+ T-cell defect for self-restricted responses [see also Clerici and Shearer, this volume].

T-cell activation via other receptors is not as strongly affected in early HIV infection as the CD3 pathway. It appears that the signals delivered via the CD28 molecule, utilizing a signal transduction pathway independent of the CD3/TCR route [16–18], result in significant enhancement of T-cell proliferation [8, 9]. Nevertheless, these observations need to be interpreted with some caution in terms of the quantitative aspects of the responses. This is because CD28 is not mitogenic on its own, and the counts measured by thymidine incorporation do not rise to normal high levels in the presence of both anti-CD3 and anti-CD28 [9]. We have also demonstrated that a subset of CD8+ cells loses its CD28 receptor and is likely to become anti-CD28 refractory during disease progression [19].

The mechanism underlying anti-CD3 unresponsiveness has been investigated but no evidence for biochemical defects in signal transduction pathways, such as Ca^{2+} flux or protein kinase C activation has been found [8]. Expression of CD25, the IL-2 receptor α chain, is normal, but IL-2 production is decreased in T cells from asymptomatic HIV-infected men [9, 20]. Proliferation can be partially restored by the addition of exogenous IL-2 but the normal levels of anti-CD3 activation are not reached [9].

Selective Loss of Memory T-Cell Function in Early HIV Infection

The decreased T-cell responsiveness in early HIV infection, summarized above, can in part be explained by an in vivo preferential loss of functional subsets and memory T cells or by a functional anergy of the remaining memory cell populations. Phenotypic analysis of T cells from early asymptomatics shows a significant decrease in CD29 expressing cells in peripheral blood [8, 9, 21–23]. The CD29 antigen, recognized by the prototype 4B4 antibody, is the integrin β chain expressed by a CD4+ subset referred to as the 'helper-inducer' cells [24]. Antigen-reactive T cells and T cells that provide helper activity for polyclonal and antigen-specific antibody synthesis are mainly in the CD29+ subset [24, 25]. The early dip in the number of this population may therefore be functionally significant. Although the CD29+, CD4+ population is not fully identical to the 'memory' type cells which express CD45RO, the 180-kD isoform of the CD45 antigens, there is an overlap between these CD4+ populations [see figure 6 in the chapter by Giorgi, this volume]. While the CD45RO+ 'memory' and CD45RA+ populations are mostly complementary, there is less complementarity between the CD29+ and the CD45RA+ T cells. It is therefore not surprising that the CD45RA+ cells are not dramatically increased in early HIV-1 infection [8, 9, 19, 21, 22, 26, 27]. Results for subset changes from seroconversion to time of diagnosis of AIDS are shown in figure 1.

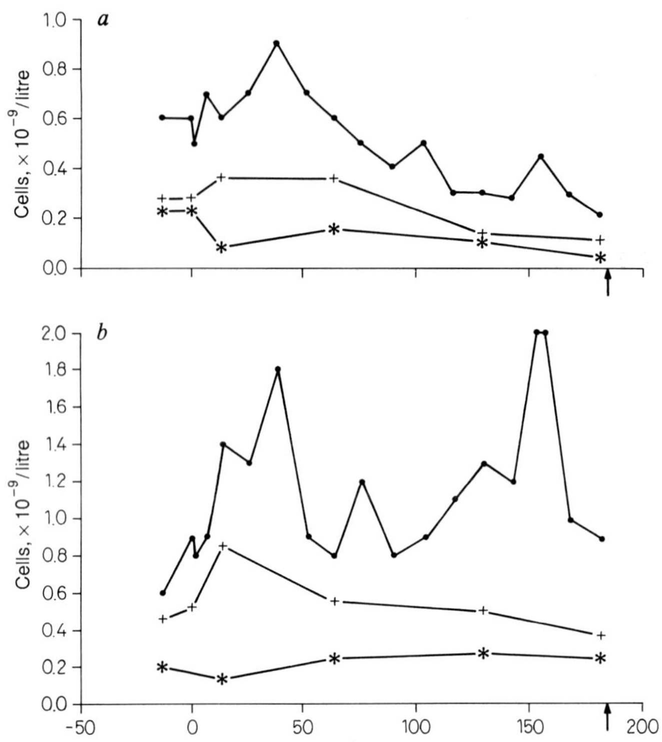

Fig. 1. Subpopulations of CD4+ cells (*a*) and CD8+ cells (*b*) in a representative patient who progressed rapidly from seroconversion to AIDS. The total (•), CD45RA+ (+) and CD29+ (*) subpopulations are shown within the CD4 and CD8 subsets. Zero denotes the first time point giving a Western blot-positive result. The arrow indicates the time of AIDS diagnosis.

The transition of CD45RA 'naive' to CD45RO 'memory' cells has been observed after T cells first encounter their specific antigen [28, 29]. The increased expression of various adhesion molecules by 'memory' cells may facilitate recognition and enhance proliferation upon secondary activation by specific antigen [25, 29]. 'Memory' cells, but not 'naive' cells, readily respond to suboptimal concentrations of anti-CD3 and anti-CD2

MAb, while both subpopulations respond well to these mitogens when administered in optimal higher concentrations, particularly when costimuli such as anti-CD28 and IL-2 are also given [8, 30]. From Clerici's data [15], it can be inferred that 'memory' cells, but not 'naive' cells, are able to respond to low-density self-restricted presentation of allogenic peptides whereas both subsets do respond to high-density allo-restricted presentation of allogenic peptides or allo-MHC molecules. During the early phases of HIV-1 infection, responses to suboptimal concentrations of antibodies [8] and to low-density self-restricted presentations are particularly deficient. These observations indicate that 'memory' cell functions are damaged.

The loss of memory cells in early infection can be explained by the preferential infection of proliferating cells by retroviruses [31] and/or by suicidal activation [reviewed by Meyaard et al. and Janossy, Borthwick et al., this volume]. These two possibilities are not mutually exclusive but the latter phenomenon is not necessarily caused by active viral replication. Upon antigenic stimulation in vivo, memory cells might be the main target that can be infected. Indeed, both in SIV and HIV infection evidence for preferential infection of memory cells has been obtained, although still only a small fraction of this subset harbours HIV [32–34]. In addition, 'naive' cells are likely to be inhibited or suboptimally activated, as a result of the defects of antigen presenting cells [reviewed by Knight et al., this volume] thus abrogating the renewal of the memory pool. Cells may subsequently die via direct cytopathic effects of the virus, virus-specific cytotoxic activities [35, 36], or by other mechanism(s) [37], as shown in this volume by Gougeon et al. and Meyaard et al.

T-Cell Function in Later Stages of HIV-1 Infection

As HIV-1 infection progresses towards clinical symptoms, CD4+ T-cell numbers decrease, CD8+ T-cell numbers increase and immune functions further deteriorate [reviewed by Autran and Giorgi, this volume]. 'Memory' cells are gradually becoming more unresponsive to stimuli, and CD8+ T cells are expanding that have lost expression of CD28 and CD11b antigens and 5′-nucleotidase activity [19, 23, 38–40]. Thus the severely decreased CD4+ helper numbers may contribute to immune dysregulation reflected in other compartments including CD8+ cells and antigen presenting cells (APC) [10, 41]. This deterioration is characterized by further

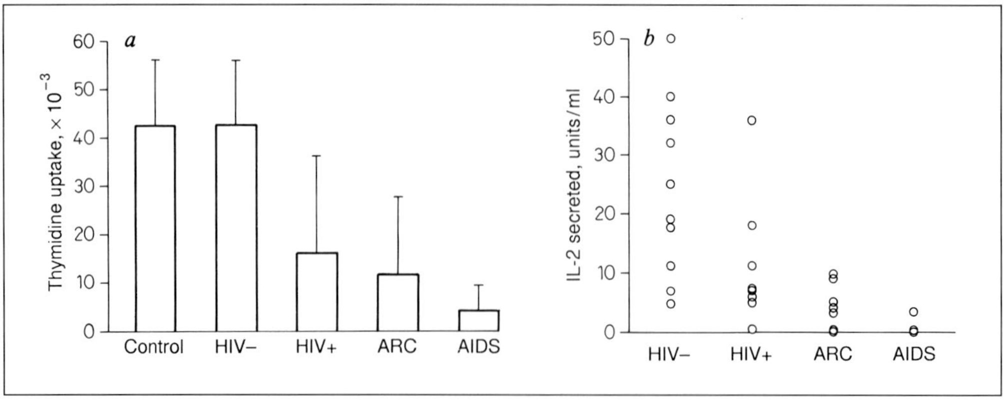

Fig. 2. The decline of anti-CD3 responsiveness (*a*) and IL-2 secretion upon anti-CD3 activation (*b*) in HIV-1-infected individuals. The anti-CD3 mitogenicity was assessed by thymidine uptake and the IL-2 secretion was expressed in units/ml culture medium following 3 days activation with suboptimal (5 µl/ml) amounts of purified CD3 antibody (CLB-T3/3).

decrease of in vitro proliferative responses and anti-CD3-induced IL-2 production [9, 42, 43], and a rapid decrease in CD4+ T-cell numbers (fig. 2).

Conclusions

T-cell reactivity in HIV infection appears to be sequentially caused through at least three different mechanisms. In early infection the damage to 'memory' cells can account for the decreased responsiveness to various stimuli in vitro. Later on in infection, both 'memory' and 'naive' T cells are functionally defective, and the dysfunction in both CD4 and CD8 cells contributes to nonresponsiveness. In the latest stages of asymptomatic infection and in symptomatic HIV-infected persons, the proportional increase of nonresponsive CD8 cells with immature or activated phenotypes may contribute to further immunological damage. The different stages can be distinguished by the sequential loss of responsiveness to nominal antigen and low-density anti-CD3 antibodies, the loss of alloantigen reactivity and finally the decreased capacity to respond to all stimuli (fig. 3).

CD4+ Nos, × $10^9/l$	0.8	0.8	0.6	0.3	0.2	0.1
T-cell response						
Nominal Ag	+	+	–	–	–	–
PWM	+	+	–	–	–	–
Anti-CD3	+	+	+	–	–	–
ALLO	+	+	+	+	+	–
PHA	+	+	+	+	+	–
B-cell function						
Ig production spontaneous	–	+	+	+	+	+
PWM-induced	+	–	–	–	–	–
APC function	+	+	+	–	–	–
Time	1–3 mo		variable timespan		6–24 months	
	↑ Infection	↑ Seroconversion		CDCII/III		ARC/AIDS

Fig. 3. The sequential loss of immune functions and the decline of CD4 cell numbers in the course of HIV infection (for explanations see text).

References

1 Revision of the CDC surveillance case definition for acquired immunodeficiency syndrome. MMWR 1987;36(suppl):1–15.

2 Fauci AS: The human immunodeficiency virus: infectivity and mechanisms of pathogenesis. Science 1988;239:617–622.

3 Clevers H, Alarcon B, Wileman T, Terhorst C: The T cell receptor/CD3 complex: a dynamic protein ensemble. Annu Rev Immunol 1988;6:629–662.

4 Tax WJ, Hermes FF, Willems RW, Capel PJ, Koene RA: Fc receptors for mouse IgG1 on human monocytes: polymorphism and role antibody-induced T cell proliferation. J Immunol 1984;133:1185–1189.

5 Miedema F, Petit AJC, Terpstra FG, Schattenkerk JKME, De Wolf F, Al BJM, Roos M, Lange JMA, Danner SA, Goudsmit J, Schellekens PTA: Immunological abnormalities in human immunodeficiency virus (HIV)-infected asymptomatic homosexual men. HIV affects the immune system before CD4+ T helper cell depletion occurs. J Clin Invest 1988;82:1908–1914.

6 Terpstra FG, Al BJM, Roos MThL, De Wolf F, Goudsmit J, Schellekens PT, Miedema F: Longitudinal study of leukocyte functions in homosexual men seroconverted

for HIV: rapid and persistent loss of B-cell function after HIV infection. Eur J Immunol 1989;19:667–673.

7 Bentin J, Tsoukas C, McCutchan JA, Spector SE, Richman DD, Vaughan JH: Impairment in T-lymphocyte responses during early infection with the human immunodeficiency virus. J Clin Immunol 1989;9:159–168.

8 Van Noesel CJM, Gruters RA, Terpstra FG, Schellekens PTA, Van Lier RAW, Miedema F: Functional and phenotypic evidence for a selective loss of memory T cells in asymptomatic HIV-infected men. J Clin Invest 1990;86:293–299.

9 Gruters RA, Terpstra FG, De Jong R, Van Noesel CJM, Van Lier RAW, Miedema F: Selective loss of T-cell functions in different stages of HIV infection. Eur J Immunol 1990;20:1039–1044.

10 Lane HC, Depper JL, Greene WC, Whalen G, Waldmann TA, Fauci AS: Qualitative analysis of immune function in patients with the acquired immunodeficiency syndrome. N Engl J Med 1985;313:79–84.

11 Roos MThL, Miedema F, Eeftink Schattenkerk JKM, De Wolf F, Goudsmit J, Lange JMA, Danner SA, Out TA, Schellekens PTA: Cellular and humoral immunity in various cohorts of male homosexuals in relation to infection with human immunodeficiency virus. Neth J Med 1989;34:132–141.

12 Shearer GM, Payne SM, Joseph LJ, Biddison WE: Functional T lymphocyte immune deficiency in a population of homosexual men who do not exhibit symptoms of acquired immune deficiency syndrome. J Clin Invest 1984;74:496–506.

13 Shearer GM, Bernstein DC, Tung KSK, Via CS, Redfield R, Salahuddin SZ, Gallo RC: A model for the selective loss of major histocompatibility complex self-restricted T cell immune responses during the development of acquired immune deficiency syndrome (AIDS). J Immunol 1986;137:2514–2521.

14 Shearer GM, Salahuddin SZ, Markham PD, Joseph LJ, Payne SM, Kriebel P, Bernstein DC, Biddison WE, Sarngadharan MG, Gallo RC: Prospective study of cytotoxic T lymphocyte responses to influenza and antibodies to human T lymphotropic virus-III in homosexual men. J Clin Invest 1985;76:1699–1704.

15 Clerici M, Via CS, Lucey DR, Roilides E, Pizzo PA, Shearer GM: Functional dichotomy of CD4+ T helper lymphocytes in asymptomatic human immunodeficiency virus infection. Eur J Immunol 1991;21:665–670.

16 Weiss A, Manger B, Imboden JB: Synergy between the T3/antigen receptor complex and Tp44 in activation of human T cells. J Immunol 1986;137:819–825.

17 Pantaleo G, Olive D, Harris D, Poggi A, Moretta L, Moretta A: Signal transducing mechanisms involved in human T cell activation via surface T44 molecules. Comparison with signals transduced via the T cell receptor complex. Eur J Immunol 1986;16:1639–1642.

18 Van Lier RAW, Brouwer M, de Groot E, Kramer IJ, Aarden LA, Verhoeven AJ: CD3/T cell receptor and CD28 use distinct intracellular signalling pathways. Eur J Immunol 1990:21:1775–1778.

19 Gruters RA, Terpstra FG, De Goede REY, Mulder JW, De Wolf F, Schellekens PTA, Van Lier RAW, Tersmette M, Miedema F: Immunological and virological markers in individuals progressing from seroconversion to AIDS. AIDS 1991;5: 837–844.

20 Hofmann B, Moller J, Langhoff E, Damgård Jakobsen K, Odum N, Dickmeiss E, Ryder LP, Thastrup O, Scharff O, Foder B, Platz P, Petersen C, Mathiesen L, Hart-

vig-Jensen T, Skinhoj P, Svejgaard A: Stimulation of AIDS lymphocytes with calcium ionophore (A23187) and phorbol ester (PMA): Studies of cytoplasmic free Ca, IL-2 receptor expression, IL-2 production, and proliferation. Cell Immunol 1989; 119:14–21.

21 Fletcher MA, Azen SP, Adelsberg B, Gjerset G, Hassett J, Kaplan J, Niland JC, Odom-Maryon T, Parker JW, Stites DP, Mosley JW, and Transfusion Safety Study Group: Immunophenotyping in a Multicentre Study: The transfusion safety study experience. Clin Immunol Immunopathol 1989;52:38–47.

22 Giorgi JV, Fahey JL, Smith DC, Hultin LE, Cheng HL, Mitsuyasu RT, Detels R: Early effects of HIV on CD4 lymphocytes in vivo. J Immunol 1987;138:3725–3730.

23 De Martini RM, Turner RR, Formenti SC, Boone DC, Bishop PC, Levine AM, Parker JW: Peripheral blood mononuclear cell abnormalities and their relationship to clinical course in homosexual men with HIV infection. Clin Immunol Immunopathol 1988;46:258–271.

24 Morimoto C, Letvin NL, Boyd AW, Hagan M, Brown HM, Kornacki MM, Schlossman SF: The isolation and characterization of the human helper inducer T cell subset. J Immunol 1985;134:3762–3769.

25 Sanders ME, Makgoba MW, Shaw S: Human naive and memory T cells: reinterpretation of helper-inducer and suppressor-inducer subsets. Immunol Today 1985;9: 195–198.

26 De Paoli P, Battistin S, Crovatto M, Modolo ML, Carbone A, Tirelli U, Santini G: Immunologic abnormalities related to antigenaemia during HIV-1 infection. Clin Exp Immunol 1988;74:317–320.

27 De Paoli P, Carbone A, Battistin S, Crovatto M; Arrighini N, Santini G: Selective depletion of the OKT4+ 4B4+ subset in lymph nodes from HIV+ patients. Immunol Lett 1987;16:71–73.

28 Tedder TF, Clement LT, Cooper MD: Human lymphocyte differentiation antigens HB-10 and HB-11. II. Differential production of B-cell growth and differentiation factors by distinct helper T cell subpopulations. J Immunol 1985;134:2989–2994.

29 Sanders ME, Makgoba MW, Sharrow SO, Stephany D, Springer TA, Young HA, Shaw S: Human memory T lymphocytes express increased levels of three cell adhesion molecules (LFA-3, CD2, and LFA-1) and three other molecules (UCHL-1, CDw29 and Pgp-1) and have enhanced IFN-gamma production. J Immunol 1988; 140:1401–1407.

30 Sanders ME, Makgoba MW, June CH, Young HA, Shaw S: Enhanced responsiveness of human T cells to CD2 and CD3 receptor mediated activation. Eur J Immunol 1989;19:803–808.

31 Gowda SD, Stein BS, Mohagheghpour N, Benike CJ, Engleman EG: Evidence that T cell activation is required for HIV-1 entry in CD4+ lymphocytes. J Immunol 1989; 142:773–780.

32 Gallatin WM, Gale MJ, Hoffman PA, Willerford DM, Draves KE, Beneviste RE, Morton WR, Clark EA: Selective replication of simian immunodeficiency virus in a subset of CD4+ lymphocytes. Proc Natl Acad Sci USA 1989;86:3301–3305.

33 Schnittman SM, Lane HC, Greenhouse J, Justement JS, Baseler M, Fauci AS: Preferential infection of CD4+ memory T cells by human immunodeficiency virus type 1: Evidence for a role in the selective T-cell functional defects observed in infected individuals. Proc Natl Acad Sci USA 1990;87:6058–6062.

34 Cayota A, Vuillier F, Scott-Algara D, Dighiero G: Preferential replication of HIV-1 in memory $CD4^+$ subpopulation. Lancet 1990;336:941–941.

35 Lanzavecchia A, Roosnek E, Gregory T, Berman P, Abrignani S: T cells can present antigen such as HIV gp120 targeted to their own surface molecules. Nature 1988; 334:530–532.

36 Plata F, Autran B, Martins LP, Wain-Hobson S, Raphael M, Mayaud C, Denis M, Guillon JM, Debre P: AIDS virus-specific cytotoxic T lymphocytes in lung disorders. Nature 1987;328:348–351.

37 Amiesen JC, Capron A: Cell dysfunction and depletion in AIDS: the programmed cell death hypothesis. Immunol Today 1991;12:102–105.

38 Joly P, Guillon J-M, Mayaud C, Plata F, Theodorou I, Denis M, Debre P, Autran B: Cell-mediated suppression of HIV-specific cytotoxic T lymphocytes. J Immunol 1989;143:2193–2201.

39 Stites DP, Moss AR, Bacchetti P, Osmond D, McHugh TM, Wang YJ, Hebert S, Colfer B: Lymphocyte subset analysis to predict progression to AIDS in a cohort of homosexual men in San Francisco. Clin Immunol Immunopathol 1989;52:96–103.

40 Pantaleo G, Koenig S, Baseler M, Lane HC, Fauci AS: Defective clonogenic potential of $CD8^+$ T lymphocytes in patients with AIDS: Expansion in vivo of a nonclonogenic $CD3^+CD8^+DR^+CD25^-$ T cell population. J Immunol 1990;144:1696–1704.

41 Fahey JL, Prince HE, Weaver M, Groopman J, Visscher B, Schwartz K, Detels R: Quantitative changes in T helper or suppressor/cytotoxic lymphocyte subsets that distinguish acquired immune deficiency syndrome from other immune subsets disorders. JAMA 1984;76:95–100.

42 Schellekens PTA, Roos MTL, De Wolf F, Lange JMA, Miedema F: Low T-cell responsiveness to activation via CD3/TCR is a prognostic marker for AIDS in HIV-1 infected men. J Clin Immunol 1990;10:121–127.

43 Pantaleo G, De Maria A, Koenig S, Butini L, Moss B, Baseler M, Lane HC, Fauci AS: $CD8^+$ T lymphocytes of patients with AIDS maintain normal broad cytolytic function despite the loss of human immunodeficiency virus-specific cytotoxicity. Proc Natl Acad Sci USA 1990;87:4818–4822.

Rob A. Gruters, UMR 103 CNRS/Biomérieux, Ecole Normale Supérieure de Lyon, 46, allée d'Italie, F–69364 Lyon Cedex 07 (France)

Janossy G, Autran B, Miedema F (eds): Immunodeficiency in HIV Infection and AIDS. EC/FERS/MRC Workshop on Immunodeficiency in HIV-1 Infections, Windsor, Surrey, 1991. Basel, Karger, 1992, pp 145–156

Inhibition of CD4+ T Lymphocyte Activation by HIV-1 Glycoproteins gp120 and gp160

Daniel Cefai[a], *Thierry Idziorek*[b], *Patrice Debré*[a], *Georges Bismuth*[a]

[a] Laboratoire d'Immunologie Cellulaire et Tissulaire, CNRS URA 186, CERVI, et [b] Laboratoire de Biologie et Génétique des Pathologies Immunitaires, CNRS URA 1463, CERVI, GH Pitié-Salpêtrière, Paris, France

HIV-1 infection results in quantitative and qualitative defects of the CD4 helper/inducer T cell function [1]. The selective infection of CD4+ T cells by HIV-1 is governed through the high affinity interaction between the viral envelope glycoprotein gp120 and the CD4 molecule. Different works showed that soluble gp120 itself could exert a suppressive effect on T cell reponses [2, 3] by binding to CD4 and competitively blocking the interaction between T cells and antigen-presenting cells (APC). gp120 was also shown to inhibit T cell activation triggered by various mitogens [4] or CD3-specific monoclonal antibodies (MAb) in the presence of APC [5]. However, the mechanism of gp120-blocking effect remains unclear.

T cells can be activated via the CD3/T cell antigen receptor (TcR) or CD2 molecules by specific MAb. This is the result of transmembrane signalling which involves early biochemical events and culminates in cell proliferation. Recent works have made it likely that the CD4 molecule may have also a signalling role in human T cells and is perhaps directly involved in signal transduction. The main evidence was that CD4-specific MAb could inhibit the T cell stimulation driven by ligands such as mitogenic lectins or CD3-specific MAb [6, 7]. Bank and Chess [8] showed that CD4 could deliver inhibitory signals to T cells, independent of MHC class II recognition. It was therefore suggested that binding of gp120 to CD4 may also transmit negative signals to T cells.

To understand how gp120/160 molecules (gps) interfere with normal T cell responses and whether they participate in the HIV-induced immune deficiency syndrome, we investigated their effect on the late cellular proliferation and early biochemical events following T cell activation. An antigen-specific T cell clone, representing a homogenous population of physiological CD4+ T cells, was stimulated with anti-CD3/TcR or anti-CD2 MAb in the absence of APC. Our results provide insights into the impairment of T cell signalling by gp120 and gp160. Moreover, correlation with internalization of these glycoproteins and modulation of the membrane CD4 level are discussed.

gp160 and gp120 Inhibit T Cell Proliferation Triggered via the CD3 TcR Molecular Complex

Antigen-specific CD4+ T cell proliferation is normally induced by recognition of the antigen on APC in the context of the major histocompatibility complex (MHC) class II molecules by the TcR together with CD4 on T cells. Binding of gps to CD4 was suggested to impair T cell responses by blocking such interaction. However, a negative direct effect of gps on T cell activation was also envisaged. To test these possibilities, a diphtheria toxoid (DT)-specific T cell clone, P28D, was used, which responds to both anti-CD3 mAb in the absence of APC and to the antigen DT presented by histocompatible APC (DR7β1-transfected murine fibroblasts). Effects of purified recombinant gp160 (Pasteur Mérieux, France) and gp120 (Transgene SA, France) was investigated. As shown in figure 1b, both glycoproteins (gps) inhibited P28D proliferation induced by UCHT1 in a dose-dependent manner (45–80% maximal inhibition at 4 μg/ml). HIV-1 glycoproteins used alone did not induce any significant P28D proliferation. The specific response to DT was also severely impaired (fig. 1a). Different CD3/TcR- or CD2-specific MAb were further tested. Stimulation by various CD3-specific MAb with different isotypes, as well as by MAb IC5Vα specific for the α chain of P28D TcR [9], was clearly impaired (not shown) whereas the CD2-induced proliferation was unaffected, indicating that gps specifically inhibit the stimulation through the CD3/TcR complex. Addition of purified soluble CD4 (sT4) reversed the inhibition in all experiments.

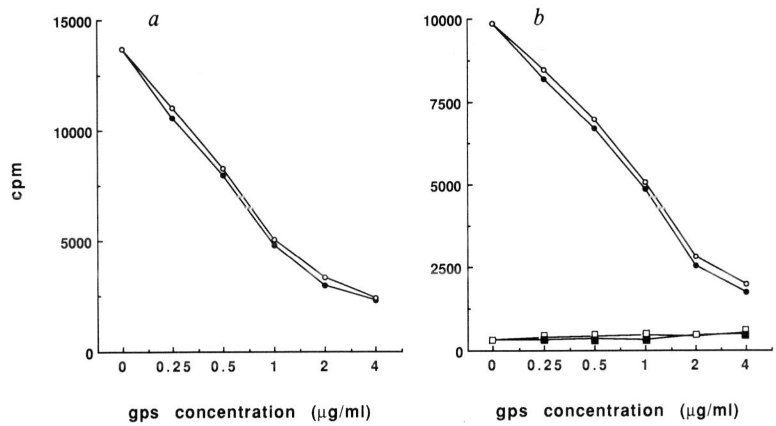

Fig. 1. Dose-dependent inhibitory effect of gp160 and gp120 on the antigen-specific and CD3-induced P28D T cell proliferation. *a* P28D T cells were stimulated in 96-well flat-bottomed plates with DT (30 μg/ml) presented by DRα:DR7β1-transfected murine fibroblasts (2.5×10^4/well) in the presence of indicated concentrations of gp160 (○) or gp120 (●). Proliferation was measured on day 3 after a 16-hour pulse with 1 μCi/well [^{3}H]thymidine. *b* P28D T cells were stimulated with the anti-CD3 MAb UCHT1 (1/400 dilution of ascitic fluid) cross-linked by a rabbit anti-mouse IgG whole molecule (1/400 final dilution) in the presence of indicated concentrations of gp160 (○) or gp120 (●). The effect of gp160 (□) and gp120 (■) used alone is also shown. Proliferation was measured as in figure 1a. Results are expressed as the mean cpm of triplicate determinations (SD $< 10\%$). Representative of five different experiments.

gp160 and gp120 Inhibit Early Biochemical Events Mediated via CD3/TcR but not via CD2

As shown in figure 2, transmembrane signalling via the CD3- or CD2-dependent T cell activation is believed to be linked to the phosphatidylinositol (PI) cycle-related process and involves calcium and pH changes. Hydrolysis of inositol phospholipids by phospholipase C (PLC) is one of the earliest biochemical events triggered by specific antigen [10] or MAb directed against the CD3/TcR complex [11] or the CD2 [12, 13]. Both an increased production of inositol phosphates (IPs) [11, 12] and diacylglycerol (DAG) are usually detected. The results presented above led us to

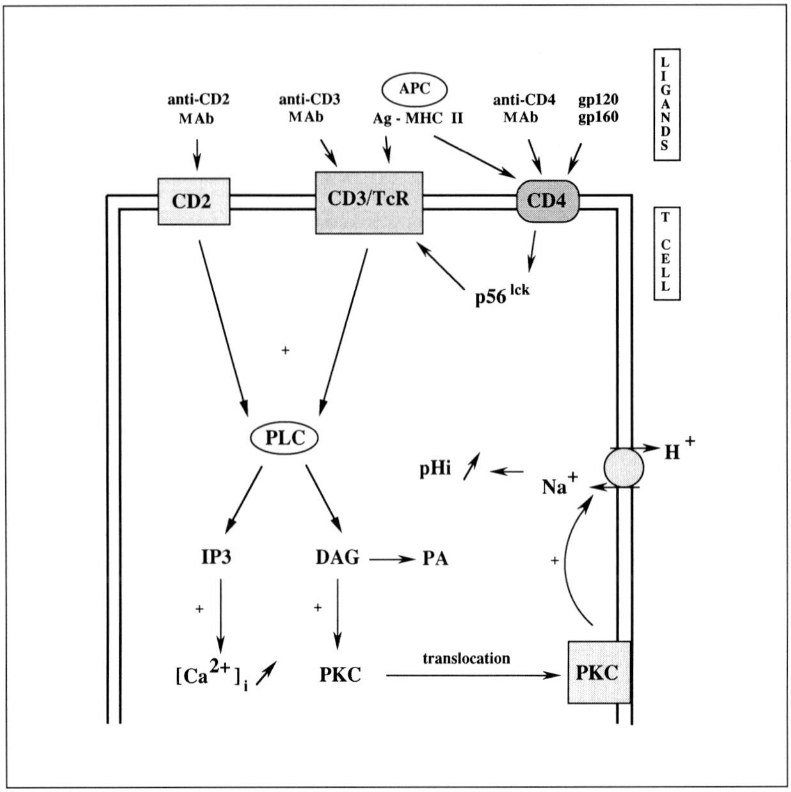

Fig. 2. Early biochemical events following CD3/TcR and CD2-mediated T cell activation (see text).

search for a possible inhibitory effect of the HIV-1 glycoproteins on the CD3/TcR mediated PLC pathway in P28D T cells.

Production of inositolphosphates (IPs) and phosphatidic acid (PA) was measured. As shown in figure 3a, phosphoinositides hydrolysis triggered via the CD3/TcR either by specific antigen (DT) or by UCHT1 was substantially decreased by a 18-hour pretreatment with 4 μg/ml gp120 (up to 60 and 45% inhibition, respectively). Phosphatidic acid (PA) labeling was investigated in parallel (fig. 3b) since its increased turnover has been reported as the direct consequence of DAG neosynthesis and phosphorylation. gp120 (4 μg/ml) reduced PA turnover in CD3-stimulated T cells (up to 50% inhibition). Identical results were obtained with gp160 in all

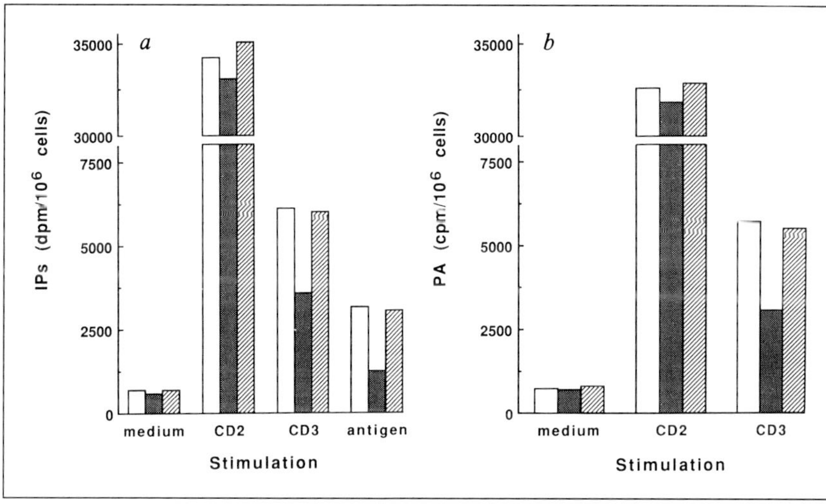

Fig. 3. gp120 inhibits PLC activity triggered via the CD3/TcR but not CD2 in P28D T cells. *a* P28D T cells were labeled to isotopic equilibrium with myo-[2-^{3}H]inositol-free medium in the absence (□) or the presence of 4 μg/ml of pg120 alone (■) or together with 4 μg/ml of sT4 (▨). Cells were then washed and incubated (10^7/ml) in Hanks' buffer containing 10 m*M* LiCl for 20 min at 37 °C. Aliquots (100 μl) were stimulated for 30 min by X11+D66 CD2-specific MAb or the CD3-specific MAb UCHT1 (1/400 dilution for each) or 5 × 10^5 DR7β1 murine transfectants prepulsed 18 h with 50 μg/ml of DT antigen. Inositolphosphates (IPs) were extracted and quantified by anion-exchange chromatography on a Dowex AG-1X8 column. Each determination was performed in triplicate. Representative of three different experiments. *b* P28D T cells were incubated for 18 h in control medium (□) or with 4 μg/ml of gp 120 alone (■) or together with 4 μg/ml of sT4 (▨). T cells were washed twice in a phosphate-free buffer containing 20 m*M* Hepes (pH 7.2), 150 m*M* NaCl, 1 m*M* $MgCl_2$, 5 m*M* KCl, 1 m*M* $CaCl_2$, 0.1% glucose, resuspended in the same buffer (10^7/ml) and incubated in carrier-free ^{32}P-orthophosphoric acid (50 μCi/ml) for 60 min at 37 °C. Aliquots (100 μl) were then stimulated with the different MAb for 30 min at 37 °C. The reaction was stopped with HCl, the ^{32}P-labeled phospholipids extracted, separated by thin-layer chromatography and phosphatidic acid (PA) quantified as described [12]. Results are expressed as the mean cpm for 10^6 cells of triplicate determinations. Representative of three experiments.

experiments. Addition of sT4 prevented the gps-induced inhibition of the CD3 pathway. In all cases, the CD2-triggered PLC activation was not affected. Note that immunofluorescence analysis of 18-hour gps-treated cells revealed that the expression of both the CD3/TcR complex and the CD2 molecule was not altered by the gps (not shown). So, the inhibitory effect of gps on the CD3/TcR pathway was not due to a decrease in the CD3/TcR membrane level.

CD3- and CD2-specific MAb induce a rapid and sustained rise of intracellular free Ca^{2+} concentration ($[Ca^{2+}]_i$) and intracytoplasmic pH (pH_i) in human T lymphocytes (fig. 2). The former is likely to result from the activation of the PLC pathway [12]. The latter is mainly due to the activation of the Na^+/H^+ membrane exchanger by protein kinase C (PKC) [14]. We further investigated whether gps could impair these phenomena. $[Ca^{2+}]_i$ measurements were performed in cells loaded with the fluorescent indicator Fura-2. Overnight incubation of P28D T cells with gps impaired by about 35% the increase in $[Ca^{2+}]_i$ induced by UCHT1 (data not shown). pH_i determinations carried out in parallel with the esterified indicator BCECF indicated that UCHT1-induced alkalinization was decreased by about 50% (data not shown). These inhibitions could be reversed by addition of sT4 (not shown). The CD2-induced increase of $[Ca^{2+}]_i$ and pH_i was, however, unaffected.

These results clearly show that, in addition to their inhibitory effect on the CD3/TcR-induced proliferation, both HIV-1 glycoproteins may also specifically impair the CD3/TcR-induced production of PI cycle-related second messengers and related activation events.

Inhibitions of T Cell Response to CD3 Correlate with Decreased CD4 Membrane Level

P28D cells were incubated at 37 °C with saturating concentrations of gps. Binding of gps was monitored at different times by immunofluorescence with 110.4 specific MAb, Membrane CD4 level was measured in parallel with OKT4 MAb which does not compete with gp120 binding to CD4. As shown in figure 4a, binding of gp120 (same results with gp160, not shown) was maximal after 1 h of incubation and then continuously decreased to 70% of its maximal value after 18 h. CD4 membrane level remained constant for 2–3 h and then decreased throughout the incubation period. Only 65–70% of initial CD4 level was detected after 18 h of incubation. The CD3-triggered IPs production was measured in parallel (fig. 4b). No significant inhibition was observed for incubations < 4 h. Maximal inhibitions were achieved after 18 h of incubation.

The parallel decrease of CD4 and gps at the P28D surface led us to investigate whether both molecules could be internalized after gps binding.

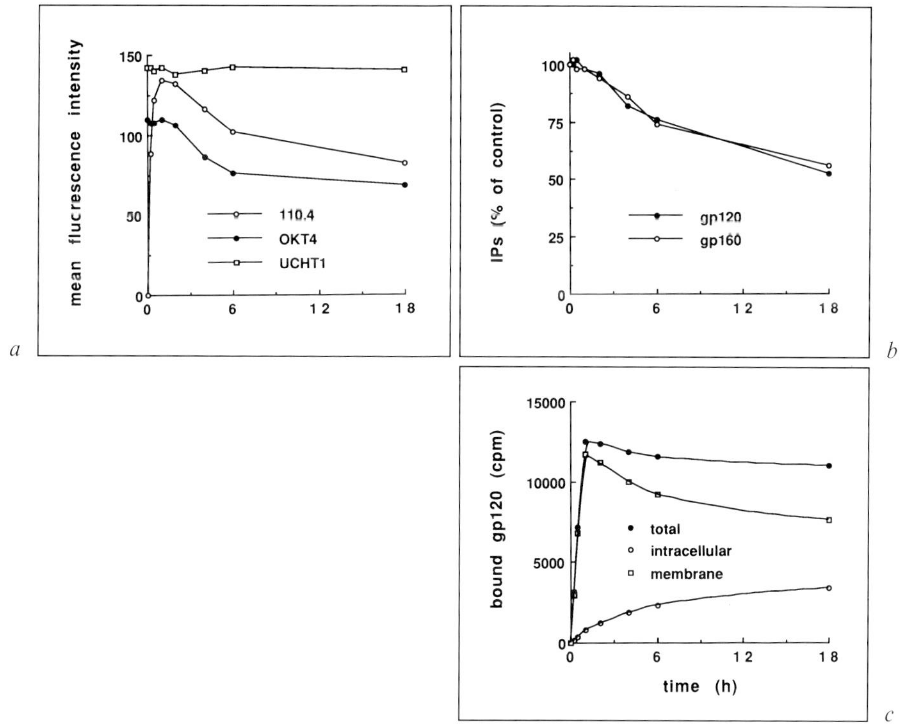

Fig. 4. Impairment of P28D T cell response to CD3 and down-modulation of CD4 membrane level are temporally related. *a* P28DT cells were incubated in culture medium wiht gp120 (3 μg/ml). At the indicated times, indirect immunofluorescence analysis was performed on a FACStar cell sorter using a FITC-conjugated goat anti-mouse IgG Fc fragment. Binding of gp120 was monitored with 110.4 MAb (1/250 dilution). CD3 and CD4 expression was analyzed with UCHT1 and OKT4 MAb, respectively. *b* P28D T cells were labeled with myo-[2-^{3}H]inositol for 24 h. During this period, cells were incubated with gp120 or gp160 for the indicated times. At the end of the labeling period, cells were washed and stimulated for 30 min with the CD3-specific MAb UCHT1 (1/400 final dilution). IPs production was measured as in figure 2a. Results are presented as percent of cpm of control cells incubated without gp120. Representative of three different determinations. *c* P28D T cells were incubated in culture medium in the presence of 10 n*M* iodinated gp120. At the indicated times, aliquots were collected. Cells were washed 3 times and total cell-associated radioactivity determined on a gamma counter. For determination of the intracellular radioactivity, membrane-bound gp120 was removed by incubating the cells for 5 min in culture medium acidified to pH 2.8 with concentrated HCl. Membrane-bound counts were determined from the difference between total and intracellular radioactivity. Representative of 4 different experiments.

For that purpose, gps were iodinated using the Bolton and Hunter technique. Cells were incubated at 37 °C with ^{125}I-gp120 (same results with gp160, not shown) and intracellular radioactivity was measured at different times after acid treatment of cells which removes surface-bound material. As shown in figure 4c, intracellular radioactivity continuously increased throughout the incubation. In contrast, total cell-associated radioactivity rapidly increased to a plateau phase and then remained constant. Confirming the results obtained with unlabelled gps, membrane-bound radioactivity rapidly increased to maximal values, probably reflecting saturation of CD4, and then continuously decreased. Internalized radioactivity, expressed as the percentage of total cell-associated radioactivity, was about 35% after 18 h of incubation (not shown). The parallel decrease of CD4 and gps membrane level and the increase of intracellular gps indicated that gps were internalized within cells and suggested further that gps and their CD4 receptor were cointernalized.

Discussion: The Significance of gp120/160 Interactions with T Cells in HIV-1 Infections

The impairment of CD4 T cell function during HIV infection is not fully understood. Among the various mechanisms which could account for such a phenomenon, an inhibitory role of gp120/160 on T cell activation has been proposed, particularly because this kind of interference could explain the defects in antigen-specific and in CD3/TcR-mediated PBL responses in HIV+ patients. However, it is not clear whether such an effect might be due to an interference of gps on T cell-APC interaction and/or to a direct immunosuppressive role on T cell signalling. We tested this latter hypothesis by assessing the effect of gps on the stimulation of a DT-specific T cell clone either by Ag and APC or by anti-CD3/TcR MAb in an APC-free cell system. Both gp120 and gp160 specifically inhibited the proliferative response triggered by APC but also by various MAb specific for the CD3/TcR complex in the absence of APC. The CD2-induced P28D activation was not affected. These results demonstrate that HIV-gps may exert a direct effect on the CD3-mediated T cell activation process.

Activation of T lymphocytes is initiated by hydrolysis of PI 4,5-biphosphate by PLC into inositol 1,4,5-triphosphate (IP_3) and DAG. The former induces the release of Ca^{2+} from intracellular stores [12, 13] while

DAG activates PKC [15]. Triggering of these early activation events may be achieved in P28D T cells by CD2 or CD3/TcR-specific MAb [12, 13]. As for proliferative responses, only the CD3/TcR-stimulated PLC pathway was inhibited by gps.

Linette et al. [16] have shown that T cell responses to CD3 and CD4 membrane expression were decreased in HIV-1-infected T cells, suggesting that CD4 should participate in the CD3-triggered T cell activation and that it may not simply act as an accessory molecule. Thus, perturbation of CD4 by specific MAb may transmit a regulatory signal to the CD3/TcR complex [8]. Ledbetter et al. [7] further demonstrated that cross-linking of CD4 by specific MAb inhibited the CD3-mediated Ca^{2+} responses. Comodulation studies also suggest that CD4 may be physically and perhaps functionally associated to the CD3/TcR complex on activated T cells [17, 18]. These findings may account for the inhibitory effects of the gps on the T cell activation process mediated through the CD3/TcR. In agreement with Weinhold et al. [5], we have also found that membrane CD4 expression was not decreased after a 2-hour exposure to gps. However, it was reduced after a further 18 h of incubation. This was temporally correlated with the time course of gps-induced inhibitions of P28D T cell responses to CD3 stimulation. This was in agreement with Kornfeld et al. [pers. commun.] who reported that a 24-hour exposure of PBL and Jurkat cells to gp120 resulted in a progressive decrease in the Ca^{2+} response. In our experiments, inhibition of the CD3/TcR-triggered IPs production was not observed for incubations with gps <4 h but was maximal after a 18-hour exposure. As CD4 was shown to enhance T cell responses to CD3 when brought into close physical proximity to the CD3/TcR complex [7], down-modulation of CD4 could partially account for the impairment of the CD3/TcR-triggered P28D T cell activation by gps.

Veillette et al. [19] and Rudd et al. [20] have reported in both murine and human T cells that CD4 is associated to the tyrosine kinase activity $p56^{lck}$ and that cross-linking of CD4 with specific MAb induces a rapid increase in this activity together with the phosphorylation of the CD3 ξ subunit [21]. $p56^{lck}$ has also been shown to be necessary for T cell response [22]. As recently demonstrated in CEM cells [23], gps induce in P28D T cells a down-modulation of $p56^{lck}$ protein level which is about 75% after 18 h of incubation and appears to be temporally related to that of membrane CD4 [Cefai unpubl. results].

Altogether, these results show that gps may alter the CD3/TcR activation process by both inhibiting the APC-T cell interaction and by directly down-regulating CD4 and the associated $p56^{lck}$. Moreover, our results make it likely that glycoproteins released from HIV-infected T cells [1] into blood may bind to circulating T cells and then are subsequently internalized. The uptake and internalization of gps by CD4 cells could at least partially explain the impairment of uninfected T cells function especially before the quantitative defect of the CD4 subset. The down-modulation of CD4 by gps would also partially account for the decline in the CD4 subset during HIV disease as was suggested by previous studies [24] and also supported by our results. However, the in vivo relevance of such inhibitions obtained herein in vitro remain to be demonstrated. Finally, these data may be of importance in late budding of the virus, and when considering HIV-1 envelope glycoproteins in vaccinal strategies.

Conclusions

Antigen-specific CD4 T lymphocyte responses are impaired during HIV infection and soluble envelope proteins gp120 and gp160 are suggested to participate in this effect. We and others have now shown that gp120/160 can interfere with the CD3/TcR-mediated stimulation of a human T cell clone both in the presence or the absence of APC. Activation was detected by late proliferation and early biochemical events following transmembrane cell signalling. Both gps clearly inhibit the T cell proliferation induced in an APC-free system by MAb specific for the CD3 molecule or the TcR α chain (up to 80% inhibition). Both gps also specifically inhibit the CD3/TcR PLC transduction pathway, without affecting the CD3/TcR cell surface expression. Thus, IPs production, PA turnover, $[Ca^{2+}]_i$ and pH_i increase are specifically impaired. By contrast, gps fail to suppress the CD2-triggered functional and biochemical P28D T cell responses. Maximal inhibitions are achieved after long-term incubations with gps (18 h) and correlate with a 35% decrease of the CD4 membrane level. We also demonstrate that gps are progressively internalized inside the cells throughout the incubation period. These results show that, in addition to their postulated role in the alteration of the interaction between CD4 on T lymphocytes and MHC class II molecules on APC, soluble HIV-1 envelope gps may directly and specifically impair the CD3/TcR-mediated activation of PLC in uninfected T cells. They further emphasize the importance of CD4 in the CD3/TcR-mediated activation pathway.

Acknowledgments

We thank H. Gouy for her expert technical assistance. We would also like to acknowledge I. Diaz and M. Ollivier (Pasteur Mérieux) for production and purification of gp160 and H. Kolbe (Transgene) for fruitful discussions about purification.

This work was supported by grants from the Institut National pour la Santé et la Recherche Médicale (INSERM), the Agence Nationale de Recherche contre le SIDA (ANRS) and the Caisse Nationale d'Assurance Maladie. D.C. is a recipient of the ANRS.

References

1 Fauci AS: The human immunodeficiency virus: Infectivity and mechanisms of pathogenesis. Science 1988;239:617–623.

2 Chirmule N, Kalyanaraman V, Oyaizu N, Pahwa S: Inhibitory influences of envelope glycoproteins of HIV-1 on normal immune responses. J AIDS 1988;1:425–430.

3 Diamond DC, Sleckman BP, Lasky LA, Greenstein JL, Burakoff SJ: Inhibition of CD4+ T cell function by the HIV envelope protein gp120. J Immunol 1989;141: 3715–3717.

4 Mann DL, Lasane F, Popovic M, Arthur LO, Robey WG, Blattner WA, Newttner MJ, Newman MJ: HTLV-III large envelope protein (gp120) suppresses PHA-induced lymphocyte blastogenesis. J Immunol 1988;138:2640–2644.

5 Weinhold KJ, Lyerly HK, Stanley SD, Austin AA, Matthews TJ, Bolognesi DP: HIV-1 gp120-mediated immune suppression and lymphocyte destruction in the absence of viral infection. J Immunol 1989;142:3091–3097.

6 Wassmer P, Chang C, Lodgberg L, Shevach EM: Role of the L3T4-antigen in T cell activation. II. Inhibition of T cell activation by monoclonal anti-L3T4 antibodies in the absence of accessory cells. J Immunol 1985;135:2237–2242.

7 Ledbetter JA, June CH, Rabinovitch PS, Grossmann A, Tsu TT; Imboden JB: Signal transduction through CD4 receptors: stimulatory vs. inhibitory activity is regulated by CD4 proximity to the CD3/T cell receptor. Eur J Immunol 1988;18:525–532.

8 Bank I, Chess L: Perturbation of the T4 molecule transmits a negative signal to T cells. J Exp Med 1985;162:1294–1301.

9 Bismuth G, Gouy G, Mariuzza R, Vaquero C, Theodorou I, Debré P: A human TcR-Ig chimeric protein used to generate a TCR chain variable region specific MAb. Mol Immunol 1990;27:1127–1136.

10 Imboden JB, Weyand C, Goronzy J: Antigen recognition by a human T cell clone leads to increases in inositol triphosphate. J Immunol 1987;138:1322–1324.

11 Imboden JB, Stobo JD: Transmembrane signalling by the T cell antigen receptor. Perturbation of the T3-antigen receptor complex generates inositol phosphates and releases calcium ions from intracellular stores. J Exp Med 1985;161:446–454.

12 Bismuth G, Theodorou I, Gouy H, Le Gouvello S, Bernard A, Debré P: Cyclic AMP-mediated alteration of the CD2 activation process in human T lymphocytes. Preferential inhibition of the phosphoinositide cycle-related pathway. Eur J Immunol 1988;18:1351–1357.

13 Pantaleo G, Olive D, Poggi A, Kozumbo WJ, Moretta L, Moretta A: Transmembrane signalling via the T11-dependent pathway of human T cell activation. Evidence for the involvement of 1,2-diacylglycerol and inositol phosphates. Eur J Immunol 1987;17:55–60.

14 Gelfand EW, Mills GB, Cheung RK, Lee JWW, Grinstein S: Transmembrane ion fluxes during activation of human T lymphocytes: role of Ca^{2+}, Na^+/H^+ exchange and phospholipid turnover. Immunol Rev 1987;95:59–64.

15 Nishizuka Y: The role of protein kinase C in cell surface transduction and tumor promotion. Nature (Lond) 1984;308:693–696.

16 Linette GP, Hartzman RJ, Ledbetter JA, June CH: HIV-1-infected T cells show a selective signaling defect after perturbation of the CD3/antigen receptor. Science (Wash) 1988;241:573–576.

17 Rivas A, Takada S, Koide J, Sonderstrup-McDevitt G, Engleman EG: CD4 molecules are associated with the antigen receptor complex on activated but not resting T cells. J Immunol 1988;140:2912–2918.

18 Saizawa K, Rojo J, Janeway CA Jr: Evidence for a physical association of CD4 and the CD3:α:β T cell receptor. Nature 1987;328:260–263.

19 Veillette A, Bookman MA, Horak EM, Bolen JB: The CD4 and CD8 T cell surface antigens are associated with the internal membrane tyrosine-protein kinase $p56^{lck}$. Cell 1988;55:301–308.

20 Rudd CE, Trevillyan JM, Dasgupta JD, Wong NL, Schlossman SF: The CD4 receptor is complexed in detergent lysates to a protein-tyrosine kinase (pp58) from human T lymphocytes. Proc Natl Acad Sci USA 1988;85:5190–5194.

21 Veillette A, Bookman MA, Horak EM, Samelson LE, Bolen JB: Signal transduction through the CD4 receptor involves the activation of the internal membrane tyrosine-protein kinase $p56^{lck}$. Nature 1989;338:257–259.

22 Glaichenhaus N, Shastri N, Littman DR, Turner JM: Requirement for association of $p56^{lck}$ with CD4 in antigen-specific signal transduction in T cells. Cell 1991;64: 511–520.

23 Juszczak RJ, Turchin H, Truneh A, Culp J, Kassis S: Effect of human immunodeficiency virus gp120 glycoprotein on the association of the protein tyrosine kinase $p56^{lck}$ with CD4 in human T lymphocytes. J Biol Chem 1991;17:11176–11183.

24 Crise B, Buonocore L, Rose JK: CD4 is retained in the endoplasmic reticulum by the human immunodeficiency virus type 1 glycoprotein precursor. J Virol 1990;64: 5585–5593.

Dr. Patrice Debré, Laboratoire d'Immunologie Cellulaire et Tissulaire,
URA CNRS 186, CERVI GH Pitié-Salpêtrière, 83, Bd. de l'Hôpital,
F–75651 Paris Cedex 13 (France)

Janossy G, Autran B, Miedema F (eds): Immunodeficiency in HIV Infection and AIDS.
EC/FERS/MRC Workshop on Immunodeficiency in HIV-1 Infections, Windsor, Surrey, 1991.
Basel, Karger, 1992, pp 157–170

Abnormalities of Interleukin-2 Receptor Chains in HIV-1 Infection

Yasmina Sahraoui[a], *Adlen Ammar*[a], *Yanto Lunardi-Iskandar*[a], *Maria Perraki*[b], *Vassilis Georgoulias*[a, b]

[a]Oncogenèse Appliquée, INSERM U268, Hôpital Paul-Brousse, Villejuif, France;
[b]National AIDS Reference Center of Crete and School of Medicine,
University of Crete, Iraklion, Crete, Greece

The cytokine interleukin-2 (IL-2) and its receptor, the IL-2 receptor (IL-2R) forms a carefully regulated system with a key role in the proliferation of normal mature T cells upon antigenic and mitogenic stimulation. The IL-2R complex displays a functional and biochemical heterogeneity as the receptor with high affinity is composed of at least two polypeptide chains. These are the 55-kd (IL-2Rα or p55) and 75-kd (IL-2Rβ or p75) molecules [1]. IL-2Rα chain has been discovered first as it is only weakly expressed on a subset of T cells with 'memory' phenotype (fig. 1) but its expression is increased upon T-cell activation. IL-2Rα is frequently referred to as the Tac antigen and it is recognized by monoclonal antibodies in the CD25 category, including the anti-Tac antibody. Nevertheless, IL-2Rα, on its own, binds IL-2 with low affinity [2, 3]. Conversely, IL-2Rβ is constitutively expressed on some T-cell subsets, albeit weakly [4–6], and its membrane expression is only moderately increased during activation (fig. 1). This β-chain alone binds IL-2 with an intermediate affinity [3] but the ligand is capable of transducing activation signals to the nucleus of the T cell.

Previous studies have already demonstrated that HIV-1 infection is associated with significant defects of the IL-2/IL-2R system, and these

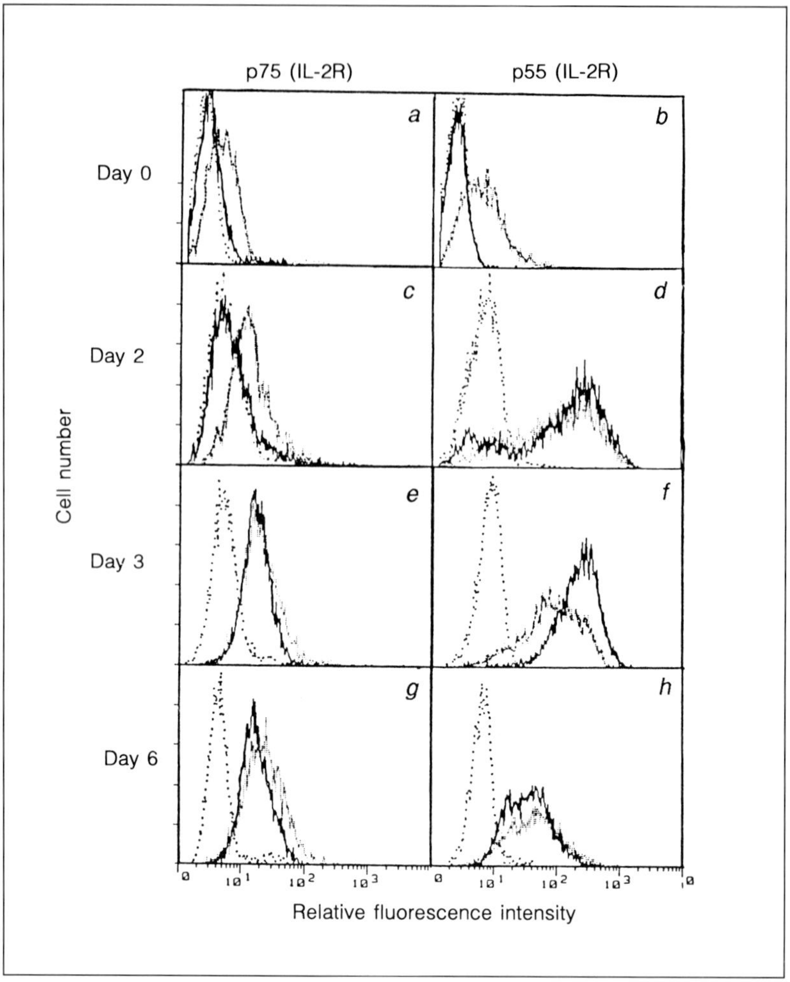

Fig. 1. The kinetics of expression of p75 (Tu27) and p55 (CD25) IL-2R chains on CD45RA+ (——) and CD45RO+ (–·–·–) T cells after PHA activation. NK cells were depleted. Negative controls were incubated with mouse Ig (······). It is shown that the intensity of expression of IL-2Rα (p55) chain after stimulation for days 2–3 (*d, f*) is higher than that of IL-2Rβ (p75) chain (*c, e*). Among resting cells CD45RO+ populations ('memory') cells express higher amounts of IL-2Rα and IL-2Rβ that the CD45RA ('virgin') T cells. [From 6, with permission.]

anomalies have been recorded even at early stages of the disease [7]. Such disorders have been documented in different HIV-1-related conditions and experimental models in vitro. It has been shown that HIV infection of normal T lymphocytes in vitro is followed by an impaired expression of IL-2Rα chain gene [8, 9], and that apparently activated T cells from HIV-1 seropositive patients can also be defective in expressing IL-2Rα. In addition, severe defects of IL-2 cytokine synthesis during T-cell activation are also part of the earliest signs recorded in asymptomatic patients (Clerici and Shearer, this volume).

There is, however, very little information concerning the modifications of IL-2Rβ chain during HIV infection. The observations about the functional consequences of the impaired expression of IL-2R chains are also scantly. One reason for this lack of fine details is the complexity of the questions involved. Apparently, different subpopulations of normal T cells express different quantities of IL-2Rα and IL-2Rβ [4–6], and during HIV-1 infection there are considerable shifts in these T-cell subsets. For example, the loss of CD4+ cells is counterbalanced by the accumulation of particular subsets of CD8+ cells (Autran and Giorgi; Hulstaert et al., this volume). For this reason the control studies need to be carefully constructed (see below). One example is that natural killer (NK) cells (CD16+) abundantly express IL-2Rβ and may alter results attributed to CD8+ populations unless removed from the suspension analyzed [4]. Thus, the observations described below are, to a great extent, preliminary but they already clearly confirm an important paradox in HIV-1 infection. This is the accumulation of lymphoid cells which, by certain criteria, appear to be activated while at the same time are deficient in their IL-2R expression.

In the present study we describe that the infection of T lymphocytes with HIV-1 in vitro induces an impaired expression of both IL-2Rα and IL-2Rβ chains on PHA-activated normal T cells. Moreover, in HIV infection the patients' cells have an abnormal constitutive expression of IL-2Rβ chain which also is associated with a decreased IL-2-induced activation of NK and LAK cells.

Technical Comments

In the investigations below, cellular and molecular techniques are complementary. Immunofluorescence studies include double labelling methods in order to analyze the expressions of the IL-2Rα, detected by the anti-IL-2Rα antibody IOT14 (CD25; Immunotech, Luminy, France) and IL–2Rβ, detected by the anti-IL-2Rβ antibody Tu27 (kindly provided by Prof. K. Sugamura, Japan [4]). The second layers include fluoresceinated goat antimouse Ig $F(ab_2)$, in combination with other lymphocyte subset reagents such as CD4, CD8 and CD16 directly conjugated with fluorochromes such as phycoerythrin. After labelling, the cells are investigated by flow cytometry. The cell proliferation assays detect the incorporation of ^{3}H-thymidine following a 72-hour incubation with various stimulants. The binding assays of ^{125}I-rIL-2 and cross-linking has been performed on the peripheral blood cells as previously described [10] with slight modification [11]. Equilibrium binding experiments have been carried out according to the design of Robb et al. [2]. The specific binding was determined by conventional competition experiments using cold rIL-2 and Scatchard analysis. The Northern analysis included the extraction of cellular RNA from both normal and patients' PHA blasts using a standard method. RNA (10 μg) was electrophoresed on agarose, blotted and hybridized with specific IL-2Rα and IL-2Rβ cDNA probes under high stringency conditions. The same filters were washed and re-hybridized with a β-actin cDNA probe serving as a semiquantitative positive control.

The investigations have been performed on blood mononuclear cells obtained from control (HIV-1 seronegative) and HIV-1-infected patients in various stages of their disease development as described below. These cells were isolated by Ficoll-Hypaque density centrifugation and the cells were resuspended in RPM1-1640 medium supplemented with 10% FCS, *L*-glutamine and antibiotics. LAK cells were also generated by incubating PBMC with 500 U/ml rIL-2 for 4 days at 37 °C in 5% CO_2 in air. NK and LAK activity was determined using ^{51}Cr-labelled K-562 and Daudi cells, respectively, the released radioactivity then being counted after a standard 4-hour release. In some experiments normal blood mononuclear cells (10^6/ml) were infected with 50 tissue culture units ($TCID_{50}$) of HIV-1. Control experiments were performed using heat-inactivated HIV. The infected cells were tested for reverse transcriptase activity, expression of IL-2Rα and IL-2Rβ chain and, finally, for expression of high affinity IL-2R in equilibrium binding experiments.

The Expression of High Affinity IL-2R Is Blocked by HIV-1 Infection

Initially, normal T cells were stimulated by PHA and infected with HIV-1 in vitro and the expression of an IL-2R chain was investigated. As soon as 48 h postinfection the expression of IL-2Rα chains was inhibited by $>$ 50% when compared to control values seen in normal T cells free of HIV-1-infection. Moreover, HIV-1-infected cells showed an almost threefold decrease in the number of high affinity binding sites/cell. Normally, on PHA-activated blasts 3,600 such receptors/cell were detected, and in HIV-1-infected cultures the observed value was 1,400 high affinity receptors/cell. When the cells were preincubated with a CD4 monoclonal antibody the inhibitory effect of HIV-1 was abolished and normal numbers of receptors were seen [12].

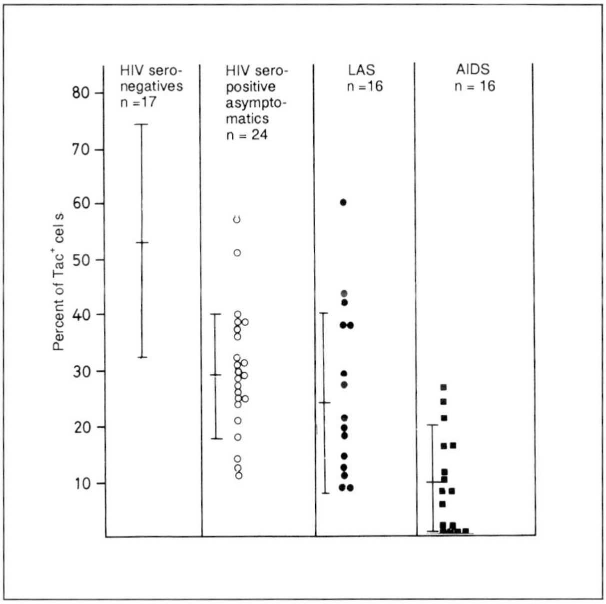

Fig. 2. Expression of IL-2Rα chain on PHA-activated blood mononuclear cells (10^6/ml) from normal blood donors and HIV-1-infected patients. The cells were stimulated with PHA-P (0.1% v/v) for 48 h and washed. The cells were stained with CD25 (IOT14) moAb using indirect immunofluorescence and analyzed by flow cytometry.

IL-2Rα Chain Expression on PHA-Stimulated Blasts from HIV-1-Infected Patients

When normal blood mononuclear cells are stimulated with strong mitogens such as PHA for 3 days, 50–60% of cells, mostly lymphoblasts, strongly express IL-2Rα detected by anti-CD25 moAbs (fig. 2). This newly generated IL-2R expression is handicapped in both HIV-1 seropositive asymptomatic individuals (CDC group II) and in patients with lymphadenopathy (CDC group III). The mean CD25+ cell numbers are 28% and 24.5%, respectively. In AIDS patients, the production of CD25+ blasts is virtually abolished (fig. 2). The impaired membrane expression of IL-2Rα is associated with a decreased accumulation of the corresponding mRNA transcript (fig. 3a). The decreased transcription of the IL-2Rα gene has also been reported by Prince et al. [9] and by our group [13]. Cross-linking experiments with

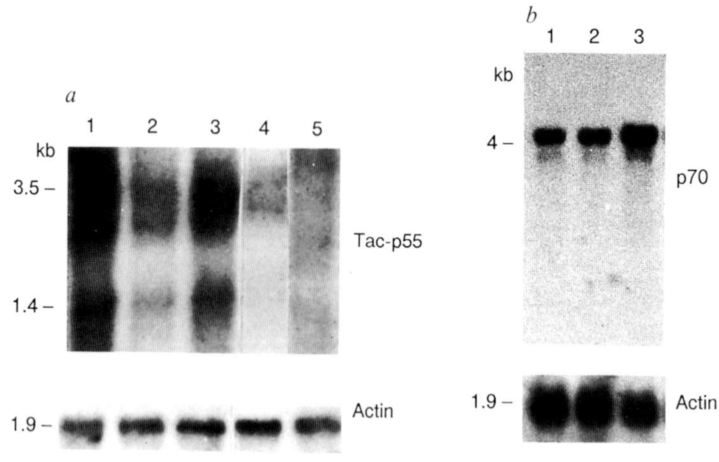

Fig. 3. Northern blot for IL-2Rα and IL2Rβ chain mRNA. Total RNA was extracted from either PHA-stimulated blood mononuclear cells taken from HIV-1 seropositive patients (*a*: lanes 2–5; *b*: lanes 1–2) or from normal blood donors (*a*: lane 1; *b*: lane 3). The blots were then hybridized with cDNA probes for IL-2Rα (*a*) and IL-2Rβ (*b*). The same filters were washed and rehybridized with a β-actin cDNA probe to give a positive control.

radiolabelled rIL-2 under low affinity conditions using PHA-activated patients' cells also indicated a decreased amount of functional receptors (fig. 4). The pattern was heterogeneous since in some cases neither IL-2Rα nor IL-2Rβ chains could be detected. However, in the majority of the studied cases, both IL-2R chains could be detected, albeit of decreased intensity in comparison to a similar cell number of normal T cells.

Analysis of IL-2Rβ Chain Expression on Lymphocytes from HIV-1-Infected Patients

In normal unstimulated lymphocytes taken from HIV-1 seronegative donors a hierarchy of IL-2Rβ subunit expression can be seen among the T-cell subsets. CD4+ cells are virtually negative and CD8+ cells express

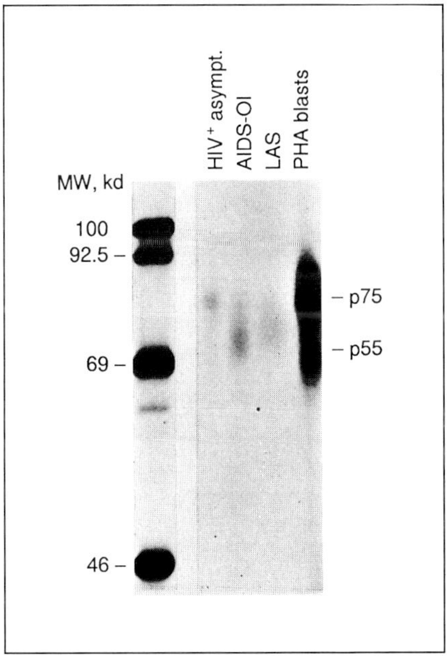

Fig. 4. Cross-linking of ^{125}I-rIL-2 on PHA blasts from HIV-infected patients at different clinical stages as described in the text. HIV$^+$ asympt. = Asymptomatic seropositive patients; AIDS-OI = AIDS patient presenting opportunistic infection; LAS = patient with persistent generalized lymphadenopathy.

detectable amounts of the β-subunit [4]. Further, within the CD8+ category, CD45RA+ ('naive' or unprimed) cells are weak expressors while CD45RO+ ('memory' or primed) cells display moderate amounts of IL-2Rβ chain (fig. 1) [5, 6]. NK cells are strongly positive [1]. Consequently, when the lymphocyte populations from a dominant CD4+ lymphocyte in the normal blood shift to CD45RO+ CD8+ cell dominance in HIV-1 infection one would expect an increase in IL-2Rβ expression, merely due to the population shift even without an increased activation. Instead, preliminary experiments shown in table 1 reveal that in the blood of HIV-1-positive patients there are decreased numbers of IL-2Rβ+ cells. This corresponds to a lower number of CD8+, IL2Rβ+ cells (table 1).

In figure 1 normal T lymphocytes from HIV-1 seronegative donors are activated by PHA and the relatively poor increase in IL-2Rβ positivity is shown as compared to the brilliant increase in IL-2Rα expression. Follow-

Table 1. Expression of IL-2Rβ chain on various T-cell subsets

Patient	Unstimulated cells		PHA-stimulated cells	
	CD4+/ TU27+	CD8+/ TU27+	CD4+/ TU27+	CD8+/ TU27+
Asympt-5	5[a]	11	24	21
Asympt-6	2	9	45	13
PGL-7	7	5	30	26
AIDS-5	5	3	9	9
AIDS-6	7	6	9	4
Normal HIV-seronegative controls (n = 6)	8±2	17±4	25±3	24±4

Both unstimulated and PHA-activated PBMC from HIV-infected subjects and normal blood donors were stained by double immunofluorescence using appropriate moAbs and cells were analyzed by flow cytometry. Numbers represent the code of each patient.
[a] Percentage of positive cells.

ing such a mitogenic stimulation of the lymphocytes from HIV-1-positive asymptomatic patients (CDC group II) and patients with lymphadenopathy (CDC group III) IL-2Rβ chain expression can be induced on both CD4 and the CD8 T cell subsets, but not on the lymphocytes from patients with AIDS (table 1). These observations have been confirmed by the study of mRNA for IL-2Rβ which showed smaller amounts of transcripts than seen in normal stimulated blasts (fig. 3b). Finally, with cross-linking by ^{125}I-rIL-2 a small amount of β-chain could be brought down in a proportion of cases (fig. 4). The different patients were heterogeneous in respect of α- and β-chain impairment, and this heterogeneity requires further analysis for correlations with other prognostic signs.

The Analysis of High Affinity IL-2R

The impaired membrane expression of both IL-2 binding polypeptides detected by flow cytometry and cross-linking analysis was corroborated with the equilibrium binding findings. The binding of ^{125}I-rIL-2 on the

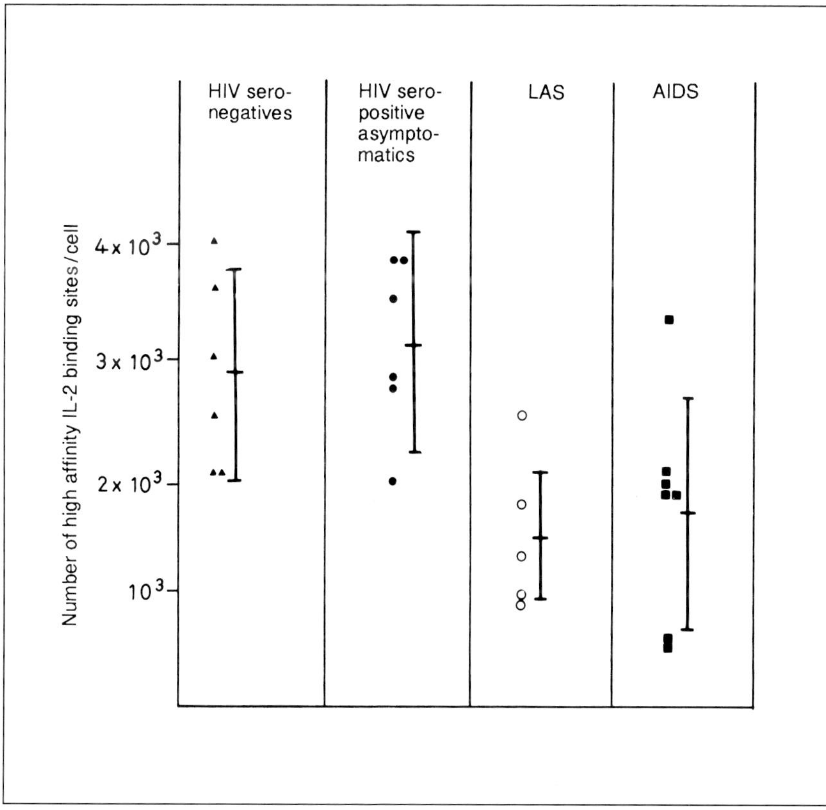

Fig. 5. Number of high affinity IL-2 binding sites: PHA-activated patients' cells for 48 h were used for equilibrium binding of ^{125}I-rIL-2 experiments as described in the text and the affinity of IL-2R was determined by Scatchard analysis. LAS = Persistent generalized lymphadenopathy patients.

patients' PHA-activated cells showed, in general, a decrease in the absolute number of high affinity binding sites/cell which was, again, more pronounced in PGL and AIDS (CDC groups III and IV, respectively) (fig. 5). In addition, in some PGL and AIDS patients a significant increase of the K_d was also observed indicating important abnormalities of the binding capacity of high affinity IL-2R expressed on these particular patients' cells (fig. 4). Moreover, in some patients (in 8 out of 10 cases tested) there was a decrease of the number of low affinity IL-2R (which corresponds to the

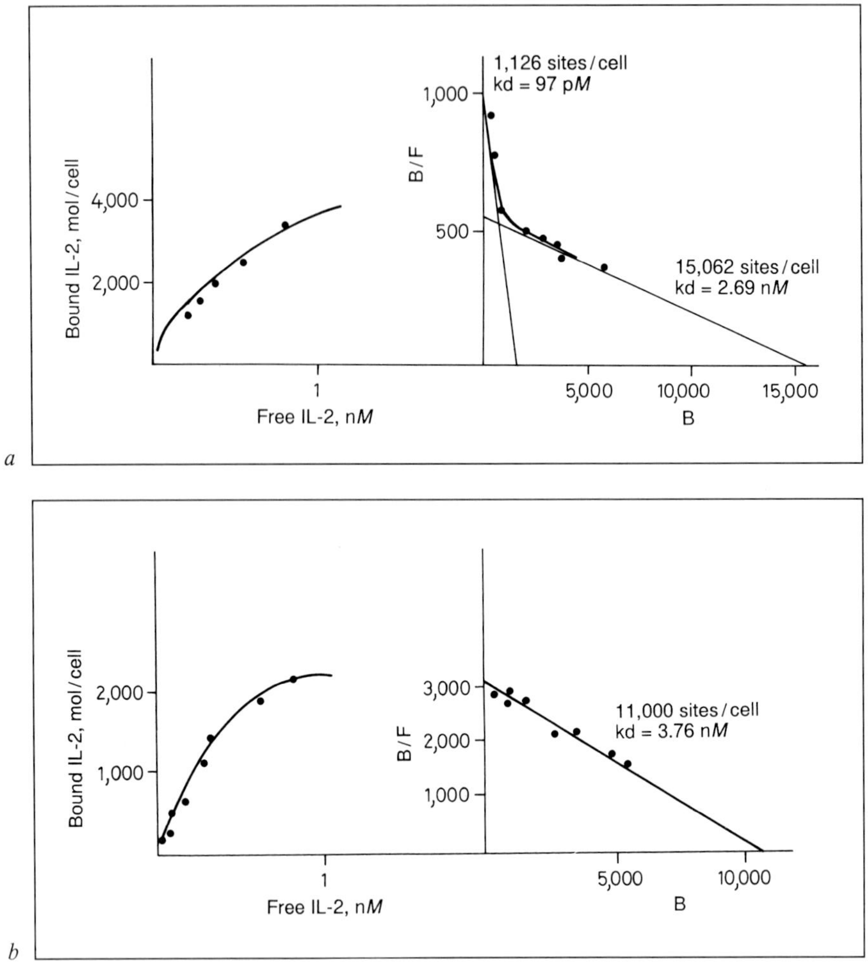

Fig. 6. Binding of ^{125}I-rIL-2 on HIV-infected patients' cells stimulated with PHA: typical binding experiments performed as described in the text, showing two types of IL-2r (*a*) or only one (*b*).

α-chain of IL-2R) (fig. 6a) whereas in others (in 3 out of 10 cases tested) only a low number of low affinity but not high affinity IL-2R could be detected (fig. 6b). These observations taken together strongly suggest that several different mechanisms are responsible for these abnormalities.

Table 2. NK and LAK activity of HIV-infected patients' PBMC

Patient	Percent of cytotoxicity	
	K-562	Daudi
Asympt-22	0[a] (0)[b]	11 (0)
Asympt-23	3 (0)	43 (10)
Asympt-24	11 (8)	0 (4)
PGL-24	67 (27)	70 (9)
PGL-25	95 (ND)	0 (0)
AIDS-23	42 (0)	0 (3)
AIDS-24	71 (18)	34 (0)
Normal seronegative controls (n = 6)	51 ± 20 (15 ± 17)	25 ± 19 (3 ± 4)

PBMC were incubated in the presence of 5,000/ml rIL-2 for 72 h, and washed cells were incubated with ^{51}Cr-labelled K-562 or Daudi target cells at different effector:target cell concentrations. Numbers represent the code of each patient. ND = Not determined.

[a] The results represent the percentage of specific lysis at an effector:target concentration of 20:1.

[b] The percentage of specific lysis before cell incubation with rIL-21 is represented in parentheses.

The Functional Consequences of Impaired IL-2R Expression

IL-2Rβ chain can transduce T cells and activate the cytolytic activity of CD16+ NK cells [14]. The simplest model to analyze the functional impairment and signal transduction of the IL-2R complex is to study NK cell function after incubation in culture medium enriched in rIL-2. These NK cells from HIV-1 seropositive patients may express some β-chain but in several patients tested high concentrations of rIL-2 failed to increase the NK activity (table 2). In addition, when rIL-2-induced cells from the HIV-1-positive patients were tested for their capacity to lyse NK-resistant Daudi cells, no induction of LAK activity could be detected in a proportion of patients (table 2). There was no correlation between the impaired rIL-2-induced NK and LAK activity in individual patients' cells. Although these functional defects could be due to the decreased constitutive expression of IL-2Rβ chain on CD8+ cells as showed by several experimental approaches, we cannot exclude that additional mechanisms might also be involved. In particular, it should be very interesting to investigate the expression of second messenger and the different pathways of cell activation since it has already been shown that impaired anti-CD3-induced cell activation of HIV-infected T cells is associated with a decreased intracytoplasmic accumulation of Ca^{2+} [15].

Conclusions

The results presented here clearly demonstrate that HIV infection induces significant abnormalities in the expression of IL-2R chains. Indeed, HIV-1 infection in vitro of normal T cells induces a decrease of both IL-2Rα and IL-2Rβ chain expression which is associated with an impaired number of high affinity binding sites/cell. A similar phenomenon is observed studying the cells taken from HIV-1-infected patients. The decreased expression of IL-2R, and its impaired function, demonstrated in model experiments with NK cells following incubation with IL-2, is particularly relevant when one considers the large increase in cells with CD45RO+ phenotype in HIV-1 infection. These cell types have a higher degree of 'constitutive' high affinity IL-2R expression than the 'naive' CD45RA+ cells and still the IL-2R profiles in the HIV-1 disease are decreased rather than increased. This phenomenon may contribute to the early cellular defects on T-cell proliferation and the preferential demise of CD45RO+ cells in cultures following powerful mitogenic, and presumably also after antigenic stimulation (see also Miedema et al., and Janossy, Borthwick et al., this volume).

The abnormalities of the IL-2R system in HIV-infected patients may have a crucial importance during the impaired T-cell function in HIV infection. Moreover, the heterogeneity of the IL-2R chain expression during the various clinical stages seems to indicate that different mechanisms should operate from these abnormalities. Several mechanisms might be involved. We have reported that HIV-infected macrophages spontaneously release a 29-kd cationic protein which inhibits mitogen-induced expression of IL-2Rα but not IL-2β chain [16, 17]. In addition, others have demonstrated that viral pg120 envelope glycoprotein induces an impaired expression of IL-2Rα chain [18]. The observation that AIDS patients' cells displayed the more profound abnormalities also raises the question whether other non-HIV-1-related secondary factors could be involved in the pathogenesis of the abnormal expression of IL-2R chains as well.

References

1 Smith KA: The structure of high affinity IL-2 receptor: α and β chains. Immunol Today 1987;8:11.
2 Robb RJ, Greene WC, Rusk CM: Low and high affinity cellular receptors for interleukin-2: Implications for the level of Tac antigen. J Exp Med 1984;160:1126.
3 Lowenthal JW, Greene WC: Contrasting interleukin-2 binding properties of the α (p55) and β (p70) protein subunits of the human high affinity interleukin-2 receptors. J Exp Med 1987;166:1156.
4 Ohashi Y, Takeshita T, Nagata K, Mori S, Sugamura K: Differential expression of the IL-2 receptor subunit, p55 and p75 on various populations of primary peripheral blood mononuclear cells J. Immunol 1989;143:3548.
5 Taga K, Kasahara Y, Yachie A, Miyawaki T, Taniguchi N: Preferential expression of IL-2 receptor subunits on memory populations whithin CD4+ and CD8+ T cells. Immunology 1991;72:15.
6 Akbar AN, Salmon M, Ivory K, Taki S, Pilling D, Janossy G: Human nalve and memory CD4+ T cells synergize in response to alloantigens. Eur J Immunol 1991;21: 2517–2522.
7 Ciobanue N, Welbe K, Kruger G, Venuta S, Gold J, Veldman SP, Wang CY, Kozimer B, Moore MAS, Safal B, Mertelsmann R: Defective T cell response to PHA and mitogenic monoclonal antibodies in male homosexuals with acquired immunodeficiency syndrome and its in vitro correction by interleukin-2. J Clin Immunol 1983;3:332.
8 Hoffmann B, Langhoff E, Linchardt BO, Odum N, Hylding-Nielsen JJ, Ryder LP, Platz P, Jakobsen BK, Bendtzen K, Jakobsen N, Lerche B, Shaffer-Nielsen C, Dickmeiss E, Ulrich K, Svejgaard A: Investigation of immunosuppressive properties of inactivated human immunodeficiency virus and possible neutralization of this effect by some patient sera. Cell Immunol 1989;121:336.
9 Prince HE, Kermani-Arab V, Fahey JK: Depressed interleukin-2 receptor expression in acquired immunodeficiency and lymphadenopathy syndromes. J Immunol 1984;133:1313.
10 Dukowitz M, Wano Y, Le Thi Bich-Thuy, Katz P, Gullen BP, Kehrl JH, Greene WC: A second human interleukin-2 binding protein that may be a component of high affinity interleukin-2 receptors. Nature 1987;327:518.
11 Allouche M, Sahraoui Y, Augery-Bourget Y, Ohashi Y, Sugamura K, Jasmin C, Georgoulias V: Presence of a p70-IL-2-binding peptide on leukemic cells from various hemopoietic lineages. J Immunol 1989;143:2223.
12 Sahraoui Y, Ammar A, Lunardi-Iskandar Y, Tsapis A, Spanakis E, N'Go N, Allouche M, Gay Bellile V, Januin C, Georgiadis V: Abnormal expression of IL-2Rβ (p70) binding polypeptide on HIV-infected patients' cells. Cell Immunol, in press.
13 Allouche M, Lunardi-Iskandar Y, Varela-Millot C, Itzaki M, Cornuet P, Meyer P, Jasmin C, Georgoulias V: Effect of phorbol myristate acetate on T cell colony formation, IL–2R expression and IL-2 production by cells from patients at all stages of HIV infection. Clin Exp Immunol 1990;81:200.
14 Ishii T, Takeshita T, Numato N, Sugamura K: Protein phosphorylation mediated by IL-2/IL-2 receptor β-chain interaction. J Immunol 1988;141:174.

15 Linette GP, Hartzman RJ, Lebdetter JA, June CH: HIV-1-infected T cells show a selective signalling defect after perturbation of CD3/antigen receptor. Science 1989; 241:573.
16 Ammar A, Cibert C, Bertoli AM, Tsilivakos V, Jasmin C, Georgoulias V: Biological and biochemical characterization of a factor produced spontaneously by adherent cells of human immunodeficiency virus-infected patients inhibiting Tac expression on normal T cells. J Clin Invest 1991;87:2048.
17 Ammar A, Bertoli AM, Sahraoui Y, Perraki M, Tsilivakos V, Meyer P, Rozenbaum W, Jasmin C, Georgoulias V: Production of a Tac inhibitory activity by adherent cells of HIV-infected subjects at different clinical stages. J AIDS, in press.
18 Diamond DC, Sleckman BP, Gregory T, Lashky L, Greenstein JL, Burakoff SL: Inhibition of CD4+ cell function by the HIV envelope protein gp120. J Immunol 1988;141:3715.

Yasmina Sahraoui, MD, Oncogenèse Appliquée, INSERM U268,
Hôpital Paul-Brousse, 14–16, av. Paul-Vaillant-Couturier,
F–94804 Villejuif (France)

Janossy G, Autran B, Miedema F (eds): Immunodeficiency in HIV Infection and AIDS.
EC/FERS/MRC Workshop on Immunodeficiency in HIV-1 Infections, Windsor, Surrey, 1991.
Basel, Karger, 1992, pp 171–184

Activated CD8+ Cells in HIV-Related Diseases

Brigitte Autran[a], *Janis V. Giorgi*[b]

[a]Laboratoire d'Immunologie Cellulaire et Tissulaire URA CNRS 0625, CERVI,
CH Pitié-Salpêtrière, Paris, France, and
[b]Cellular Immunology and Cytometry, UCLA School of Medicine,
Los Angeles, Calif., USA

HIV infection generates a strong cellular immune response in the host [1–3] together with profound changes in the immune system which are characterized by a major decline of CD4 cells and the activation of CD8 lymphocytes [4–6]. Indeed, CD8 cell numbers increases at the time of seroconversion in much the same way as observed in CMV or EBV infections [7, 8]. However, the CD8 lymphocytosis resolves in a few weeks or a few months after EBV and CMV infection, respectively, while it persists throughout HIV disease progression. A tissue infiltration of CD8+ cells is also frequently observed [9, 10] that can culminate in a condition referred to as diffuse infiltrative CD8 lymphocytosis syndrome (DILS) [11].

The CD8 cell subset contains mature T lymphocytes which have differentiated within the thymus. The distinct steps of the T-cell differentiation pathway are characterized by a sequential acquisition and loss of cell surface molecules (table 1). When leaving the thymus, CD8 cells recirculate as resting T cells. CD8+ resting precursors of CTL are triggered by an interaction of their T-cell receptor with viral peptides bound to major histocompatibility complex (MHC) antigens on the membrane of virus-infected cells [12]. This recognition event induces a controlled program of T-cell activation leading to the transient re-expression of some surface molecules that were previously expressed within the thymus, and also referred to as T-cell activation markers: i.e., CD25, CD38, as well as HLA-DR (table 1). CD8 cell activation induces differentiation into memory cells and effector CTL engaged in lysis of target cells (table 1).

Table. 1. Heterogeneity of human CD8+ cells when analyzed with commonly used antibodies

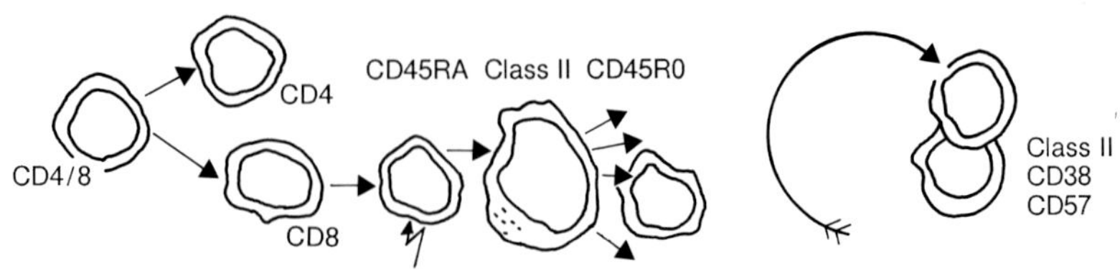

	A. Immaturity		B. Activation			C. (Irregular?) regeneration	
	Thymic differentiation (thymic acute leukemia, convoluted lymphoma)		Antigen-driven immune response (mitogenic stimulation in vitro, acute virus infections, e.g. EBV)			After immunosuppression (bone marrow or solid organ transplants), chronic virus infections, e.g. CMV and HIV-1, chronic GvH disease	
Markers	Cortical	Medullary	'Naive'	Transit blasts	'Memory'	Combinations of markers that define cells which survive poorly in vitro	
TdT[a] (nuclear enzyme)	+	–					
CD1[a] (NA1/34)	+	–				–	
CD38[b] (T10, Leu17)	++	+	+	+	–	++	
CD45RA[c] (2H4)	–	–/+	+	±	–	–	
CD45R0[c] (UCHL1)	+	+/–	–	±	+	+	
Class II[d] HLA-DR	–	–	–	+	–	++	late HIV: ±
CD25 (IL2R)[e] Tac	–	–	–	++	±	±/–	very low in HIV
Leu8[f] LAM-1	–	+	+	+	–	–	
CD29[g] (4B4)	+/–	±	±	+	+	+	not the same as CD45R0
CD57[h] (LEU-7)	–	–	–	–	– (+NK)	–/++	population heterogeneity

In humans and animals, CD8 lymphocyte-mediated antiviral activity, especially MHC class I restricted cytotoxicity, develops following infection with many viruses. This activity is instrumental in mediating recovery from primary viral infection and accelerating recovery from subsequent infections [13–16]. CD8+ CTL specific for most of the HIV gene products, including *env, gag, pol, nef* and *vif,* have been described in HIV-infected individuals at early stages of the infection [1, 2, 17–21]. The ability to detect HIV-specific effector CTL that have differentiated in vivo is a striking feature of the cellular immune response to HIV (see also Rowland-Jones and McMichael, this volume). This phenomenon contrasts with the need to re-stimulate CTL in vitro in order to detect cytotoxicity specific for other viruses such as EBV or influenza. The persistent replication of HIV may be responsible for a permanent stimulation and differentiation of HIV-specific CTL, therefore inducing high frequencies of activated effector CTL as observed at early stages of the disease [20, 21].

The protective efficiency of HIV-specific CTL is suggested by the capacity of CD8+ lymphocytes to inhibit the in vitro replication of HIV or SIV in CD4+ lymphocytes by both cytotoxic and nonlytic mechanisms [22–24]. In contrast, the CD8+ CTL may be deleterious for the host by clearing immunocompetent cells expressing HIV antigens, and thus participate to the progressive CD4+ cell lymphopenia [25]. Similarly, HIV-specific CD8+ CTL that are recruited in tissues may have a role in the pathogenesis of some HIV-related disorders by inducing functional alterations [2, 21, 25–27].

Footnote to table 1.

---

[a] Early stages of intrathymic development (thymic immaturity).
[b] Thymic immaturity plus activation.
[c] This pair of CD45R reagents are complementary in defining 'naive' and 'memory' cells.
[d] Activation marker, frequently transiently expressed on T blasts.
[e] Activation marker which is normally retained in small amounts on 'memory' T cells.
[f] Human leukocyte adhesion molecule-1.
[g] The antigen is the integrin β_1-chain, an adhesion molecule which is widely distributed in tissues. On CD4 T cells, the CD45RO+ and the CD29+ populations are largely overlapping, and referred to as memory cells. On CD8 T cells, the CD45RO+ cells and the CD29+ are not equivalent; CD45RO+ CD8+ cells are memory cells, but the function of the CD29+ CD8+ population is unknown.
[h] An NK cell-associated marker which on CD3+ CD8+ cells designates a non. NK subset with suppressive action as well as cytotoxic potential (see Sadat-Sowti et al., this volume).

Table 2. Selective increase in CD8+ subsets in HIV infection

Study group[1]	n[2]	Total CD8+ mm[3]	Percent proportion of total CD8+ with select phenotype			
			CD38+	HLA-DR+	Leu8–	CD57+
Hetero	50	541	19	5	42	19
MR-neg	86	621	21	8	49	35
HR-neg	25	591	15	5	53	34
ST-pos	34	826	31	15	62	32
LT-pos	83	882	43	29	64	34
AIDS	32	891	69	20	76	40

[1] Hetero: Heterosexual seronegative; MR-neg: medium risk seronegative homosexual; HR-neg: high risk seronegative homosexual; ST-pos: short-term seropositive; LT-pos: long-term seropositive.
[2] The minimum numbers of samples studied for each marker are shown. For HLA-DR+, CD38+ and Leu8– higher numbers of ST-pos cases were analyzed (n = 47).

There is a body of evidence indicating a loss of CTL responses along with progression towards AIDS; however, mechanisms responsible for CTL dysfunction are still poorly understood. In AIDS patients, CD8+ cells expressing T-cell activation markers [6] have been shown to lose both their cytolytic and clonogenic potential [28] (see Pantaleo et al., this volume). Similarly, CD8+ T cells expressing the CD57 antigen are expanded in stage IV patients and release a 20–30 kD glycoprotein inhibiting cytolytic T and NK cell activities [21, 29] (see Sadat-Sowti et al., this volume).

For these reasons, the question is raised whether such an activation of CD8 cells may both contribute to protection towards AIDS and participate to the pathogenesis of the HIV-related disorders and immune dysfunction.

CD8 Lymphocyte Activation in HIV Infection

The CD8+ cells are increased in every subgroup of HIV-infected patients including recently diagnosed AIDS patients as shown in table 2 (see also figure 2 in Janossy, Jiang, et al., this volume). Advanced disease is accompanied by somewhat lower absolute numbers of CD8 cells which is

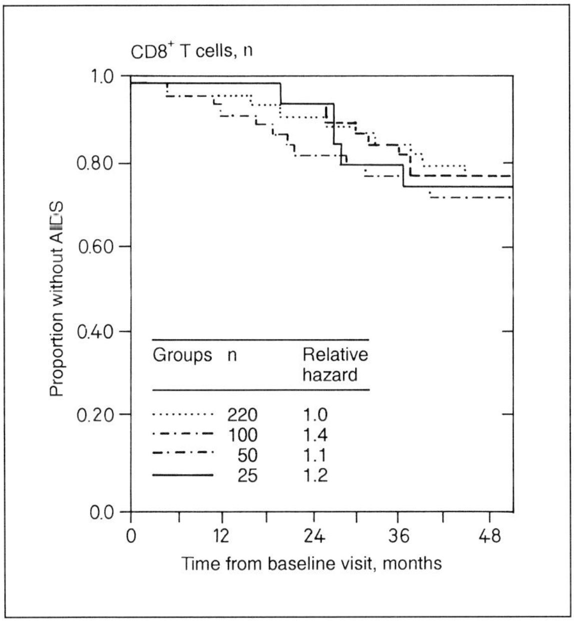

Fig. 1. Lack of prognostic significance of total CD8+ cell level in HIV infection: Kaplan-Meier plots of the proportion of individuals without AIDS of those men studied in Fahey et al. [30] and shown in figure 7 (Giorgi, this volume) based on the CD8+ measurements.

due to the development of pan-leukopenia at this stage. Total CD8 values are not prognostic for development of AIDS as shown in the Kaplan-Meier plots for survival in figure 1.

The immunophenotype of the CD8+ cells in the circulation of HIV-infected people reflects the activation of their immune systems. Dual and triple color immunofluorescence studies have demonstrated that distinct CD8+ subsets are elevated in HIV-1 disease [6, 31–38]. The activation markers appear on CD8+ cell surface at the time of seroconversion and remain throughout HIV disease progression. Table 2 shows elevations in several of the CD8+ subsets which we have examined. CD8+ populations evaluated in our studies include those defined by the leukocyte differentiation antigens CD38 [39–41], HLA-DR (MHC class II), Leu8 [42, 43] and CD57 [21, 29]. The type of CD8+ cells present in the circulation of HIV-1-infected individuals reflects the stage of the individual's disease. Early

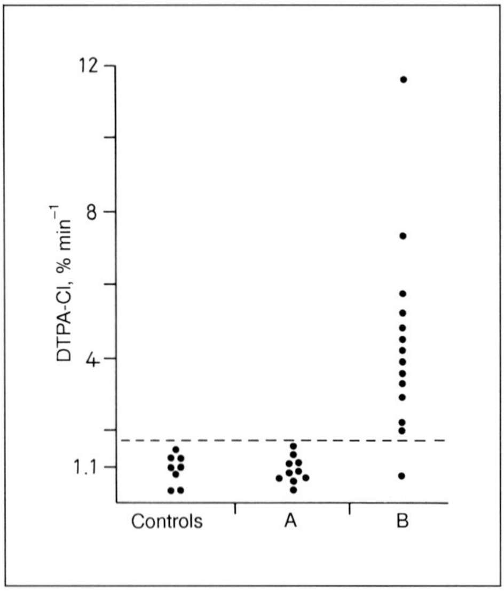

Fig. 2. Correlation between the CTL activity of alveolar lymphocytes and alveolocapillary membrane permeability. Clearance of inhaled 99Tc^{m} DTPA in HIV-infected patients without (controls) or with (A and B) a lymphocytic alveolitis. Fresh alveolar lymphocytes from group A patients did not lyse autologous alveolar macrophages, whereas a cytolytic activity was observed with lymphocytes from group B patients (range 9–60% at a lymphocyte to macrophage ratio of 20:1).

after seroconversion, the expanded CD8+ subsets have the phenotypes Leu8–, CD45RA–, CD38+ and HLA-DR+. Subsequently, there is a continued rise in the number of CD38+ cells and this subset ultimately is highly associated with advanced HIV disease [6, 36, 44]. Increased proportions of CD8+ cells expressing CD38 are correlated with several of the activation markers (sIL-2R, β_2-microglobulin, neopterin and soluble CD8 (sCD8) [45]) which are elevated in HIV-1-infected persons.

We have performed three color immunofluorescence studies to clarify the relationship between the coexpression of different activation antigens on CD8+ cells. It appears that the CD8+ cells that are HLA-DR+ but CD38– (i.e., CD8+ HLA-DR+ CD38–) are more abundant in individuals who have stable HIV disease. In contrast. individuals with very advanced HIV disease have many CD38+ cells (i.e., CD8+ CD38+ HLA-DR+ or CD8+ CD38+ HLA-DR–) but almost no CD8 cells which express HLA-

DR but no CD38 (i.e., CD8+ CD38− HLA-DR+). Studies are underway in our laboratory to determine the prognostic significance of elevations in distinct CD8+ subsets and to test which CD8+ subsets have anti-HIV activity in assays of control of viral replication and cytotoxicity.

Contribution to Disease Pathogenesis of Activated CD8 Cytotoxic T Lymphocytes Specific for HIV

A lung alveolitis composed of CD8+ lymphocytes is frequently observed in up to 60% of HIV-infected patients irrespective of their clinical staging [10]. In stage II or III patients, this massive CD8+ cell infiltration contains cells with cytotoxic potential as suggested by their D44-mAb staining [46]. Up to 35% of the alveolar CD8+ cells coexpress the HLA-DR and CD38+ antigens but lack the IL-2 receptors and transferrin receptors (table 3). In these patients, the alveolar CD8+ cells mediate spontaneous cytotoxic activity against MHC compatible alveolar macrophages or cells expressing the HIV-1 *env, gag* and *nef* antigens [2, 20, 27]. These data indicate that the CD8+ cells detected in the HIV-related lung alveolitis of stage II and III patients are activated and differentiated in vivo into HIV-specific CTL.

The physiopathological consequences of the activation of HIV-specific CTL in vivo have been analyzed by radioclinical studies and pulmonary function tests performed in parallel to CTL assays in vitro. Figure 2 shows the relationship between the alveolar CTL activity and the increase of the alveolocapillary membrane permeability, as assessed by the clearance of technetium-labelled diethylenetriamine pentaacetate (DTPA) [47]. These clinical studies and pulmonary function tests indicate a strong correlation between the presence of HIV-specific CTL in lungs and the appearance of clinical abnormalities. Similarly, the transfer of syngeneic CTL specific for the respiratory syncytial virus has been shown to induce a lethal pneumonia in infected mice [48]. Therefore, the local immune conflict between differentiated CD8+ cytotoxic T cells specific for HIV and alveolar macrophages could be responsible for tissue injuries such as lesions of the lung epithelium.

Cytotoxic T cells specific for HIV have also been demonstrated in the cerebrospinal fluid from patients with HIV-related neurological disorders [26]. Moreover, activated CD8+ cells infiltrate the germinal centers of lymphoid organs, releasing various cytokines such as γ-interferon [49]. In

Table 3. Comparative analysis of CD8 cell surface phenotypes in four HIV-related disorders and after BMT

Patients	Cells evaluated	CD3+	CD4+	CD8+	CD8+ CD57+	CD8+ CD25+	CD8+ CD71+	CD8+ DR+	CD8+ CD38+
Alveolar lymphocytosis	alveolar lymphocytes	87±6	10±8	75±15	15±9	4±4	2±1	30±17	ND
CD8+ hyperplasia (>3×10⁹/l) with splenomegaly	PBL	75±13	5±2	70±15	38±8	0.5±5	2.5±2	23±13	41±11
CD8+ hyperplasia (>1.5×10⁹/l) no splenomegaly	PBL	73±13	19±7	50±18	24±10	0.5±5	1±9	18±16	27±17
ITP	spleen cells	56±9	8±8	41±6	9±1.5	1±0.1	ND	12±3	15.5±4
BMT	PBL	57±21	21±9	48±20	9±9	0.5±5	ND	21±20	ND

The numbers of CD8+ cell evaluations in alveolar lymphocytosis, CD8+ hyperlymphocytosis with splenomegaly, or without splenomegaly, ITP, or in the first 3 months after allogeneic bone marrow transplantation (BMT) were 15, 3, 8 and 10 respectively.
Results are expressed as geometric mean and standard deviations of percent of positive lymphocytes in the overall lymphocyte population studied.

part, these CD8 cells represent differentiated CTL directed against the *env, gag* and *nef* HIV proteins [27] which could indeed play a role in the progressive lymphoid depletion observed in AIDS.

Taken together, these observations strongly suggest that as a result of their activation HIV-specific CTL develop into effector killer cells that are capable of recognizing host cells expressing HIV antigens at a time when immune functions are still intact. These virus-specific activated CD8+ CTL may thereby contribute to the clinical expression of HIV-related disorders.

HIV-Related CD8+ Hyperplasia

A persistent CD8+ hyperlymphocytosis can be observed in some HIV-infected patients. Indeed, in Pitié-Salpêtrière Hospital in Paris, a stable CD8+ hyperlymphocytosis (over 1,500/mm^3 for more than 2 months) was present in 10% of HIV seropositive patients [50]. This group of 131 patients was composed of 55% stage II or III patients (CD8+ cells 1,699/ mm^3, mean CD4+ cells = 410/mm^3) and 45% stage IV patients (mean CD8+ cells = 1,887/mm^3; CD4+ cells = 311/mm^3) indicating that this profound CD8 hyperplasia was not related to the stage of the disease. In these patients, signs of a diffuse lymphoid cell infiltration similar to that seen in DILS were absent. CD8+ hyperlymphocytosis was associated with good prognosis since the onset of AIDS in those patients with sustained CD8+ hyperplasia was significantly delayed when analyzed 18 months later compared with patients in whom CD8+ cells had fallen below 1,500/mm^3 [50].

Extreme cases of such CD8+ hyperplasia have also been reported in 3 stage III patients with a massive hyper-CD8+ lymphocytosis ($> 3\times10^9$/liter) and pseudotumoral splenomegaly. The histopathological analysis of lymphoid organs showed no B or T cell clonal proliferation but an intense polyclonal CD8+ lymphoid hyperplasia with an immunoblastic cell infiltration [51, and E. Oksenhendler, in preparation].

The phenotypes of the CD8+ PBL and spleen cells from stage III patients with CD8+ hyperlymphocytosis, regardless of whether or not they had splenomegaly, were similar to those observed in spleen cells from HIV+ patients splenectomized for idiopathic thrombopenic purpura (ITP) (table 3). The same CD8 phenotypes were also observed in early regenerating PBL after allogeneic bone marrow transplantation. Thus, CD8+ cells in all these cases had similar expression of T-cell activation markers with high levels of CD8+ cells coexpressing the HLA-DR and CD38 but not the CD25 or CD71 antigens. However, we observed a higher expression of CD38 than of HLA-DR antigens on the CD8+ cells from patients with pseudotumoral CD8+ hyperlymphocytosis. The numbers of CD8+ CD57+ cells in patients with CD8+ hyperlymphocytosis were also higher as compared with the values observed in CD8+ alveolar lymphocytosis or in ITP.

The recent report of CD8+ DILS suggested that CD8+ cells could be expanded from cytotoxic CD8+ cells specific for HIV proteins. The question of whether this pseudotumoral proliferation of CD8+ cells contained high levels of CTL specific for HIV proteins was therefore investigated. However, neither effector CTL nor precursor cells of CTL specific for the *env* and *gag* HIV-1-BRU proteins could be detected in the CD8+ PBL or in spleen cells from these patients [51]. These results differed from those of studies performed in parallel on spleen cells from patients splenectomized for ITP in which we detected effector CTL specific for most of the HIV proteins [27]. In contrast, the CD8+ CD57+ lymphocytes contained in the CD8+ hyperlymphocytosis inhibited in vitro NK and CTL cytolytic effector cell activities as described in Sadat-Sowti et al., this volume. Furthermore, after depleting the CD8+ CD57+ cells from the suspension, HIV-specific CTL responses still remained low in this CD8+ cell hyperplasia, suggesting that inhibitory effects from CD8+ CD57+ cells were not responsible for the apparent CD8 cell anergy. Overall, these data indicate that the HIV-related pseudotumoral expansion of partially activated, polyclonal CD8+ cells does not result in the accumulation of functional CTL clones specific for HIV.

Conclusions

Activation of CD8+ cells is a hallmark of the HIV infection and activated CD8+ cells are detectable throughout the infection in the peripheral blood and in the tissues. Various HIV-related diseases including idiopathic CD8+ lymphocytic alveolitis, ITP or pseudotumoral splenomegaly are characterized by an expansion or an infiltration of CD8+ cells. The activated status of the CD8+ cells in HIV disease is manifest by a somewhat dissociated cell surface expression of several T-cell activation markers. A critical issue is whether this expression of markers, classically referred to as T-cell activation markers, reflects a true, antigen-induced CD8+ T-cell activation or only a local or general virus-induced production of cytokines which lead to indiscriminate activation of irrelevant CD8+ clones with no apparent antiviral specificity.

The 'activated' phenotype of CD8+ cells reflects diverse functional cell types. Indeed, at early stages of the disease, CD8+ cells are activated in vivo and differentiate into cytotoxic effector cells exhibiting high frequencies of HIV-specific CTL. These CTL recognize host cells expressing HIV antigens in organs such as lungs and lymphoid organs at a time where immune functions are still intact. These activated virus-specific CD8+ CTL may therefore be responsible for tissue injuries, thus contributing both to the control of HIV replication and to the clinical expression of HIV-related diseases.

In contrast, HIV-specific CTL are not detectable within the CD8+ HLA-DR+ cells from AIDS patients or from pseudotumoral CD8+ cell proliferation. The mechanisms underlying this loss of CD8+ cell function remain poorly understood. Several hypotheses have been proposed involving a viral escape mutant mechanism (see Rowland-Jones and McMichael, this volume) or a state of anergy which could be related either to the loss of T-helper cell function, or to a nonrandom loss of T-cell repertoire. Finally, the continuous activation of the immune system and mainly of CD8+ cells may also play a major role in the CD8+ cell dysfunction and in the loss of the immune control of HIV replication.

References

1 Walker BD, Chakrabarti S, Moss B, Paradis TJ, Flynn T, Durno AG, Blumberg RS, Kaplan JC, Hirsch MS, Schooley RT: HIV-specific cytotoxic T lymphocytes in seropositive individuals. Nature 1987;328:345.

2 Plata F, Autran B, Martins LP, Wain-Hobson S, Raphael M, Mayaud C, Denis M, Guillon JM, Debré P: AIDS virus-specific cytotoxic T lymphocytes in lung disorders. Nature 1987;328:348–351.

3 Autran B, Plata F, Debré P: MHC restricted cytotoxicity against HIV. J AIDS 1991; 4:361–367.

4 Munoz A, Carey V, Saah AJ, Phair JP, Kingsley LA, Fahey JL, Ginzburg HM, Polk BF, and the Multicenter AIDS Cohort Study: Predictors of decline in CD4 lymphocytes in a cohort of homosexual men infected with human immunodeficiency virus. J AIDS 1988;1:396–404.

5 Fahey JL, Prince H, Weaver M, Groopman J, Visscher B, Schwartz K, Detels R: Quantitative changes in T helper or T suppressor/cytotoxic lymphocyte subsets that distinguish acquired immune deficiency syndrome from other immune subset disorders. Am J Med 1984;76:95–100.

6 Giorgi JV, Detels R: T cell subset alterations in HIV-infected homosexual men: NIAID multicenter AIDS cohort study. Clin Immunol Immunopathol 1989;52:10–18.

7 Schooley RT, Hirsch MS, Colvin RB, Cosimi AB, Tolkoff-Rubin NE, McCluskey RT, Burton RC, Russell PS, Herrin JT, Delmonico FL, Giorgi JV, Henle W, Rubin RH: Association of herpesvirus infections with T lymphocyte subset alterations, glomerulopathy, and opportunistic infections after renal transplantation. N Engl J Med 1983;308:307–313.

8 Crawford DH, Brickell P, Tidman N, McConnell I, Hoffbrand AV, Janossy G: Increased numbers of cells with suppressor T cell phenotype in the peripheral blood of patients of infections mononucleosis. Clin Exp Immunol 1981;43:291–297.

9 Guillon JM, Fouret P, Mayaud C, Picard F, Raphael M, Touboul JL, Chaunu MP, Hauw JJ, Akoun G: Extensive T8-positive lymphocytic visceral infiltration in a homosexual man. Am J Med 1987;82:655–661.

10 Guillon JM, Autran B, Denis M, Fournet P, Plata F, Mayaud C, Akoun G: HIV-related lymphocytic alveolitis. Chest 1988;94:1264–1270.

11 Itescu S, Brancato LJ, Buxbaum J, Gregersen PK, Rizk CC, Croxson TS, Solomon GE, Winchester R: A diffuse infiltrative CD8 lymphocytosis syndrome in human

immunodeficiency virus infection: a host immune response associated with HLA-DR5. Ann Intern Med 1990;112:3–10.
12 Paul WE: Fundamental Immunology, ed 2. New York, Raven Press, 1989.
13 Yap KL, Ada GL, McKenzie IFC: Transfer of specific cytotoxic T lymphocytes protects mice inoculated with influenza virus. Nature 1978;273:238–240.
14 Quinnan GV, Kirmani N, Rook AH, Manischewitz JF, Jackson L, Moreschi G, Santos GW, Saral R, Burns WH: Cytotoxic T cells in cytomegalovirus infection. N Engl J Med 1982;307:7–13.
15 McMichael AJ, Gotch FM, Noble GR, Beare PAS: Cytotoxic T-cell immunity to influenza. N Engl J Med 1983:309:13–17.
16 Oldstone MBA, Blount P, Southern PJ, Lampert PW: Cytoimmunotherapy for persistent virus infection reveals a unique clearance pattern from the central nervous system. Nature 1986;321:239.
17 Nixon DF, Townsend ARM, Elvin JG, Rizza CR, Gallwey JJ, McMichael AJ: HIV-1 gag-specific cytotoxic T lymphocytes defined with recombinant vaccinia virus and synthetic peptides. Nature 1988;85:484–487.
18 Rivière Y, Tanneau-Salvadori F, Régnault A, Lopez O, Sansonetti P, Guy B, Kieny MP, Fournel JJ, Montagnier L: Human immunodeficiency virus-specific cytotoxic responses of seropositive individuals: distinct types of effector cells mediate killing of targets expressing gag and env proteins. Virology 1989;63:2270–2077.
19 Koenig S, Earl P, Pawell D, Pantaleo G, Merli S, Moss B, Fauci AS: Group-specific, MHC class I restricted cytotoxic responses to HIV-1 envelope proteins by cloned peripheral blood T cells from an HIV-1 infected individual. Proc Natl Acad Sci USA 1988;85:8638–8642.
20 Hoffenbach A, Langlade-Demoyen P, Vilmer E, Dadaglio G, Michel F, Mayaud C, Autran B, Plata F: Very high frequencies of HIV-specific cytotoxic T lymphocytes in humans. J Immunol 1989;142:452–456.
21 Joly P, Guillon JM, Mayaud C, Plata F, Theodorou I, Denis M, Debré P, Autran B: T cell mediated suppression of HIV-specific cytotoxic T lymphocytes. J Immunol 1989;143:2193–2201.
22 Walker CM, Moody DJ, Stites DP, Levy JA: $CD8^+$ lymphocytes can control HIV infection in vitro by suppressing virus replication. Science 1986;234:1563.
23 Kannagi M, Chalifoux LV, Lord CI, Letvin NL: Suppression of simian immunodeficiency virus replication in vitro by $CD8^+$ lymphocytes. J Immunol 1988;140:2237–2242.
24 Brinchmann JE, Gaudernack G, Vartdal F: $CD8^+$ T cells inhibit HIV replication in naturally infected $CD4^+$ T cells. J Immunol 1990;144:2961–2966.
25 Lanzavecchia A, Roosnek E, Gregory T, Berman P, Abrignani S: T cells can present antigens such as HIV gp120 targeted to their own surface molecules. Nature 1988; 334:530–533.
26 Sethi KK, Naher H, Stroehman N: Phenotypic heterogeneity to cerebrospinal fluid derived from HIV-specific and HLA-restricted cytotoxic T cell clones. Nature 1988; 335:171–181.
27 Hadida F, Parrot A, Kieny MP, Sadat-Sowti B, Mayaud C, Debré P, Autran B: C-terminal and central regions of HIV-1 nef recognized by CTL from lymphoid organs: an in vitro limiting dilution analysis. J Clin Invest 1992;89:53–60.

28 Pantaleo G, Koenig S, Baseler M, Lane HC, Fauci AS: Defective clonogenic potential of $CD8^+$ T lymphocytes in patients with AIDS. J Immunol 1990;144:1696–1701.

29 Sadat-Sowti B, Debré P, Idziorek T, Guillon JM, Hadida F, Oksenhendler E, Katlama C, Mayaud C, Autran B: A lectin-binding soluble factor released by $CD8^+$ $CD57^+$ lymphocytes from AIDS patients inhibits T cell cytotoxicity. Eur J Immunol 1991;21:737–741.

30 Fahey JL, Taylor JMG, Detels R, Hofmann B, Melmed R, Nishanian P, Giorgi JV: The prognostic value of cellular and serologic markers in infection with human immunodeficiency virus type 1. N Engl J Med 1990;322:166–172.

31 Prince HE, Arens L, Kleinman SH: CD4 and CD8 subsets defined by dual-color cytofluorometry which distinguish symptomatic from asymptomatic blood donors seropositive for human immunodeficiency virus. Diagn Clin Immunol 1987;5:188–193.

32 Lewis DE, Puck JM, Babcock GF, Rich RR: Disproportionate expansion of a minor T cell subset in patients with lymphadenopathy syndrome and acquired immunodeficiency syndrome. J Infect Dis 1985;151:555–559.

33 Ziegler-Heltbrock HWL, Schramm W, Stachel D, Rumpold H, Kraft D, Wernicke D, Von der Helm K, Eberle J, Deinhardt F, Rieber EP, Riethmuller G: Expansion of a minor subpopulation of peripheral blood lymphocytes ($T8^+$/$Leu\text{-}7^+$) in patients with haemophilia. Clin Exp Immunol 1985;61:633–641.

34 Prince HE, Kreiss JK, Kasper CK, Kleinman S, Saunders AM, Waldbeser L, Mandigo G, Kaplan HS: Distinctive lymphocyte subpopulation abnormalities in patients with congenital coagulation disorders who exhibit lymph node enlargement. Blood 1985;66:64–68.

35 Stites DP, Casavant CH, McHugh TM, Moss AR, Beal SL, Ziegler JL, Saunders AM, Warner NL: Flow cytometric analysis of lymphocyte phenotypes in AIDS using monoclonal antibodies and simultaneous dual immunofluorescence. Clin Immunol Immunopathol 1986;38:161–177.

36 Salazar-Gonzalez JF, Moody DJ, Giorgi JV, Martinez-Maza O, Mitsuyasu RT, Fahey JL: Reduced ecto-5′-nucleotidase activity and enhanced OKT10 and HLA-DR expression on CD8 (T suppressor/cytotoxic) lymphocytes in the acquired immune deficiency syndrome: Evidence of CD8 cell immaturity. J Immunol 1985;135:1778–1785.

37 Tedder TF, Boyd AW, Freedman AS, Nadler LM, Schlossman SF: The B cell surface molecule B1 is functionally linked with B cell activation and differentiation. J Immunol 1985;153:973–979.

38 Giorgi JV, Nishanian PG, Schmid I, Hultin LE, Cheng HL, Detels R: Selective alterations in immuno-regulatory lymphocyte subsets in early HIV infection. J Clin Immunol 1987;7:140–150.

39 Jackson DG, Bell JI: Isolation of a cDNA encoding the human CD38 (T10) molecule, a cell surface glycoprotein with an unusual discontinuous pattern of expression during lymphocyte differentiation. J Immunol 1990;144:2811–2815.

40 Alessio M, Roggero S, Funaro A, De Monte LB, Peruzzi L, Geuna M, Malavasi F: CD38 molecule: Structural and biochemical analysis on human T lymphocytes, thymocytes, and plasma cells. J Immunol 1990;145:878–884.

41 Funaro A, Spagnoli GC, Ausiello CM, Alessio M, Roggero S, Delia D, Zaccolo M, Malavasi F: Involvement of the multilineage CD38 molecule in a unique pathway of cell activation and proliferation. J Immunol 1990;145:2390–2396.
42 Kansas GS, Engleman EG: Phenotypic identification of suppressor-effector, suppressor-amplifier and suppressor-inducer T cells of B cell differentiation in man. Eur J Immunol 1987;17:453–457.
43 Camerini D, James SP, Stamenkovic I, Seed B: Leu-8/TQ1 is the human equivalent of the Mel-14 lymph node homing receptor. Nature 1989;342:78–82.
44 Taylor J, Afrasiabi R, Fahey JL, Korns E, Weaver M, Mitsuyasu R: Prognostically significant classification of immune changes in AIDS with Kaposi's sarcoma. Blood 1986;67:666–671.
45 Prince HE, Kleinman S, Czaplicki C, John J, Williams AE: Interrelationships between serologic markers of immune activation and T lymphocyte subsets in HIV infection. J AIDS 1990;3:525–530.
46 Autran B, Mayaud C, Raphael M, Plata F, Denis M, Gourguin A, Guillon JM, Debré P, Akoun G: Evidence for a cytotoxic T-lymphocyte alveolitis in human immunodeficiency virus-infected patients. AIDS 1988;2:179–183.
47 Meignan M, Guillon JM, Denis M, Joly P, Rosso J, Carrette MF, Baud L, Parquin F, Plata F, Debré P, Akoun G, Autran B, Mayaud C: Increased lung epithelial permeability in T-lymphocytic alveolitis. Am Rev Resp Dis 1989;40:65–91.
48 Cannon MJ, Openshaw PJM, Askonas BA: Cytotoxic T cells clear virus but augment lung pathology in mice infected with respiratory syncytial virus. J Exp Med 1988; 168:1163–1168.
49 Devergne O, Peuchmaur M, Crevon MC, Trapani JA, Maillot MC, Galanaud P, Emilie D: Activation of cytotoxic cells in hyperplastic lymph nodes from HIV-infected patients. AIDS 1991;5:1071–1079.
50 Valentin MA, Autran B, Berlureau P, Bellanger F, Rosenheim M, Katlama C: CD8 lymphocytosis and HIV infection progression. Abstr WB2438, VIIth Int Conf on AIDS, Florence 1991.
51 Oksenhendler E, Autran B, Malbec D, Wolf M, Rabian C, Clauvel JP: Hyper-CD8 lymphocytosis and pseudo-tumoral splenomegally in 3 HIV-infected patients. Abstr WB85, VIIth Int Conf on AIDS, Florence 1991.

Dr. Brigitte Autran, Laboratoire d'Immunologie Cellulaire et Tissulaire, URA CNRS 0186, CERVI, CH Pitié-Salpêtrière, 83, Bd. de l'Hôpital, F–75651 Paris Cedex 13 (France)

Janossy G, Autran B, Miedema F (eds): Immunodeficiency in HIV Infection and AIDS. EC/FERS/MRC Workshop on Immunodeficiency in HIV-1 Infections, Windsor, Surrey, 1991. Basel, Karger, 1992, pp 185–194

The Staging and Prognostic Value of Subset Markers on CD8 Cells in HIV Disease

Frank Hulstaert[a], *Kenneth Strauss*[a], *Maryse Levacher*[b], *Guido Vanham*[c], *Luc Kestens*[c], *Bruce Allan Bach*[d]

[a] Becton Dickinson Immunocytometry Systems, Erembodegem, Belgium; [b] Hôpital Claude Bernard, Paris, France; [c] Institute of Tropical Medicine, Antwerp, Belgium; [d] Becton Dickinson Immunocytometry Systems, San Jose, Calif., USA

Progressive CD4 lymphocyte depletion is well documented in human immunodeficiency virus (HIV) infection and is clearly associated with disease progression. However, profound changes of immune cellular activation also occur in HIV disease, particularly in the CD8 T cell compartment. The expansion of this cell population and increased expression of activation antigens on CD8 cells are now as clearly documented as are the declines in the CD4 cell cohort. Indeed, changes in CD8 cells at the time of HIV infection often precede declines in the CD4 compartment; they persist inexorably throughout the disease course, and actually increase in patients near death from AIDS.

Numerous activation markers on CD8 cells, including CD38, HLA-DR and CD69, increase significantly after HIV seroconversion. CD57 is also elevated on CD8 cells although there is less agreement that it is an activation marker. There is evidence that the proportional increase in these surface antigens may serve as a very sensitive measure of host response to the disease [1]. We summarize here the results of several studies performed at centers in Europe and the USA designed to evaluate the degree of expression of activation markers at various stages of HIV infection and to determine the value of these markers in early identification of patients likely to have rapidly progressive disease.

Technical Requirements for These Studies

Since the investigations are focussed on the CD8 populations, a minimal requirement is double staining with CD8 and another antibody. Optimized Simultest brand combinations of murine monoclonal antibodies, directly conjugated with the fluorochromes fluorescein (FITC, green) and phycoerythrin (PE, orange) were used to stain peripheral blood. A typical panel used for 2-color phenotyping, with appropriate CD designations, is listed in table 1.

In many areas, however, triple marker combinations are required to further dissect CD8 heterogeneity. The third marker utilized is either a biotin-conjugated antibody detected by streptavidin second layer conjugated to duochrome (DC, red), or more recently an antibody directly conjugated to peridinyn-chlorophyll-A protein (PerCP, red). The use of directly labelled antibodies has greatly simplified sample preparation and has brought immunophenotyping to the routine laboratory.

Another major requirement for immunophenotyping studies is the use of high resolution flow cytometry. In these studies analysis was performed using a FACScan brand instrument, calibrated with CaliBRITE brand beads and AutoCOMP brand software.

A lysed whole blood procedure, including Simultest reagents and SimulSET software, was used in a large multicenter study to obtain reference ranges for the major lymphocyte subsets [2]. Using this approach, matching reference ranges were obtained in all participating centers. In addition, using the same methods, the CD4 and CD8 percent measures

Table 1. Reagents used in the study

Reagent	CD numbers	Utility/target cells
LeucoGATE	CD45/CD14	quality assured lymphocyte gating
Isotype control	IgG1/IgG2	control for nonspecific staining
Leu 3/2	CD4/CD8	CD4+ T cells and CD8 lymphocytes
Leu 2/DR	CD8/HLA-DR	activated (MHC Class II expressing) CD8 cells
Leu 2/17	CD8/CD38	activated/immature (CD38+) CD8 cells
Leu 2/8	CD8/Leu 8	CD8 cells expressing LECAM-1, associated with cellular adhesion
Leu 7/2	CD57/CD8	NK and T cell differentiation marker, expression is associated with certain viral infections

were reported in a multicenter study to be associated with a coefficient of variation (CV) of 3% due to sample preparation and a CV of 6% when sequential samples of the same subject were measured [3].

This methodology was also applied consistently in several clinical trials, focussing on different aspects of the HIV disease.

One hundred microliters of peripheral blood, collected in EDTA, was gently vortexed with 20 μl of each monoclonal reagent combination and then incubated for 15 min in the dark at room temperature. 2 ml of (1×) FACS brand lysing solution was added to lyse the red blood cells over 10 min, also at room temperature in the dark. The samples were then centrifuged at 250 *g* for 5 min, the supernate aspirated and the pellet resuspended in 2 ml phosphate-buffered saline (PBS) containing 1% sodium azide. This was followed by another centrifugation step at 250 *g* for 5 min. Cells were fixed in 0.7 ml PBS with 1% paraformaldehyde. These samples were stored at 4 °C and analyzed within 24 h by a FACScan flow cytometer using SimulSET software only after at least 1 h of fixation.

Quality Assured Analysis

A minimum of 10,000 total events was acquired using SimulSET brand software to define a lymphocyte scattergate using both the lightscatter and fluorescence information from the LeucoGATE brand reagent (CD45/CD14). This analysis gate was used to analyse each tube of the reagent panel.

The analysis gate was automatically optimized to include at least 95% of the lymphocytes and no more than 10% of nonlymphocytes in the sample. Where possible a 'large gate' was set, excluding as few lymphocytes as possible from the analysis [4]. For each tube a minimum of 2,000 lymphocytes were acquired and analyzed using the same scattergate. For three-color analysis a lymphocyte gate was defined manually using the LYSIS II brand software.

Markers for determining cells positive and negative for any reagent were set by the SimulSET software using directly conjugated antibodies of irrelevant specificity (Simultest control) as negative controls.

Dotplots and quadrant statistics from two-color analysis were generated by SimulSET software. All dotplots were reviewed using a set of written criteria [2, 3]. Dimly staining CD4+ cells, proven not to be lymphocytes by backgating, were excluded by repositioning the quadrant boundaries in an unambiguous and predefined manner. Results were transferred electronically to database systems for further statistical analysis.

Multiple determinations of the CD4 and CD8 lymphocyte values were obtained on the same patient sample in all cases. Comparison of these values permitted the application of additional internal consistency checks. Where possible, it was verified whether the sum of percentage results for T, B and NK cells accounted for 100% of all lymphocytes [5].

Immunophenotype results are expressed as absolute count/mm^3, as percentages of total lymphocytes or as proportions of CD3, CD4 or CD8 lymphocytes. The notation used in this paper will be 'minor subset per major subset' (e.g. CD38/CD8) for in-proportion data. In-proportion analysis allows the normalization of results to the denominator of a larger CD3+, CD4+ or CD8+ lineage and allows comparison between groups whose overall numbers may differ.

Patients

HIV-1 seropositive subjects were grouped into approximately equal numbers of asymptomatic (ASY; CDC Group II), lymphadenopathy syndrome (LAS; CDC Group III), AIDS-related complex (ARC; CDC Group IV A, B, C2, E) and AIDS patients (CDC Group IV C1, D) [6]. They were evaluated clinically and by lymphocyte immunophenotypic analysis, and compared to healthy, age-matched seronegative controls.

Changes in Major Lymphocyte Populations

The ASY and LAS groups have nearly identical immunophenotypes and can thus be considered as a single group. Lymphocytes in ARC and AIDS patients are also comparable, with the exception of lower CD4 cell number and a higher proportion of CD38 on CD8 cells in AIDS. Marked and reproducible phenotypic changes are noted between the control and ASY/LAS groups for most lymphocyte populations and between the ASY/LAS and ARC/AIDS patients for a more limited number of subsets. Compared to controls, ASY patients show a loss of approximately 50% in the absolute number of CD4 cells, while the CD8 counts rise by nearly 25%. The fall in CD4 numbers cannot be attributed to a selective decrease in its subsets, as defined by CD45RA or Leu-8 expression, nor do these subsets correlate with disease outcome. In later stage AIDS patients there is often an overall decline in all lymphocyte population counts as a result of generalized pancytopenia.

Percentages of CD8 cells increase by 30–50% at seroconversion and continue their progressive increase throughout all stages of HIV infection. These changes, coupled with the steady fall in CD4 percentages, serve to maintain a relatively constant overall percentage of total T cells while simultaneously lowering the CD4:CD8 ratio. B and NK cell populations show minor declines as HIV disease worsens. Within the B cell compartment there is a shift from the Leu 8+ to the Leu 8– phenotype on nearly a third of the cells, suggesting changes consistent with an increased activation state.

The most common two-color combination used to evaluate HIV patients is CD4/CD8. This reagent pair does not necessarily exclude monocytes, which are dimly CD4+ but CD3–, or cells dimly CD8+ and CD3–, which belong to the NK lineage. There is evidence that a mean of 20% of total CD8 lymphocytes are NK cells. Thus, evaluation of HIV patients by means of the two-color combination of CD3/CD4 and CD3/CD8 may provide a more accurate T cell helper:suppressor ratio (H/S), as it allows for the elimination of monocyte and NK cell contamination. The true T cell H/S ratio values are higher than the conventional H/S ratio derived from the CD4/CD8 combination, thereby providing a potentially more sensitive threshold for discerning early T cell subset changes associated with HIV infection.

Activated T Cells in HIV Infection

Even more dramatic than the expansion of the CD8 compartment is the increased expression of numerous activation antigens on these cells. The most striking change occurs in HLA-DR expression on CD8 cells. The percentage of CD8 cells expressing CD38, CD57 and CD69 also rises 40–60% at seroconversion and continues a progressive rise as disease stage worsens (fig. 1). The absolute count of activated CD8 cells also increases with advancing clinical stage, reaches a plateau, and then declines in late-stage AIDS, usually as a result of generalized lymphopenia. Distinct from the above markers, the loss of Leu-8 expression on CD8 cells generally signals cellular activation. The percentages and absolute counts of Leu 8– CD8 cells increased in a pattern similar to the other activation markers.

One marker of activation, CD38 expression on CD8 cells, shows particular promise as a surrogate marker for progressive HIV infection. The simultaneous increase in CD38+ CD8 cells coupled with the decrease in CD38– CD8 cells makes the proportion of total CD8 cells expressing CD38 a powerful discriminator between ASY/LAS and ARC/AIDS patients, comparable in clinical staging value to the CD4 percentage, absolute count or CD4:CD8 ratio [7]. CD38 expression on CD8 cells increases in direct proportion to disease severity. Its distribution within the CD8 cellular compartment is greatest in AIDS patients with *Pneumocystis carinii* pneumonia (PCP), yet is significantly lower in AIDS patients with Kaposi's sarcoma (KS).

CD8 and CD38 are expressed on both T lymphocytes and NK cells. By simultaneously and independently performing both two- and three-color

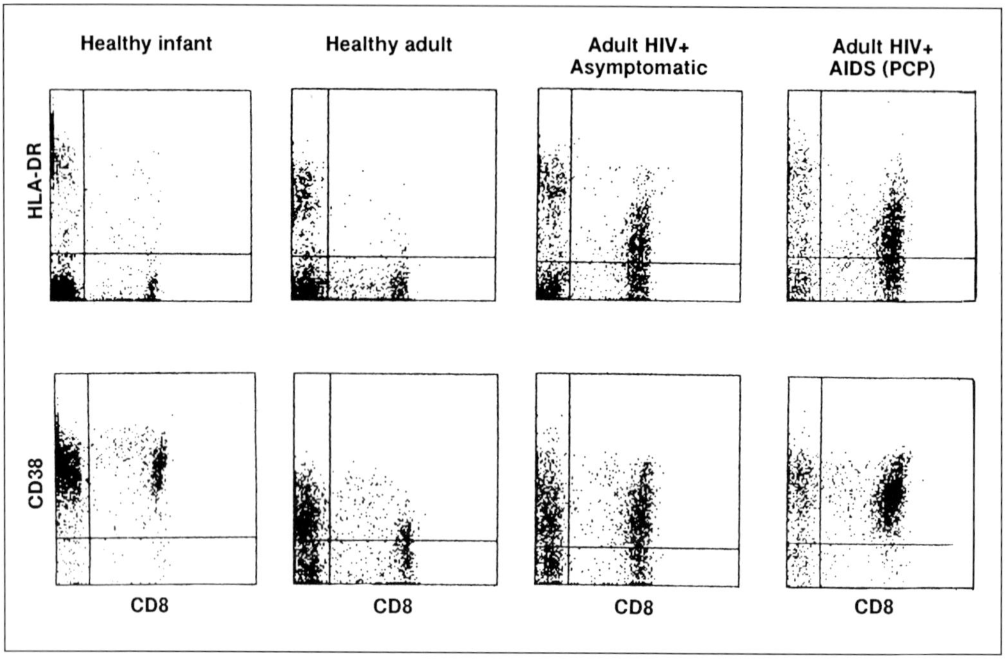

Fig. 1. Typical 2-color flow cytometry dotplots illustrating CD8 subsets defined using HLA-DR and CD38 expression in healthy infants and adults, as well as in asymptomatic and symptomatic HIV-1 seropositives.

flow cytometric analysis in well-controlled studies, we observed that T cells which exhibit high density coexpression of CD38 and CD8 increase in HIV infection whereas the low-density CD8+ NK compartment does not expand. Further three-color work has suggested that in HIV disease HLA-DR expression on CD8 cells increases at seroconversion and continues to rise during the LAS and ARC stages. At that time CD8 cells progressively increase their expression of CD38. In AIDS, CD38 expression reaches its maximum, while HLA-DR expression may decline slightly (fig. 2).

Therapy with zidovudine has been reported [8] to significantly reduce the expression of HLA-DR on T cells when measured at 12 and 24 weeks after treatment initiation in ARC patients. Also affected were p24 antigen levels, which decreased significantly, and CD4 cell percentages, which showed a slight, transient rise. This study requires confirmation and extension to include the full range of activation markers.

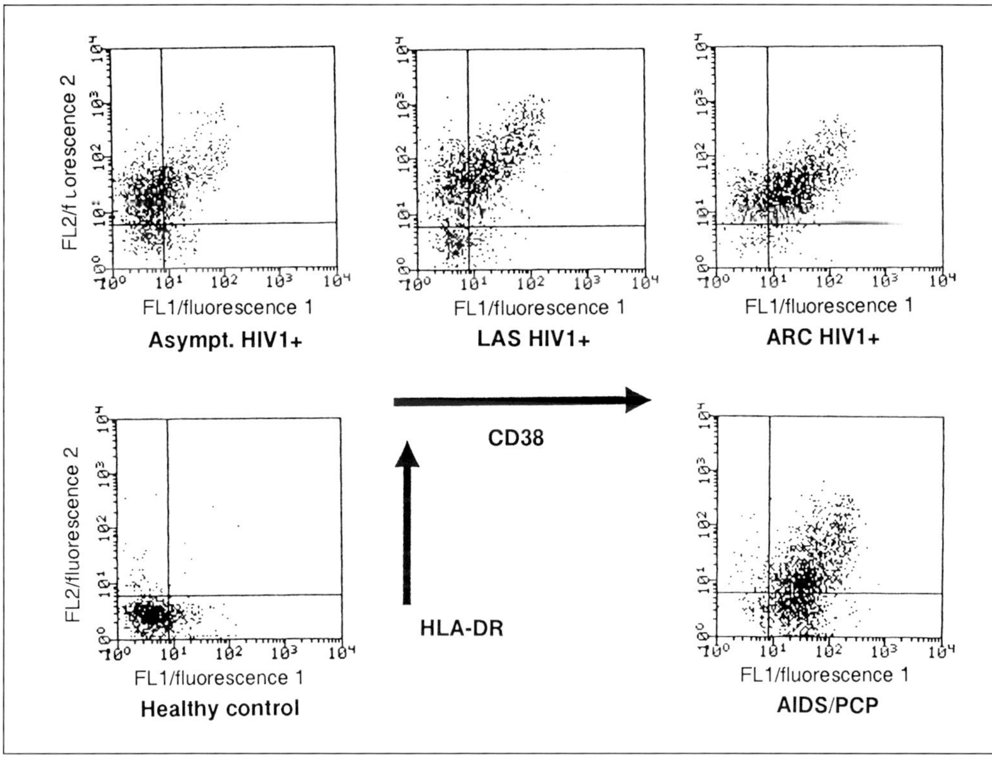

Fig. 2. Flow cytometry dotplots using 3-color immunophenotyping on lymphocytes. Cells brightly positive for CD8 (fluorescence 3) were gated (not shown), and HLA-DR expression (fluorescence 2) versus CD38 expression (fluorescence 1) in a healthy adult and in HIV-1 seropositives in various clinical stages of HIV-disease are depicted. This pattern illustrates the sequential shifts in CD8 phenotype typically observed in HIV infection, dissecting the contributions of HLA-DR and CD38 positive subsets.

Not all activation markers demonstrate increased cellular expression in HIV disease. There is a consistent and reproducible decline in interleukin 2 receptor (IL2R) expression across all HIV clinical stages [9], whereas soluble serum IL2R levels are reportedly increased in HIV disease.

AIDS patients demonstrate elevations in CD38+ CD8 cells that are strikingly similar to those seen in healthy newborns and infants. However, the intense expression of CD57 and HLA-DR on CD8 cells in AIDS patients clearly distinguishes them from normal infants who have little or no expression of these markers (fig. 1) (see also Janossy, Jiang et al., this volume).

Prognostic Significance of Activation Markers

Besides its utility in accurately staging HIV disease the CD38 marker on CD8 cells is valuable as a predictor of disease outcome. Adult ASY and LAS patients who did not progress to ARC or AIDS in a 2-year follow-up study had a significantly lower percent expression of CD38 on the CD8 cells compared to those who did progress (45 vs. 66%, $p = 0.003$). Patients with ARC who did not progress to AIDS or death also had significantly lower percentages than those who did progress (66 vs. 86%, $p = 0.014$). When the percent CD38 on CD8, percent HLA-DR on CD8 and the absolute CD4 counts were compared for their correlation to disease progression, the variable with greatest independent predictive power for progression of HIV disease was the percent CD38 on CD8 cells. This marker could predict outcome as early as the asymptomatic period and had equal or greater prognostic value than the CD4 count. Its rise persists until the very end stage of disease, with PCP patients expressing CD38 on nearly 100% of their CD8 cells [7, 8].

Functional Role of Activated T Cells

The impact of HIV infection on the human immune system clearly includes not only destruction of certain cell cohorts, principally the CD4+ T cells, but the concomitant activation of other cell lines, among which are CD8-bearing T cells. It is unclear if this activation plays a role in retarding disease progression or actually facilitates it, or whether both effects may occur during the disease course. Also unclear is the functional significance of each of these markers.

The CD38 molecule has been implicated as modulator of the expression of other cell surface activation markers via a unique proliferation pathway not used in CD2 or CD3 activator systems [10]. HLA-DR is a MHC class II antigen that shows increased expression on CD8 cells during graft-rejection episodes and in viral infections other than HIV [11]. CD57 is also elevated in other viral infections, particularly cytomegalovirus [12]. Its expression, like that of HLA-DR, often increases early in HIV disease and then remains stable across several disease stages. CD57 values have also been found to be elevated in seronegative homosexuals [1]. Leu-8 (LECAM-1) is an adhesion molecule [13], and its loss from CD8 cells in HIV disease is associated with cellular activation.

The cellular activation implied by increased expression of CD38, HLA-DR, CD57 and loss of Leu-8 does not appear to result in functional advantage to the CD8 cells. On the contrary, these activated cells seem less able to respond in vitro to mitogens, foreign antigens and monoclonal antibodies directed against them. Their life span in culture appears shortened as well [14]. However, it is still unclear whether results in vitro, or even events detected in vivo in the peripheral blood, reflect actual pathogenic processes in the host.

It is likely that variability in the asymptomatic incubation time to HIV seroconversion and to the development of signs of acquired immunodeficiency are related to elements in the host response and to the virulence of the infecting viral strain. It is interesting to speculate that the HIV virus, after initial destruction of the host's key line of defense against it, may develop a protected niche in the immune system by handicapping and eventually eliminating the cytotoxic cell activated against it.

Conclusions

Despite these unanswered questions about the true meaning and consequence of cellular activation in HIV disease, it is clear that the increased expression of activation markers on CD8 cells provides a sensitive tool for both staging disease and predicting outcome. Still to be determined are the exact levels of these markers at which certain clinical events are likely to occur, as well as the most reproducible, standardized method for measuring them. Also to be explored is the exact response of activation markers to anti-retroviral therapy and the confidence with which such a response can be used to assess efficacy of treatment and prediction of outcome. As the process of finding answers to these questions unfolds, it will become clear that the measurement of CD8 cell activation can provide information crucial for the staging, treatment and prognosis of HIV patients.

Acknowledgements

We thank Sylvain Tallet, Francis Vanlangendonck and Gloria Young for their skillful technical assistance. We also thank Louise Klinkenbergh and Heidi De Gols for their assistance in generating the manuscript.

References

1 Giorgi J, Detels R: T-cell subset alteration in HIV-infected homosexual men: NIAD multicenter AIDS cohort study. Clin Immunol Immunopathol 1989;52:10–18.
2 Reichert T, De Bruyere M, Deneys V, et al: Lymphocyte subset reference ranges in adult Caucasians. Clin Immunol Immunopathol 1991;60:190–208.
3 Fink P, Burrichter H, Kaehler H, Trendelenburg C, Hulstaert F: Comparison of variations for peripheral blood lymphocyte subsets using flow cytometry. XV ISAC Congr, Bergen, Norway. Cytometry 1991(suppl 5): abstr 457C.
4 Loken M, Brosnan J, Bach B, Ault K: Establishing optimal lymphocyte gates for immunophenotyping by flow cytometry. Cytometry 1990;11:453–459.
5 NCCLS Document H42-P: Clinical Applications of Flow Cytometry: Quality Assurance and Immunophenotyping of Peripheral Blood Lymphocytes.
6 Classification System for Human T-Lymphotropic Virus Type III/Lymphadenopathy-Associated Virus Infections: Centers for Disease Control, US Department of Health and Human Services, Atlanta, Georgia. Ann Intern Med 1986;105:234–237.
7 Levacher M, Pocidalo J, Tallet S, Hulstaert F, Lowder J: Two-color immunofluorescence and flow cytometric analysis of peripheral blood lymphocyte subsets in clinical stages of HIV-1 infection. The importance of CD8 subsets; in Osterhaus A, Uytdehaag F (eds): Idiotype Networks in Biology and Medicine. Proc Congr Idiotype Networks in Biology and Medicine, Gennep, The Netherlands, April 17–20, 1989.
8 Levacher M, Tallet S, Dazza M, Dournon E, Rouveix B, Pocidalo J: T activation marker evaluation in ARC patients treated with AZT. Comparison with CD4+ lymphocyte count in non-progressors and progressors towards AIDS. Clin Exp Immunol 1990;81:177–182.
9 Zola H, Koh L, Mantzioriz B, Rhodez D: Patients with HIV infection have a reduced proportion of lymphocytes expressing the IL-2 receptor p55 chain (TAC, CD25). Clin Immunol Immunopathol 1991;59:16–25.
10 Funaro A, Spagnoli GC, Ausiello CM, et al: Involvement of the multilineage CD38 molecule in a unique pathway of cell activation and proliferation. J Immunol 1990; 145:2390–2396.
11 Van Es A, Baldwin WM, Oljans PJ, Tanke HJ, Ploem JS, Van Es LA: Expression of HLA-Dr on T lymphocytes following renal transplantation, and association with graft-rejection episodes and cytomegalovirus infection. Transplantation 1984;37: 65–69.
12 Gratama JW, Kluin-Nelemans HC, Langelaar RA, et al: Flow cytometric and morphologic studies of HNK1+ (Leu 7+). Clin Exp Immunol 1988;74:190–195.
13 Picker LJ, Terstappen LWMM, Rott LS, Streeter PR, Stein H, Butcher EC: Differential expression of homing-associated adhesion molecules by T cell subsets in man. J Immunol 1990;145:3247–3255.
14 Prince HE, Czaplick CD: Preferential loss of Leu 8-, CD45R-, HLA-DR+ CD8 cell subsets during in vitro culture of mononuclear cells from human immunodeficiency virus Type I (HIV)-seropositive former blood donors. J Clin Immunol 1989;9:421–428.

Dr. Frank Hulstaert, Becton Dickinson Immunocytometry Systems, PO Box 13, Denderstraat 24, B–9320 Erembodegem (Belgium)

Janossy G, Autran B, Miedema F (eds): Immunodeficiency in HIV Infection and AIDS. EC/FERS/MRC Workshop on Immunodeficiency in HIV-1 Infections, Windsor, Surrey, 1991. Basel, Karger, 1992, pp 195–210

Activated CD45RO, CD8+ T Lymphocytes in HIV-1 Infection[1]

George Janossy[a], *Jian Dong Jiang*[b], *Julia P. Roboz*[b], *Margarita Bofill*[a], *J. George Bekesi*[b]

[a] Department of Clinical Immunology, Royal Free Hospital School of Medicine, London, UK, and [b] Department of Neoplastic Diseases and the T.J. Martell Laboratory for Leukemia, Cancer and AIDS Research, The Mount Sinai School of Medicine, New York, N.Y., USA

Several laboratories investigating the function and phenotype of human peripheral T lymphocytes have now reached a unified view of subset heterogeneity about CD4/CD8/NK cell trichotomy and the developmental steps within these pathways in terms of the expression of CD45 isoforms on these cells [1–3]. This consensus suggests that among both CD4 and CD8 cells some express high molecular weight isoforms (200–220 kD) defined by CD45RA monoclonal antibodies (MAbs). These are regarded 'naive' or unprimed T cells which after activation by antigens or mitogens switch to a display of low molecular weight isoforms. One of these (180 kD) is detected by CD45RO MAbs [2, 3], and the corresponding cell populations are referred to as 'memory' or primed cells. Only CD3+ T cells with a functional T-cell receptor switch their CD45 isoforms. By contrast, CD3– NK-like cells remain CD45RA+/RO–, i.e. a 'naive' population of cells, as might be expected from the fact that these cells lack functional T-cell receptors and fail to undergo a proliferative expansion under the influence of foreign antigens [4]. This scheme has become particularly interesting when it was documented that 'memory' T cells express many

[1] This work was supported by the MRC AIDS Directed Programme of Great Britain (to G.J. & M.B.), NIH/NIAID contracts NO1-AI-72660 and by the T.J. Martell Foundation for Leukaemia, Cancer and AIDS Research (to J.G.B.). We are grateful to Ms. Nicola Borthwick and Eva Medina for excellent technical help.

adhesion molecules including ICAM-1 (CD54) and LFA-3 (CD58) more richly than the 'naive' populations [1, 5], and also differ in their cytokine synthetic capacity (table 1) [reviewed in 6, 7]. In addition, CD45RO+ cells recirculate in the tissues more readily than CD45RA+ 'naive' cells [reviewed in 8]. A further point is that the 'memory' T cells also show at least

Table 1. Properties of 'naive' and 'memory' T cells[a]

Features	Naive cells[b]	Memory cells
Phenotype		
CD45RA (e.g. 2H4)	+++	–
CD45RO	–	+++
CD7[c]	+++/++	+/–
CD44 (Pgp-1)	++	+++
CD29 (e.g. 4B4)[d]	++	+++
CD18/CD11a (LFA-1)	+	++
CD2	++	+++
CD58 (LFA-3)	+	++
CD54 (ICAM-1)[c]	–	+
CD25 (IL-2Rα)[c, e]	–	+
MHC Class II (DR)[c]	–	+
Cytokine synthetic capacity		
IL-2	++++	++
IL-4	–	++
IL-6	–	++
IFNγ	–	++
Other features		
Size	smaller	larger
Probable turnover	slower	faster
'Surveillance' of tissue sites	minimal	active

[a] Data from various reviews [8–10].
[b] NK cells ($CD16^+$, $CD3^-$, TCR^-) including cells weakly expressing CD8 show features of the 'naive' cells.
[c] These antigens are further increased upon T-cell activation.
[d] The antibodies in this cluster detect integrin β chain which is recognizing a subset of CD4 cells with helper-inducing function. This subset is not the same as the CD45RO+ population. Among CD8+ cells CD29 is not differentiating between CD45RA+ and CD45RO+ populations.
[e] The IL-2Rβ chain shows a more complicated relationship to 'naive' and 'memory' populations which is different in CD4 and CD8 subsets [see Sahraoui et al., this volume].

a weak expression of IL-2 receptor (IL-2R) α chain (CD25) and Class II molecules while resting 'naive' cells are negative. Taken together, these observations strongly suggest that the 'memory' type cells are not a truly resting lymphoid population. The number of CD45RO+ cells might be maintained by a continuous, albeit perhaps slow, cell cycling under the influence of (cross-reacting) antigens [6]; these cells are then poised to increase their IL-2R and many other adherence molecules particularly swiftly under antigenic influence.

This area of investigation is important in terms of AIDS pathogenesis for a number of reasons. First, phenotypic and functional studies in HIV-1 infection have to consider the 'naive' to 'memory' shifts in populations of cells in order to find and interpret truly pathological changes. For example, it is relevant to consider that in HIV-1 infection there is an accumulation of 'memory' type CD8 cells [9–11], and normally this should lead to an increased rather than decreased display of IL-2Rα chains [reviewed by Sahraoui et al., this volume]. Second, as the recent views suggest, the maintenance of immunological memory is a far more *dynamic* process than merely a long-term survival of a cohort of quiescent ('inert' or 'spore-like') population of cells. If a constant level of background stimulation is required to maintain memory, then this process (a) may deteriorate when an infectious virus rapidly mutates [reviewed by Rowland-Jones and McMichael, this volume], and (b) it may remain eminently susceptible to severe dysregulating influences. It then becomes conceivable that, for example, in the absence of adequate IL-2 synthesis a most efficient cohort of weakly IL-2R+ memory cells dwindle – and other poorly adjusted subsets with less stringent IL-2 requirement, and perhaps poorer HLA restriction and meagre specificity, emerge as a dominant, numerically impressive and functionally disappointing CD8+ cohorts of cells. Consequently, these issues place the turnover of various CD8+ subsets and the dynamics of these cells' repopulation into a new focus of attention.

This review paper therefore serves two aims. First, we wish to quantify both the changes of CD45 isoforms and the other frequently utilized phenotypic features referred to as markers of 'immaturity' and 'activation' in HIV-1 infection. Our investigation is an extension of the analysis presented by Autran and Giorgi, and Hulstaert et al. [this volume], and it represents the combined experience of two laboratories in London and New York on different populations of HIV-1-infected persons. Second, we confirm that the most suitable technology for these investigations is the use of specially designed triple marker combinations on flow cytometer which bring together three features: (a) a subset marker (e.g. CD8); (b) CD45 isoform (RA or RO), and (c) a functional, 'immaturity' or activation marker. This is an open-ended plan which allows the investigation of interesting MAbs to prognostically significant (CD38), or functional antigens (CD28) or other cytoplasmic molecules, e.g. associated with apoptosis [12], cytotoxic function etc., in the precise context of CD8 subsets.

By the criteria applied in this study, HIV-1 infection appears to lead to a secondary disorder of poorly adjusted 'memory' CD8+ cells rather than NK-like cells. Similar events are also recorded in other conditions, such as complications in organ and bone marrow transplants, when an extreme degree of CD8 activation is accompanied by immunosuppression [13–16].

Comments About the Methods and Quality Controls Used

Our techniques were based on the quality-assured analysis introduced in the Becton-Dickinson (BD) protocols with high resolution flow cytometry [17, 18]. The set of double labelling probes used were commercially available as listed in table 1 of the paper by Hulstaert et al. [this volume]. Triple marker methods were also standardized. The CD45 associated antibodies were CD45RA (Sn130, cf. Akbar et al. [2]) and CD45RO (UCHL1, cf. Beverley [3]). The CD38 antibody used is referred to as RFT10 [19] and it is equivalent to OKT10 [16] and Leu17 of BD range. A common feature of these antibodies is that they detect differentiation antigens which are expressed in differing amounts on functionally distinct populations of cells rather than displayed in an 'all-or-none' pattern. For this reason the gating during the triple assays requires three considerations: (a) the appropriate reagent control and the avoidance of overconjugation when directly labelled reagents are used; (b) careful colour compensation for the three channels: fluorescein (FITC; green), phycoerythrin (PE; orange) and Duochrome (DC; red), which we have used here. Recently, this compensation has been made easier by utilizing perdinyn-chlorophyll PerCP [20] instead of DC as a red marker, and (c) biological controls to establish the gating on the basis of the known characteristics of lymphocyte subsets in the thymus, fetal tissues, cord blood and the peripheral blood of normal adults [21, 22].

In our laboratory we use three such 'bio-controls' during the gating on FACS (fig. 1a–c). CD45RA and CD45RO are expressed on two largely

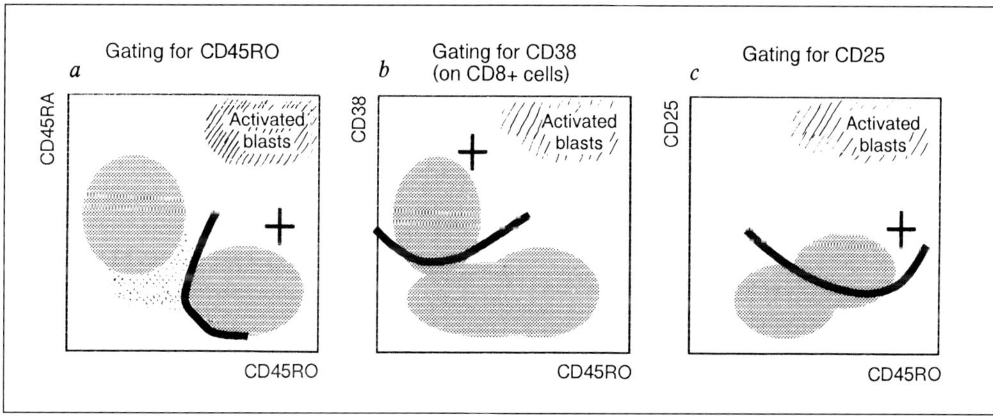

Fig. 1. The concept of gating on the basis of cell populations rather than single parameters of staining intensity. The three main normal blood controls are shown to document the populations gated positive for CD45RO+ cells ('memory' type) (*a*), immature CD38+ cells (*b*) and weakly CD25+ cells (*c*). T cells activated in cultures express these markers more intensively as shown (▨). In routine work this analysis is greatly facilitated by the Paint-a-Gate programme of the FACScan (Becton-Dickinson).

exclusive subsets (fig. 1a) [see also figure 6a by Giorgi, this volume] with some intermediate weaker staining, which are arbitrarily referred to as negatives; these cells represent 10–20% among CD8+ lymphocytes in cord blood. On the other hand, activated CD45RA+ blasts stimulated by CD3, PHA or in mixed lymphocyte cultures express high levels of both antigens while they 'switch' their phenotype to CD45RO: these are doubly positive [2]. The association between the relative immaturity of CD38+, CD8+ cells in cord and normal blood samples can be shown by the relatively strong CD38 positivity, referred to as positive, on 'naive' CD45RA+, CD45RO– cells (fig. 1b) and the weaker expression, referred to as 'negative', on 'memory' CD45RO+ cells. This corresponds to previously documented data about OKT10 expression during thymocyte and T-cell development [22]. Again, activated T blasts express CD38 at a higher level than fresh blood T cells. Finally, CD25 (IL-2Rα chain) is negative on CD45RA+ and very weakly expressed on CD45RO+ cells [5], and then greatly increased on activated blasts, as expected (fig. 1c). These three controls are routinely required for setting the gates on blood samples taken from HIV-1-infected patients.

Changes of CD8+ Subsets in HIV-1 Infections

The means of the absolute lymphocyte counts in the controls and in the HIV seropositive groups are shown in figure 2, and the corresponding percent values with the standard deviations are documented in table 2. The counts in the same groups studied in the collaborating laboratories showed no significant differences, and therefore some of the observations could be pooled.

The total CD8 counts in individuals with HIV infection (CDC2) were higher than the values seen in HIV seronegative homosexuals and in heterosexual controls, although the lymphopenia seen in drug abusers and in AIDS has brought the values down again. Within the CD8+ populations phenotypic shifts were detected. Of these, the slightly increased numbers of CD57+ lymphocytes were not attributable to HIV-1 infections because the values were already raised in the healthy HIV-1 seronegative homosexuals. The significance of the increased Class II expression was ambiguous for two reasons. Albeit the Class II+, CD8+ cells in all HIV-infected groups were higher (>400/mm^3) than in the control (80/mm^3) and the homosex-

Table 2. Changes in CD8+ lymphocyte populations during HIV-1 infection[a]

Patient groups	CDC category	n	Total absolute counts/mm^3		
			lymphocytes	CD4+	CD8+
Adults					
Heterosexual control	0	91	1,977 ± 350[c]	852	597 ± 135
HIV-1– homosexual males	0	75	1,940 ± 401	762	688 ± 196
HIV-1+ homosexuals	CDC2	197	1,810 ± 367	506	910 ± 211
HIV-1+ homosexuals with PGL	CDC3	17	1,605 ± 200	391	820 ± 140
HIV-1+ i.v. drug abusers	CDC2	40	1,431 ± 241[d]	260	727 ± 243
ARC and AIDS	CDC4	72	1,295 ± 343	159	797 ± 171
Pediatric					
Control	0		2,180 ± 261	876	530 ± 102
HIV-1+ children to DA mothers	CDC2		3,172 ± 379	390	1,548 ± 253

[a] The observations in the two centres were pooled as no significant differences were seen.

[b] In this table the percentages shown are within the total lymphocyte population. The corresponding absolute values in blood are depicted in figure 2.

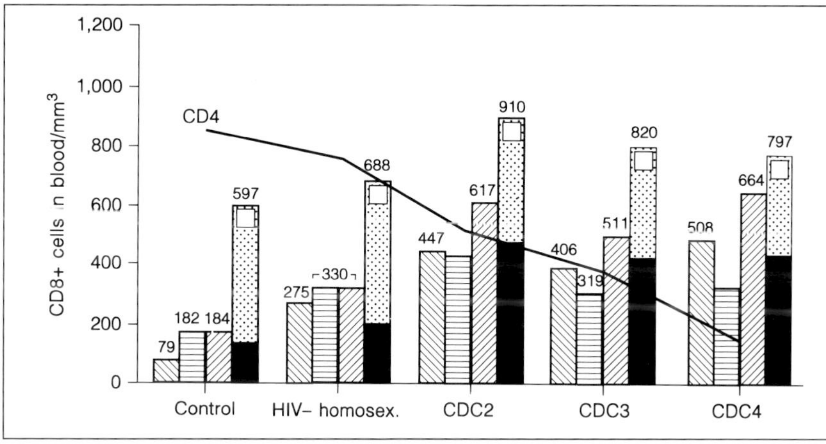

Fig. 2. The absolute numbers of CD8+ subsets in heterosexual control, HIV-1 sero-negative homosexuals and in HIV-1+ individuals classified by CDC criteria. The Class II (▧), Leu7 (▤) and CD38 (▨) values are separately shown, while the CD45RA (▦) and CD45RO (■) values are shown within the bar of the total CD8 numbers. A few CD8+ cells are negative for both CD45RA/RO (empty area in CD8 bar).

Percent within lymphocytes [b]					
total CD8+	CD8+ CD38+	CD8+ HLA-DR+	CD8+ CD57+	CD8+ CD45RA+	CD8+ CD45RO+
30.2 ± 6.6	9.3 ± 1.8	4.0 ± 0.6	9.1 ± 1.2	17.4 ± 1.4	7.0 ± 1.4
35.5 ± 8.7	17.0 ± 5.9	14.2 ± 4.1	17.0 ± 5.9	20.9 ± 5.9	10.1 ± 3.1
49.8 ± 14.5	34.1 ± 7.6	24.7 ± 4.8	24.4 ± 8.7	22.0 ± 7.1	23.7 ± 7.3
51.1 ± 11.3	n.t.	n.t.	n.t.	20.5 ± 8.3	28.2 ± 10.1
50.8 ± 17.0	37.5 ± 9.3	25.7 ± 5.1	18.4 ± 6.6	n.t.	n.t.
61.5 ± 16.1	51.3 ± 12.6	39.3 ± 10.6	27.1 ± 7.6	23.2 ± 12.1	35.8 ± 14.1
21.8 ± 2.6	11.3 ± 0.9	5.3 ± 1.6	7.8 ± 0.8	13.6 ± 2.1	16 [e]
48.8 ± 5.1	43.1 ± 3.5	32.7 ± 4.1	31.0 ± 4.5	16.9 ± 2.9	71 [e]

[c] Means ± SD.
[d] Value significantly different from controls.
[e] From Froebel et al. [10].

ual control groups (276/mm^3), in our studies we utilized very sensitive methods and many of these cells only weakly expressed Class II antigens. Such a minimally increased Class II expression is partly attributable to the increased numbers of CD45RO+ cells which normally have slightly raised Class II (see below). In addition, in some individuals we have seen dramatic but temporary increases in both the number and intensity of Class II expression, a phenomenon associated with acute infectious, mostly viral, episodes.

The increased numbers of CD38+ and CD45RO+ cells were the most consistent alterations of CD8+ cells in HIV-1 infection. Both the proportions and absolute numbers of the CD38+ and CD45RO+ subsets of the CD8+ populations were 2–3 times higher in all HIV+ groups than in the corresponding control groups. As a result, both the normally dominant CD38–, CD8+ and CD45RA+, CD8+ populations have switched their features to dominant CD38+, CD8+ and CD45RO+, CD8+ types during HIV infection. The prognostic significance of the increased CD38 positivity is reviewed by Hulstaert et al. [this volume]. The changes in both CD38 positivity and switching from the CD45RA to the CD45RO subset was also apparent in HIV-1+ children born to drug abuser mothers (table 2) [10].

Triple Colour Analysis of CD8+ Cells Expressing CD38 Antigens

During the next phase it was studied whether the combination of CD38 and CD45RA/RO analysis on CD8+ cells by utilizing triple markers (fig. 3) may further increase the analytical power of the method. The two combinations used most frequently were CD8/CD38/CD45RA (fig. 3a, c, e) and CD8/CD38/CD45RO (fig. 3b, d, f). It was first confirmed that in the cord blood the majority of CD8+ lymphocytes expressed CD38 and

---

Fig. 3. Flow cytometric analysis utilizing triple labelling with antibodies CD8 (FITC), CD38 (PE) and CD45 antibodies (biotinylated and labelled with Duochrome-conjugated streptavidin). *a, b* Cord blood with CD8+, CD38+, CD45RA+ cells. *c, d* Normal blood with CD8+ cells of heterogeneous phenotypes. *e, f* HIV-1+ patient (CDC4) with CD8+, CD38+, CD45RO+ cells. Two triples are shown: (*a, c, e*) CD8/CD38/CD45RA and (*b, d, f*) CD8/CD38/CD45RO. Only cells gated on the CD8 positivity are shown and the weakly CD8+ and CD8– cells are ignored. The large numbers designate the major subpopulations of CD8+ cells: (1) 38+, RA–, RO–; (2) 38+, RA+, RO–; (3) 38–, RA+, RO–; (4) 38–, RA–, RO+, and (5) 38+, RA–, RO+.

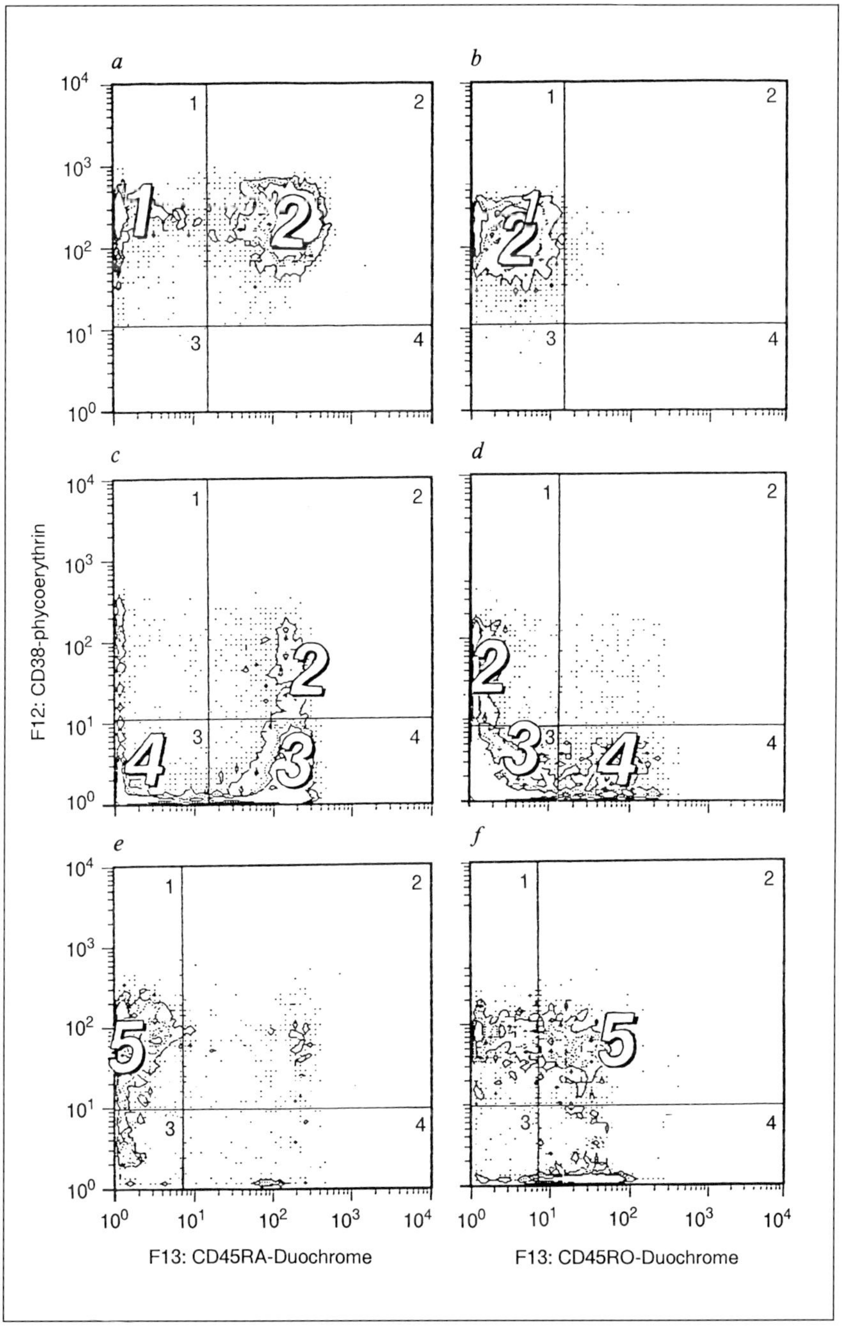
a
b
c
d
e
f
1
2
3
4
5
F12: CD38-phycoerythrin
F13: CD45RA-Duochrome
F13: CD45RO-Duochrome
10^0
10^1
10^2
10^3
10^4

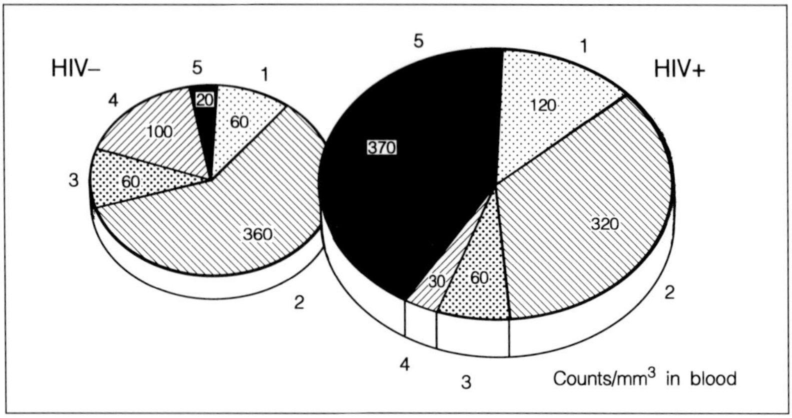

Fig. 4. The calculated absolute numbers of various CD8+ subsets in the normal HIV-1 seronegative and in the HIV-1 seropositive blood. There is a considerable increase in the CD45RO+, CD38+ subset during HIV-1 infection. The numbers around the edges refer to those shown in figure 3.

CD45RA but were CD45RO– (group '2' in fig. 3a, b) although some cells were CD45RA–/RO– double negatives (group '1'; probably the most recent emigrants from the thymus). An important observation shown in the normal blood is that the CD38+ cells still expressed CD45RA, as group '2' of immature T cells. These, together with CD38–, CD45RA+/RO– 'naive' cells (group '3') were the dominant CD8+ cells (60–70%), while CD8+ cells of 'memory' type were CD38– (group '4' in fig. 3c, d). In conclusion, in normal unstimulated blood the CD38+ and CD45RO+ cells are largely nonoverlapping populations.

It can already be seen in figure 2 that in HIV-1+ samples among CD8+ cells, CD38+ and CD45RO+ populations overlap. This point has been directly demonstrated in figure 3e, f where large proportions (45–55%) of CD8+ cells could be mapped into the CD38+, CD45RO+/RA– (group '5') category. The numerical changes in HIV-1+ blood are therefore summarized in figure 4. In essence, there is a 5- to 15-fold increase in the CD8+, CD45RO+ population – and the triple staining detects these cells more discriminately than the double staining methods. This phenotypic feature also reveals that these group '5' cells are activated T cells despite the expression of CD57 antigens on some of them, and not NK-like cells which would show a 'naive' CD45RA+/RO– phenotype. This finding is con-

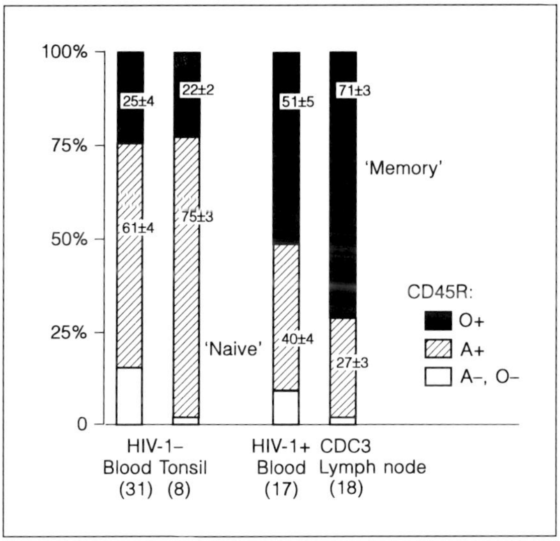

Fig. 5. The proportions of CD45RA+ (A+) and CD45RO+ (O+) cells within the CD8+ lymph node populations. The frozen histological sections were labelled by double CD8/CD45 staining and only CD8+ cells were counted. Some CD8+ cells appear to be double negative for CD45 isoforms (A–, O–). The controls are tonsils and the HIV-1 samples are from lymph nodes with persistent generalized lymphadenopathy.

firmed by CD3/CD8/CD45RO or CD38 triples which demonstrate the CD3 positivity of CD45RO+/CD38+ cells in patients with HIV-1 [23]. Similar observations have been reported by Prince and Jensen [11] using other triple combinations. An analysis of CD57+ cells among the various categories shows that these peculiar CD8+ populations, which are not at all attributable to or associated with HIV-1 infections (see above), are distributed all over the various '2', '3', '4' and '5' categories of figure 3 and cannot be placed exclusively to group '5'.

The Wide Distribution and Early Appearance of CD45RO+, CD8+ Populations in HIV Infection

The inversion of the proportions of CD45RA+ versus CD45RO+ cells within the CD8+ populations is one of the most consistent changes in the lymphoid tissues of HIV-1-infected individuals (fig. 5). In 18 samples from

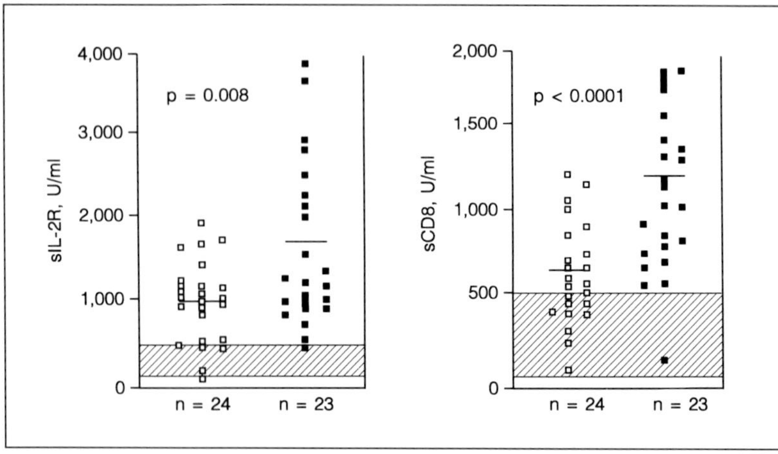

Fig. 6. Serum values of soluble IL-2R (cIL-2R, α chain; CD25 antigen) and soluble CD8 (sCD8) at first detection of seropositivity in 47 HIV-1-infected patients grouped according to subsequent clinical outcome through a 32-month mean follow-up period. □ = Patients without disease progression; ■ = patients with disease progression. Shaded areas are the range of control values in HIV-1 controls. From Pizzolo et al. [31] with permission.

patients with persistent generalized lymphadenopathy (CDC3) the percentages of CD45RO+ cells within the CD8+ populations were even higher (71%) than in the blood (51%), and the germinal centre infiltrating CD8+ cells were almost exclusively CD45RO+. In normal lymph nodes CD8+ cells are mostly CD45RA+. In addition, these CD8+ cells of 'memory' type have appeared in the lymph nodes early during HIV-1 infections, 4–5 years before the onset of clinical symptoms [9]. In HIV-infected children the switch to CD45RO positivity by the CD8+ cells is also seen early in the disease [10].

Similar observations have been made by analysing the CD38+, CD8+ subset (table 2). In a longitudinal cohort study of homosexual men at high risk of developing AIDS, significant increase was demonstrated in the CD8+, CD38+ subpopulations concomitant with positive polymerase chain reaction to the *gag, LTR* and *env* genes of HIV-1, but prior to detectable HIV-1 antibodies investigated by Western blot. This early and marked increase in the CD8+ and CD8+, CD38+ populations consistently appeared to be the first suggestive sign of impending seroconversion throughout the study [24].

These observations about the wide distribution and early appearance of particular CD8+ subsets may facilitate the detection of soluble CD8 (sCD8) molecules in the serum of HIV seropositive individuals as a useful assay for CD8 activation for the following reasons. It has been reported that sCD8 is released predominantly by Class II+, CD8+ cells [25]. Similarly, soluble IL-2R is released by activated T cells [26]. As there are indications that immunological hyperreactivity at the time of seroconversion is related to a more rapid clinical course leading to AIDS [27], it is of particular interest to investigate these parameters early in the disease. The extent of CD8+ lymphocyte activation can be assessed by the sCD8 assay [28, 29] and the soluble IL-2R (α chain) test is a measure of activations in general [30]. Recent results shown in figure 5 reveal that in serum samples taken at first detection of seropositivity the sCD8 values > 1,300 U/ml herald rapid progression and < 550 U/ml indicate slow progression while the intermediate value range 550–1,300 is not discriminatory [31]. When compared to CD4 lymphocyte counts, the assays for sCD8 have additive predictive value in terms of disease progression [32].

Concluding Remarks

One aspect of the poor prognostic significance of sCD8 in the serum that such a finding indeed suggests that certain aspects of immune activation by the CD8 lineage are deleterious. Nevertheless, observations above need to be assessed carefully. For example, the early accumulation of CD8+, CD45RO+ cells in children appeared to be the part of the powerful immune activation by the HIV-1 virus and was not associated with rapid progression [10]. Only phenotypic studies with triple marker combinations are likely to be sensitive enough to clarify the problems of CD8 subset disorders. Despite the promising start in terms of discrimination, critical investigations with the CD8/CD38/CD45RO triple are still missing: cell sorting experiments are required, similar to those described by Pantaleo et al. [this volume], to investigate the antiviral cytotoxicity in these populations. Furthermore, in other studies the short survival and poor responses to mitogens by these CD8+ cell populations have been documented – and it will be necessary to see the expression of apoptosis associated markers [12] and the display or the lack of auxiliary receptors promoting T-cell activation, such as the β chain of the IL-2R [33] and CD28 molecule [34] within these subsets.

References

1 Sanders ME, Makgoba MW, Sharrow SO, Stephany D, Springer TA, Young HA, Shaw S: Human memory T lymphocytes express increased levels of three cell adhesion molecules (LFA-3, CD2, and LFA-1) and three other molecules (UCHL1, CDw29, and Pgp-1) and have enhanced IFN-gamma production. J Immunol 1988; 140:1401–1407.

2 Akbar AN, Terry L, Timms A, Beverley PCL, Janossy G: Loss of CD45R and gain of UCHL1 reactivity is a feature of primed T cells. J Immunol 1988;140:2171–2178.

3 Beverley PCL: Minireview: human T cell subsets. Immunol Lett 1987;14:263–267.

4 Lanier LL, Le AM, Civin CI, Loken MR, Phillips JH: The relationship of CD16 (Leu-11) and Leu-19 (NKH-1) antigen expression on human peripheral blood NK cells and cytotoxic T lymphocytes. J Immunol 1986;136:4480–4486.4.

5 Buckle AM, Hogg N: Human memory T cells express intercellular adhesion molecule-1 which can be increased by interleukin-2 and interferon-γ. Eur J Immunol 1991;20:337–341.

6 Beverley PCL: Is T cell memory maintained by cross-reactive stimulation? Immunol Today 1990;11:203.

7 Akbar AN, Salmon M, Janossy G: The synergy between naive and memory T cells during activation. Immunol Today 1991;12:184–188.

8 Mackay CR: T cell memory: the connection between function, phenotype and migration pathways. Immunol Today 1991;12:189–192.

9 Janossy G, Bofill M, Racz P: Double immunofluorescence analysis of lymphoid tissues during HIV-1 infection; in Racz P, Haase AT, Gluckman JC (eds): Modern Pathology of AIDS and Other Retroviral Infections. Basel, Karger, 1990, pp 201–221.

10 Froebel KS, Doherty KV, Whitelaw JA, Hague RA, Mok JY, Bird AG: Increased expression of the CD45RO (memory) antigen on T cells in HIV-infected children. AIDS 1991;5:97–99.

11 Prince HE, Jensen ER: Three-color cytofluorometric analysis of CD8 cell subsets in HIV-1 infection. J Acquir Immune Defic Syndr 1991;4:1227–1232.

12 Piacentini M, Autuori F, Dini L, Farrace MG, Ghibelli L, Piredda L, Fesus L: Tissue transaminase is specifically expressed in neonatal rat liver cells undergoing apoptosis upon epidermal growth factor stimulation. Cell Exp Res 1991;263:227–235.

13 Maher P, O'Toole CM, Wreghitt TG, Spiegelhalter DJ, English TA: Cytomegalovirus infection in cardiac transplant recipients associated with chronic T cell subset ratio inversion with expansion of a Leu-7+ TS-C+ subset. Clin Exp Immunol 1985; 62:515–524.

14 Janossy G, Prentice HG, Grob JP, Ivory K, Tidman N, Grundy J, Favrot M, Brenner MK, Campana D, Blacklock HA: T lymphocyte regeneration after transplantation of T cell depleted allogeneic bone marrow. Clin Exp Immunol 1986;63:577.

15 Moller J, Dickmeiss E, Ryder LP, Jacobsen N, Svejgaard A: Increased frequencies of the CD29 and CD57 markers and decreased frequency of CD45RA within CD4+ and CD8+ subsets after allogeneic bone marrow transplantation in man. Scand J Immunol 1991;33:499–504.

16 Gratama JW, Wursch AM, Nissen C, Gratwohl A, D'Amaro J, de Gast GC, van Rood JJ, Speck B: Influence of graft-versus-host disease prophylaxis on early T-

lymphocyte regeneration following allogeneic bone marrow transplantation. Br J Haematol 1986;62:355–365.

17 Loken M, Brosnan J, Bach B, Ault K: Establishing optimal lymphocyte gates for immunophenotyping by flow cytometry. Cytometry 1990;11:453–459.

18 Reichert T, De Bruyere M, Deneys V, et al: Lymphocyte reference ranges in adult Caucasians. Clin Immunol Immunopathol 1991;60:190–208.

19 Gobbi M, Caligaris-Cappio F, Janossy G: Normal equivalent cells of B cell malignancies: analysis with monoclonal antibodies. Br J Haematol 1983;54:393–403.

20 Axberg I, Gale MJ Jr, Afar B, Clark EA: Characterization of T-cell subsets and T-cell receptor subgroups in pigtailed macaques using two- and three-color flow cytometry. J Clin Immunol 1991;11:193–204.

21 Janossy G, Tidman N, Papageorgiou ES, Kung PC, Goldstein G: Distribution of T lymphocyte subsets in the human bone marrow and thymus – an analysis with monoclonal antibodies. J Immunol 1981;126:1608–1613.

22 Janossy G, Bofill M, Rowe D, Muir J, Beverley PCL: The tissue distribution of T lymphocytes expressing different CD45 polypeptides. Immunology 1989;66:517–525.

23 Cambell D, Bach BA, Jackson AL, Ismail E, Fahey JL: CD38 expression on T and NK cells in HIV-infected individuals using three color flow cytometry. Int Conf AIDS 1990;6:218.

24 Yagi MJ, Joesten ME, Wallace J, Roboz JP, Bekesi JG: Human immunodeficiency virus type 1 (HIV-1) genomic sequences and distinct changes in CD8+ lymphocytes precede detectable levels of HIV-1 antibodies in high-risk homosexuals. J Infect Dis 1991;164:183–188.

25 Agostini C, Pizzolo G, Zambello R, et al: Shedding of soluble form of the CD8 complex by CD8+ HLA-DR+ cells in HIV-1-infected patients. AIDS 1991;5:813–819.

26 Prince HE, Kleinman SH, Maino VC, Jackson AL: In vitro activation of T lymphocytes from human immunodeficiency virus (HIV)-seropositive blood donors. I. Soluble interleukin-2 receptor (IL-2R) production parallels cellular Il-2R expression and DNA synthesis. J Clin Immunol 1988;8:114–120.

27 Simmonds P, Beatson D, Cuthbert RJ, Watson H, Reynolds B, Peutherer JF, Parry JV, Ludlam CA, Steel CM: Determinants of HIV disease progression: 6-year longitudinal study in the Edinburgh haemophilia/HIV cohort. Lancet 1991;338:1159–1163.

28 Reddy MM, Lange M, Grieco MH: Elevated soluble CD8 levels in sera of human immunodeficiency virus-infected populations. J Clin Microbiol 1989;27:257–260.

29 Agostini C, Semenzato G, Vinante F, Sinicco A, Trentin L, Zambello R, Zuppini B, Zanotti R, Siviero F, Veneri D, Foa R, Pizzolo G: Increased levels of soluble CD8 molecule in the serum of patients with acquired immunodeficiency syndrome (AIDS) and AIDS-related disorders. Clin Immunol Immunopathol 1989;50:146–153.

30 Lang JM, Coumaros G, Levy S, Falkenrodt A, Steckmeyer M, Partisani M, Aleksijevic A, Lehr L, Koehl C: Elevated serum levels of soluble interleukin-2 receptors in HIV infection: correlation studies with markers of cell activation. Immunol Lett 1988;19:99–102.

31 Pizzolo G, Vinante F, Morosato L, de Sabata D, Sinicco A, Raiteri R, Agostini C, Semenzato G: Determinants of HIV progression. Lancet 1992;339:130.
32 Pizzolo G, Vinante F, Morosato L, de Sabate D, Sinicco A, Raiteri R, Agostini C, Semenzato G: Prognostic significance of soluble CD8 serum levels in HIV-1 infection. AIDS 1991;6:133–134.
33 Ohashi Y, Takeshita T, Nagata K, Mori S, Sugamura K: Differential expression of the IL-2 receptor subunit, p55 and p75 on various populations of primary peripheral blood mononuclear cells. J Immunol 1989;143:3548–3557.
34 Gruters RA, Terpstra FG, De Goede RE, Mulder JW, De Wolf F, Schellekens PT, Van Lier RA, Tersmette M, Miedema F: Immunological and virological markers in individuals progressing from seroconversion to AIDS. AIDS 1991;5:837–844.

Prof. G. Janossy, Department of Clinical Immunology, The Royal Free Hospital, School of Medicine, Pond Street, GB–London NW3 2QG (UK)

Janossy G, Autran B, Miedema F (eds): Immunodeficiency in HIV Infection and AIDS. EC/FERS/MRC Workshop on Immunodeficiency in HIV-1 Infections, Windsor, Surrey, 1991. Basel, Karger, 1992, pp 211–218

Role of CD8+ T Lymphocytes in the Pathogenesis of HIV Infection

Giuseppe Pantaleo, Cecilia Graziosi, Luca Butini, Anthony S. Fauci

Laboratory of Immunoregulation, National Institute of Allergy and Infectious Diseases, National Institutes of Health, Bethesda, Md., USA

Human immunodeficiency virus (HIV) induces an immune response in infected individuals. In fact, the development of an antibody response to the various components of HIV within 2–3 weeks from the onset of primary HIV infection represents the most reliable diagnostic tool to determine if a person has been infected with the virus. In addition to the HIV-specific humoral response, an HIV-specific cell-mediated immune response has been described in a significant proportion of HIV-infected individuals as indicated by HIV-specific T cell proliferation and cytotoxicity [1]. Furthermore, on the basis of the findings in other viral infections, one may speculate that the immune system of the host has several other means to mediate antiviral activity. Some of these antiviral mechanisms are listed in table 1. Despite the fact that an HIV-specific immune response is found in all HIV-infected individuals and that different immunocompetent cells potentially contribute to the antiviral immunity, the systemic nature of HIV infection and the progression to disease in the majority of seropositive individuals indicate that the immune response to HIV is inadequate.

CD8+ T Lymphocytes Are a Major Component of the HIV-Specific Cell-Mediated Immune Response

CD8+ T lymphocytes may play a crucial role in the control of viral spreading. Major histocompatibility class I-restricted cytotoxic T lymphocytes (CTL) have been documented in the peripheral blood (PB), lymph

Table 1. Putative protective mechanisms against HIV

Neutralizing antibodies for virions or cytolytic antibodies for infected cells
CTL that kill HIV-infected cells through the recognition of a viral peptide product in association with MHC protein
NK cells lytic for HIV-infected cells
NK cells or macrophages that bind antibodies with specificity for HIV and lyse infected target cells (ADCC)
Secretion of antiviral cytokines (e.g. IFN-α) from activated T lymphocytes

nodes, cerebrospinal fluid and bronchoalveolar lavage fluid of infected individuals [2–6]. The majority of these CTL are class-I restricted and belong to the CD8+ T cell subset [6]. A unique feature of HIV-specific cytolytic activity is that it can be readily detected in fresh PB mononuclear cells (MC) indicating that a high frequency of HIV-specific CTL is associated with this chronic viral disease. Targets of HIV-specific CTL include both structural and regulatory proteins such as *env, gag,* reverse transcriptase, *nef* and *vif* gene products [2–8] and HIV-specific cytolytic activity is detected in 25–40% of HIV-infected individuals in the early stages of disease [S. Koenig, pers. commun.]. Further evidence that CD8+ T lymphocytes are an important component of the specific immune response against HIV has been the demonstration that CD8+ T cells may suppress virus replication in vitro. It has been suggested that CD8+ cell-mediated inhibition of HIV replication may occur both by lytic and nonlytic mechanisms [9, 10]. It is still unclear, however, whether this inhibition occurs in a MHC-restricted or -unrestricted manner [9, 10].

Mechanisms Responsible for the Dysfunction of CD8+ T Lymphocytes in HIV-Infected Individuals

Given the potential primary role played by CD8+ T lymphocytes in anti-HIV immunity, a precise elucidation of the dysfunction of CD8+ T cells associated with HIV infection may add new insights for a better understanding of the mechanisms of immune defense against HIV in the host.

In contrast to the refractoriness of CD8+ T lymphocytes to infection with HIV, functional impairments of CD8+ T cells have been described in HIV-infected individuals. These include: (i) reduced potential to proliferate in vitro in response to a variety of stimuli and decreased clonogenic potential [11–13]; (ii) progressive loss of HIV-specific cytolytic activity [4, 13]. These defects of CD8+ T lymphocytes are consequent to the expansion in the circulation of a subset of CD8+ T cells phenotypically characterized by the surface expression of major histocompatibility complex (MHC) class II (HLA-DR) molecules [12]. Phenotypic analysis of PBMC demonstrated that HLA-DR antigen was expressed on 25% of total CD8+ T lymphocytes in a representative seropositive individual (fig. 1a–c). Interestingly enough, expression of HLA-DR antigen was not associated with other markers of T cell activation such as the CD25 molecule (the receptor for IL-2) or the transferrin receptor [12]. In addition, when sorted CD8+DR+ and CD8+DR– cell subsets were analyzed for their potential to proliferate in vitro after stimulation with anti-CD3 mAb plus IL-2 or IL-4, the proliferative defect was predominantly confined to the CD8+DR+ cell subset (fig. 1d, e). Similar results were obtained after stimulation of CD8+DR+ T cells with other stimuli, including mitogens (PHA and PMA) and anti-CD2 mAb used alone or in combination with anti-CD28 mAb. Furthermore, we have recently performed phenotypic and functional analyses of CD8+ T lymphocytes in anatomic sites other than PB (i.e. lymphoid organs), since PB lymphocytes represent only 2% of the total lymphocyte pool [14]. We have demonstrated that the expansion of the CD8+DR+CD25– cell population was far greater in the lymphoid organs (lymph nodes, adenoids and tonsils) compared to PB and that the replicative capabilities were severely impaired in these CD8+DR+CD25– cells isolated from lymphoid organs [G. Pantaleo et al., pers. commun.]. Taken together, these results indicate that a large proportion (50%) of CD8+ T lymphocytes belong to the non-replicating CD8+DR+CD25– cell population.

As mentioned above, HIV-specific cytolytic activity may be readily detected in fresh PBMC in 25–40% of HIV-infected individuals. The detection of cytolytic T cells in the circulation indicates an ongoing HIV-specific cell-mediated immune response. On the basis of animal studies and of experimental evidence in vitro, the detection of HIV-specific CTL reflects effective immunity against HIV and protection against viral dissemination. However, there are insufficient data to implicate a relationship between the presence of HIV-specific cytolytic activity and clinical status. In this regard, preliminary data obtained in our laboratory seem to

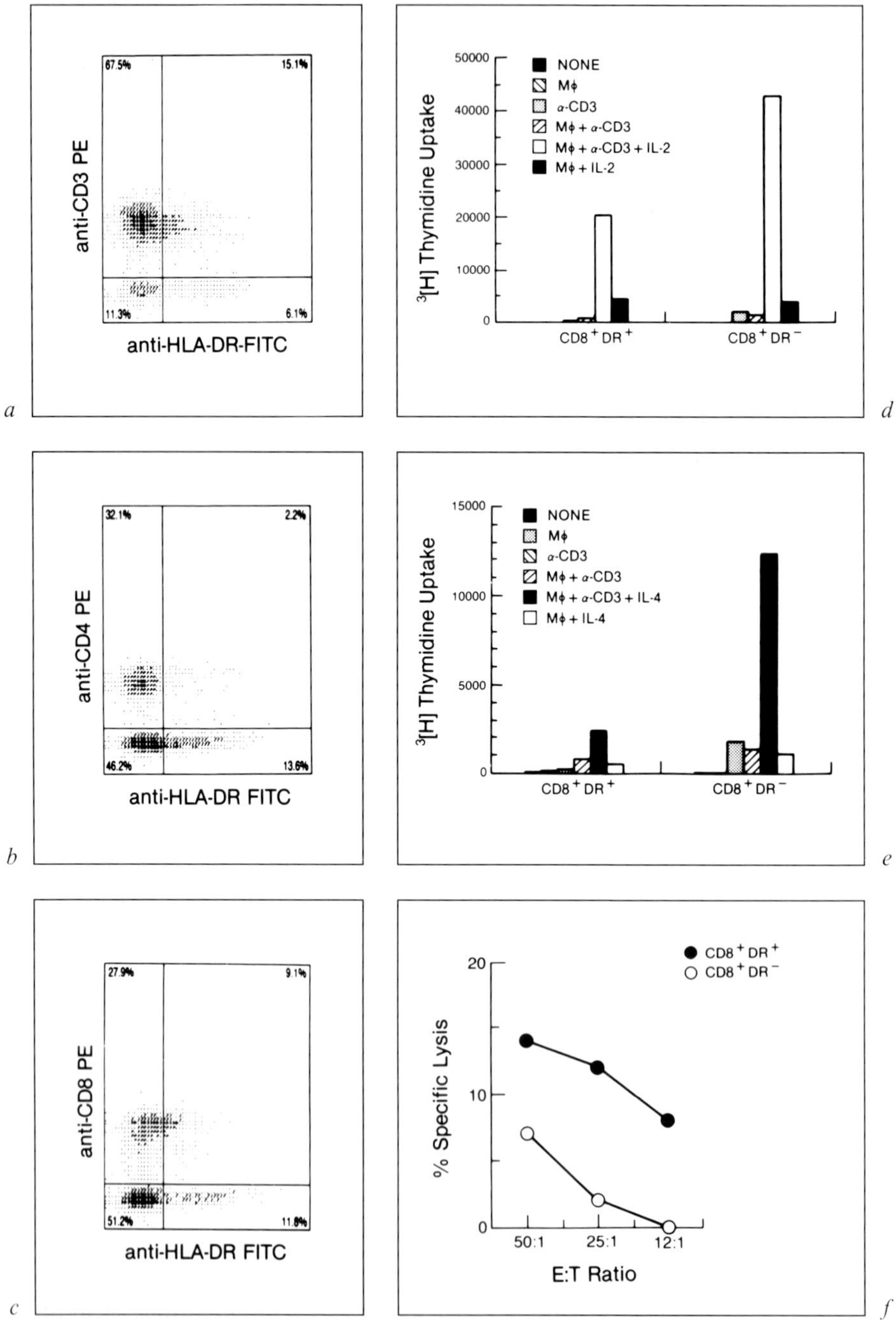
a
67.5%
15.1%
11.3%
6.1%
anti-CD3 PE
anti-HLA-DR-FITC
d
50000
40000
30000
20000
10000
0
NONE
Mϕ
α-CD3
Mϕ + α-CD3
Mϕ + α-CD3 + IL-2
Mϕ + IL-2
^{3}[H] Thymidine Uptake
CD8^{+} DR^{+}
CD8^{+} DR^{-}
b
32.1%
2.2%
46.2%
13.6%
anti-CD4 PE
anti-HLA-DR FITC
e
15000
10000
5000
0
NONE
Mϕ
α-CD3
Mϕ + α-CD3
Mϕ + α-CD3 + IL-4
Mϕ + IL-4
^{3}[H] Thymidine Uptake
CD8^{+} DR^{+}
CD8^{+} DR^{-}
c
27.9%
9.1%
51.2%
11.8%
anti-CD8 PE
anti-HLA-DR FITC
f
CD8^{+} DR^{+}
CD8^{+} DR^{-}
20
10
0
% Specific Lysis
50:1
25:1
12:1
E:T Ratio

indicate that the presence of HIV-specific cytolytic activity does not correlate with levels of virus replication in PB (i.e. plasma viremia), with the ability to isolate virus in vitro and more importantly with clinical status [G. Pantaleo et al., pers. commun.]. One cannot exclude the possibility that the presence of HIV-specific cytolytic activity is related to high levels of viral antigen expression, which may reflect active virus replication. In this regard, it is noteworthy that appropriate CTL activation cannot occur without high levels of virus expression. Therefore, it should be taken in consideration that persistent detection of CTL activity may also reflect continuing active disease in the host.

It has been observed that a progressive loss of HIV-specific cytolytic activity occurs during HIV infection [4, 15]. This decline in HIV-specific CTL activity has been associated with clinical progression of disease and thus a relationship to progression to AIDS has been suggested; however, due to the limited number of patients investigated, more studies are necessary to confirm this hypothesis. Nonetheless, the putative mechanisms responsible for the progressive decrease and ultimately the loss of HIV-specific cytolytic activity have been extensively studied. The HIV-mediated depletion of CD4+ cells and the consequent impairment of CD4+ helper T cell-dependent HIV-specific effector cell responses may contribute significantly to the loss of HIV-specific cytolytic activity [16]. In addition, rapid changes in the primary structures of individual isolates of HIV occur in vivo [17]. It is possible that with the progression of HIV infection the immune system in the host is not able to generate immune responses against these new variants. These viral variants may not be able to appropriately activate the preexistent CTL. The lack of appropriate activation of preexistent CTL will result in the observed loss of HIV-specific cytolytic

---

Fig. 1. a–c Cytofluorometric analysis of CD3, CD4, CD8 vs. HLA-DR antigen expression in PBMC of a HIV-infected individual. *d, e* Sorted CD8+DR+ and CD+DR− cell populations (5×10^4), isolated from the PBMC whose phenotypic analysis is shown in figure 1a–c, were stimulated with anti-CD3 mAb in the presence of irradiated (4,000 rad) allogeneic exogenous monocytes (10^4) and IL-2 or IL-4 were added 24 h after initiating the cultures. At the end of the period of stimulation (72 h) cell proliferation was evaluated by [^{3}H]TdR uptake. *f* Cytolytic activity against target cells expressing HIV-1_{IIIB} *env, gag, pol* and *nef* proteins by fresh sorted CD8+DR+ (●) and CD8+DR− (○). Control target cells were infected with vaccinia virus vector VSC8. Data are expressed as percentage-specific lysis after subtraction of background levels.

activity in vitro. Loss of HIV-specific cytolytic activity may also be dependent on a suppressor mechanism [18, 19]. In HIV-seropositive individuals a variable proportion of CD8+ T cells express CD57 antigen. These CD8+CD57+ cells have been shown to suppress HIV-specific cytolytic activity in vitro by the release of a soluble factor. However, HIV-specific cytolytic activity can also be detected in HIV-infected patients with expansion in vivo of the CD8+CD57+ subset. For this reason, there is a dilemma about the suppressive versus potential cytotoxic activity of the CD8+CD57+ subset, as discussed further by Sadat-Sowti et al., this volume.

Finally, a progressive reduction in the frequency of HIV-specific CTL may determine the loss of HIV-specific cytolytic activity. In this regard, we have recently shown [15] that HIV-specific CTL are predominantly contained within the poorly replicating CD8+DR+CD25– T cell subset (fig. 1f). These results suggest that the decline in HIV-specific cytolytic activity may be explained, at least in part, by a defective in vivo clonal expansion of the precursors of HIV-specific CTL. Of note is the fact that the loss of HIV-specific cytolytic activity is selective, since CD8+ T lymphocytes maintain normal cytolytic potential even in patients with AIDS [15].

Conclusions

Given that the experimental evidence seems to indicate that CD8+ T lymphocytes may exert protection against HIV infection, the emergence of the functional impairment of CD8+ T cells may explain in part why the immune system in the host does not prevent dissemination of HIV and progression to AIDS. It is likely that multiple mechanisms, working independently or in concert with each other, are responsible for the dysfunction of CD8+ T lymphocytes in HIV-infected patients. However, a key event in the emergence of the functional impairment of CD8+ T lymphocytes seems to be played by the expansion of the CD8+DR+ T cell subset. It is likely that the expansion of this CD8+ cell subset reflects an important immunopathogenic event that, interestingly enough, does not seem to be specific for HIV infection. In fact, expansion of CD8+DR+ T cells is a finding common to other viral diseases (Epstein-Barr virus and cytomegalovirus) as well as bone marrow transplantation and chronic inflammatory diseases (systemic lupus erythematosus and rheumatoid arthritis). In these pathologic conditions the defects in CD4+ T lymphocytes are neither qualitatively nor quantitatively comparable to those observed in HIV infection, suggesting that the expansion of the CD8+DR+ T cell subset may not be related directly to the depletion of CD4+ T lymphocytes. In contrast, a common feature of HIV infection and these other pathologic conditions is the chronic nature of the disease.

In conclusion, it is highly likely that CD8+ T lymphocytes play a fundamental role in the control of HIV infection. However, it is also likely that CD8+ T lymphocytes as well as other components of HIV-specific immunity have the greatest chance of clearing HIV infection and preventing dissemination and clinical progression during the earliest periods after infection. Furthermore, as a consequence of both the depletion in CD4+ T lymphocytes and the functional impairment of the other immunocompetent cells, HIV-specific immune responses are rendered ineffective in controlling viral spreading and progression to AIDS.

References

1 Salk J: The WHO virus approach to immunization. Development and evaluation of a vaccine for human immunodeficiency virus (HIV) infection. Ann Intern Med 1989;110:373–385.

2 Plata F, Autran B, Pedroza-Martins L, Wain-Hobson S, Raphael M, Mayaud C, Denis M, Guillon JM, Debre P: AIDS virus-specific cytotoxic T lymphocytes in lung disorders. Nature 1987;328:348–351.

3 Walker BD, Chakrabarti S, Moss B, Paradis TJ, Flynn T, Durno AG, Richard S, Blumberg RS, Kaplan JC, Hirsch MS, Schooley RT: HIV-specific cytotoxic T lymphocytes in seropositive individuals. Nature 1987;328:345–348.

4 Hoffenbach A, Langlade-Demoyen P, Dadaglio G, Vilmer E, Michel F, Mayaud C, Autran B, Plata F: Usually high frequencies of HIV-specific cytotoxic T lymphocytes in humans. J Immunol 1989;142:452–462.

5 Sethi KK, Naher H, Stroehmann I: Phenotypic heterogeneity of cerebrospinal fluid-derived HIV-specific and HLA-restricted cytotoxic T-cell clones. Nature 1988;335: 178–181.

6 Koenig S, Earl P, Powell D, Pantaleo G, Merli S, Moss B, Fauci AS: Group-specific, major histocompatibility complex class I-restricted cytotoxic responses to human immunodeficiency virus 1 (HIV-1) envelope proteins by cloned peripheral blood T cells from an HIV-1-infected individual. Proc Natl Acad Sci USA 1988;85:8638–8642.

7 Walker BD, Flexner C, Paradis TJ, Fuller TC, Hirsch MS, Schooley RT, Moss B: HIV-1 reverse transcriptase is a traget for cytotoxic T lymphocytes in infected individuals. Science 1988;240:64–66.

8 Koenig S, Fuerst TR, Wood LV, et al: Mapping the fine specificity of a cytolytic T cell response to HIV-1 nef protein. J Immunol 1990;145:127–135.

9 Walker CM, Moody DJ, Stites DP, Levy JA: CD8+ T lymphocyte control of HIV replication in cultured CD4+ cells varies among infected individuals. Cell Immunol 1989;119:470–475.
10 Tsubota H, Lord CI, Watkins DI, Morimoto C, Letvin NL: A cytotoxic T lymphocyte inhibits acquired immunodeficiency syndrome virus replication in peripheral blood lymphocytes. J Exp Med 1989;169:1421–1434.
11 Margolick JB, Volkman DJ, Lane HC, Fauci AS: Clonal analysis of T lymphocytes in the acquired immunodeficiency syndrome. Evidence for an abnormality affecting individual helper and suppressor T cells. J Clin Invest 1985;76:709–715.
12 Pantaleo G, Koenig S, Baseler M, Lane HC, Fauci AS: Defective clonogenic potential of CD8+ T lymphocytes in patients with AIDS. Expansion in vivo of a nonclonogenic CD3+CD8+DR+CD25– T cell population. J Immunol 1990;144:1696–1704.
13 Gruters RA, Terpstra FG, De Jong R, Van Noesel CJM, Van Lier RAW, Miedema F: Selective loss of T cell functions in different stages of HIV infection. Eur J Immunol 1990;20:1039–1044.
14 Westermann J, Pabst R: Lymphocyte subsets in the blood: a diagnostic window on the lymphoid system? Immunol Today 1990;11:406–410.
15 Pantaleo G, De Maria A, Koenig S, et al: CD8+ T lymphocytes of patients with AIDS maintain broad cytolytic function despite the loss of human immunodeficiency virus-specific cytotoxicity. Proc Natl Acad Sci USA 1990;87:4818–4822.
16 Fauci AS: The human immunodeficiency virus: infectivity and mechanisms of pathogenesis. Science 1988;239:617–622.
17 Tersmette M, Gruters RA, de Wolf F, et al: Evidence for a role of virulent human immunodeficiency virus (HIV) variants in the pathogenesis of acquired immunodeficiency syndrome: studies on sequential HIV isolates. J Virol 1989;63:2118–2125.
18 Joly P, Guillon J-M, Mayaud C, Plata F, Theodorou I, Denis M, Debre P, Autran B: Cell-mediated suppression of HIV-specific cytotoxic T lymphocytes. J Immunol 1989;143:2193–2201.
19 Sadat-Sowti B, Debre P, Idziorek T, Guillon J-M, Hadida F, Okzenhendler E, Katlama C, Mayaud C, Autran B: A lectin-binding soluble factor released by CD8+CD57+ lymphocytes from AIDS patients inhibits T cell cytotoxicity. Eur J Immunol 1991;21:734–741.

Dr. Giuseppe Pantaleo, Laboratory of Immunoregulation, National Institute of Allergy and Infectious Diseases, National Institutes of Health, Building 10, Room 11B-13, Bethesda, MD 20892 (USA)

Janossy G, Autran B, Miedema F (eds): Immunodeficiency in HIV Infection and AIDS.
EC/FERS/MRC Workshop on Immunodeficiency in HIV-1 Infections, Windsor, Surrey, 1991.
Basel, Karger, 1992, pp 219–227

The Role of Cytotoxic Lymphocytes in HIV Infection

Sarah Rowland-Jones, Andrew McMichael

Molecular Immunology Group, Institute of Molecular Medicine,
John Radcliffe Hospital, Oxford, UK

Cytotoxic T lymphocytes (CTL) carry the CD8 surface marker and recognize antigen presented by MHC class I molecules. They were first recognized as being important in allo reactions in vitro [1] and were later shown to be responsible for lysing virus-infected cells in vitro [2]. Zinkernagel and Doherty [3] showed that virus-specific CTL recognized virus antigen in the context of self MHC molecules. Townsend et al. [4, 5] showed that virus-specific CTL recognized peptide fragments of virus proteins presented by class I MHC molecules, and Bjorkman et al. [6, 7] showed that crystallized class I molecules contained peptides within a cleft on the outer surface of the molecule. Thus the T cell receptors on CTL react with foreign peptide bound in class I molecules and different class I molecules bind different peptides [8]. This means that MHC class I molecules have a pivotal role in CTL function and this must be intimately associated with the genetic polymorphism of the MHC molecules. It is highly probable that differential immune function has played a role in selecting particular class I alleles and that, for any given virus infection, different class I MHC types should offer greater or lesser degrees of immune protection in the context of the CTL response. In acute virus infections, particularly influenza, it has been shown that CTL have a protective role [9, 10]. They are activated early in infection [11] and have been shown to be responsible for clearing influenza virus [9] and respiratory syncytial virus from infected lungs [12]. In persistent virus infections, there is also evidence that they play a role in containing virus, for instance in Epstein-Barr virus infection [13, 14]. In certain circumstances, CTL may contribute to pathology, for instance, in lymphocytic choriomeningitis virus (LCMV) infection [15] and respiratory syncytial virus infection [12]. In the latter example, an infusion of a large number of cloned CTL to infected mice caused severe pneumonitis, whereas smaller numbers of CTL cleared the infection. Briefly summarized, therefore, the evidence from other virus infections is that CTL are a major component of the immune response to virus infection and usually contribute to recovery but sometimes to immunopathology.

CTL and HIV Infection

HIV-specific CTL were first demonstrated by Plata et al. [16] and Walker et al. [17, 18]. Surprisingly, in both examples they were demonstrated without any attempt to restimulate the response in vitro before measuring lytic activity, as is required for virtually all other viruses studied. Plata et al. [16] found effector CTL present in broncho-alveolar lavage material and Walker et al. [18] demonstrated CTL in freshly prepared peripheral blood lymphocytes. In both cases, envelope-specific, and later *gag*- and *pol*-specific CTL were demonstrated and HLA restriction was observed [19, 20]. Subsequently, other groups have demonstrated HIV-specific CTL, mainly directed at *gag*, *pol* and *env*, but also at epitopes in regulatory proteins such as *nef* and *tat* [21]. Nixon et al. [19] identified a peptide epitope in HIV *gag* and subsequently a large number of epitope peptides have been identified [22]. In our hands, CTL activity is most reliably demonstrated by restimulating peripheral blood cultures with autologous activated (by phytohaemagglutinin) T lymphoblasts. This protocol gives measurable CTL responses in more than 80% of patients. We have also found fresh blood CTL responses, but only in about 25% of patients; other groups find responses in a higher proportion of patients and this may relate to the interval between infection and measurements, which in our case has been around 7–8 years.

In patients with fresh blood CTL responses, it has been calculated that the level of CTL must be very high, up to 1% of all lymphocytes for a single epitope [23]. A number of measurements have been made on precursor (i.e. memory) CTL frequencies. These have generally come out at between 1 in 5,000 and 1 in 15,000 [23, 24] which are high levels compared to other virus infections (usually less than 1 in 20,000) but are well under 1%. It has been suggested that this implies a terminal differentiation of CTL to end-stage effector cells with possible depletion of memory CTL pools [23].

The loss of CTL activity as patients progress to AIDS has also been documented [25, 26], and levels measured in patients with AIDS are generally very low. In contrast, it has not been possible to determine whether CTL activity is detectable before seroconversion, but by analogy with other virus infections this could be the case.

Function of CTL in HIV Infection

Studies have shown that CTL in vitro inhibit virus replication [17, 27, 28]. This was initially shown by Walker et al. [17] in culture who showed that CD8 T cells were activated during culture of lymphocytes from patients with HIV infection and that ultimately virus replication diminished. Brinchmann et al. [27] and Kannagi et al. [28] have found that CTL can terminate HIV replication in vitro. In two of the studies contact between CTL effector and target has been shown not to be essential because inhibition would occur across a membrane. On the other hand, Kannagi et al. [28] demonstrated that there was an important role for the LFA1 adhesion molecules, implying direct contact. It is possible, therefore, that CD8 CTL may release cytokines which inhibit HIV replication and this effect may be important in controlling infection in vitro, although the direct lytic effect should also be a powerful inhibitor of virus replication. These experiments argue for an important role for CTL in control of HIV infection.

Examination of the epitopes recognized by CTL shows that many of these in *gag* or *pol* are quite conserved between different HIV-sequenced isolates. Thus, the HLA B27 epitope described by Nixon et al. [19] was fairly well conserved between HIV-1 and HIV-2, and HIV-1-specific CTL could cross-react with HIV-2 *gag* expressed by recombinant vaccinia virus and by added peptides [29]. This is in line with much previous evidence from other viruses where CTL tend to recognize conserved virus proteins and cross-react with different virus strains [30, 31]. On the other hand, if CTL are important in controlling the virus infection, they might exert selective pressure on virus to change and escape. This phenomenon is well recognized for virus escape from antibody control in HIV and other viruses [32], but is largely confined to surface glycoproteins where antibody can neutralize infectivity. Escape mutation in HIV might be possible where very low copy numbers of provirus exist within infected cells so that mutation in provirus sequence can allow that cell to replicate large numbers of escape mutants; in contrast with influenza virus infection for example, cells may be infected with many thousand virus particles and they would all have to change for virus to escape CTL recognition, an unlikely event.

We have observed in 3 patients that fluctuating CTL response to three epitopes restricted by HLA B8 has occurred [33]. In these patients the dominant epitope has changed over time and responses to some peptides have appeared to disappear completely. We therefore sequenced, after PCR amplification, around these three epitopes at different time points in these patients. A surprising degree of sequence variability was observed (fig. 1). When peptides corresponding to these sequences were made, it was found that CTL for each of the three epitopes could discriminate between these naturally occurring variants [33]. This meant that sequences were present in the provirus in these patients that could not elicit CTL responses. This strongly implies that CTL were driving escape mutation. In 1 of the 3 patients observed, this variability and appearance of apparent escape mutants coincided with the patient developing CDC stage 3 disease and complete loss of the CTL response.

In contrast to the above findings, 3 patients that have HLA B27 have not lost their ability to respond to this epitope [33]. There has been limited sequence variation in this epitope but CTL have cross-reacted fully. Similar findings have been made by Meyerhans et al. [34] for this epitope where variability was observed but all peptides made were fully recognized by patients' CTL. Thus, the occurrence of escape mutants may be dependent on the HLA type. This could be because certain HLA types, such as B8, may be more fastidious in their requirements for peptide binding, in respect of peptide sequence, than other HLA types such as B27. However, it is known that at least three positions in the peptide sequence in HLA B27-restricted epitopes are critical [35, 36], but these do not change in virus sequences obtained so far. Therefore, another component may be the function of the epitope region of the *gag* molecule for HIV. If a critical region cannot change without impairing *gag* function, escape mutation may be much less likely. By these two mechanisms, HLA type should have a profound effect on the ability of escape mutations to occur. Perhaps the simultaneous use of three epitopes in association with HLA B8 partially compensates for escape mutation occurring at one or more of these; in contrast, all B27+ patients seem to be relying on a single peptide epitope.

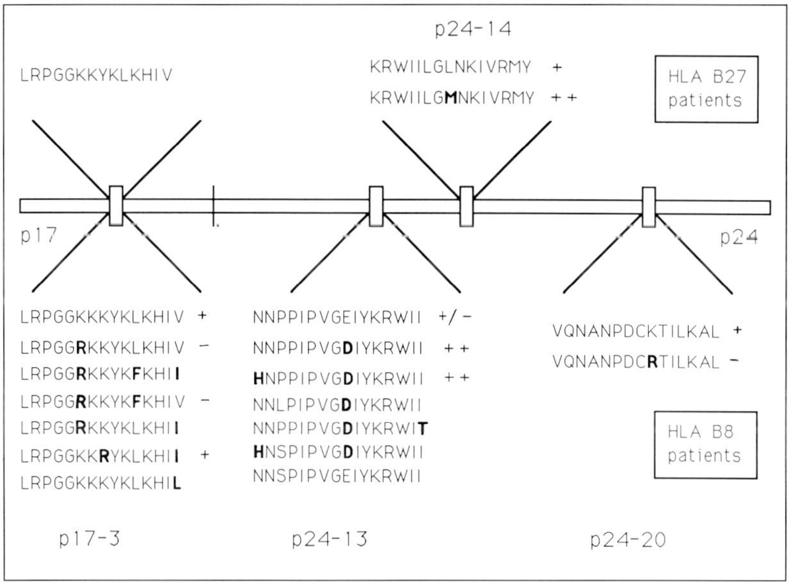

Fig. 1. Variation in the sequences of CTL epitopes in HIV *gag* in patients with HLA B8 and HLA B27. Sequence variants are shown in the single letter amino acid code with the index sequence at the top. The variants are grouped as those occurring in B27-positive and B8-positive patients. + indicates that patient CTL responds to the peptide variant; – indicates no response; +/– indicates that earlier CTL lines responded but later ones did not. Sequences that are present in patients which do not elicit CTL responses probably arise as escape mutants. [For details of experiments, see 33.]

However, HLA B8, as part of the A1 B8-DR3 haplotype, has been associated with rapid disease progression [37, 38], loss of CD4 T cells and poor outcome in three studies and this might in part be due to the contribution of the B8 towards the CTL response in these patients.

A question that arises from these studies is whether partial loss of one epitope by mutation is sufficient to give the virus survival advantage if there are other epitopes being presented by the same or other class I MHC molecules. In our view, this is likely to be the case because even small selective advantages may be sufficient to give a particular virus quasispecies the edge. One feature of HIV infection is that it is not localized to a single tissue or even area within the tissue, and that virus-infected cells are scattered throughout the body. This may present special problems for the immune system and there may be sites where the CTL response cannot be concentrated and, therefore, is of borderline efficiency. In these circumstances, loss of the single epitope out of this small number to which the immune system is reacting, could easily give the virus survival advantage. However, survival advantage in terms of immune escape may be modified by the replicative properties of the virus quasi-species and that virus may not necessarily immediately become rampant and completely dominant. Thus, an escape mutant may not necessarily account for a high percentage of identified provirus sequences. The original susceptible sequences may remain in high number because provirus is a form in which virus can escape CTL recognition altogether until activated. Therefore, the pattern of sequences identified may be complex, and a simple takeover of a single virus species with escape sequences present should not be expected and is not observed. This contrasts with the only other case of reported virus escape from CTL responses where mice transgenic for a T cell receptor specific for a single epitope in LCMV were infected with that virus and an escape mutant with a single amino acid change in the epitope sequence emerged [40].

The occurrence of escape mutants in HIV infection implies an important role for both antibody and CTL in attempted control of the infection. Both are probably making major contributions to the long latent period. However, escape mutants may contribute to the eventual breakdown of the immunity to HIV. Other factors, such as the continuous driving of CTL to terminal differentiation and overstimulation of the antibody response, may also be relevant.

Conclusions

HIV clearly presents special problems to the immune system. Initially, although the virus infects immunocytes, the total effect on the effectiveness of the immune response is likely to be small and the immune system reacts in a normal manner. However, a very potent CTL response is observed. This is unusually strong although similar levels of response have been observed in HTLV-1 infection [C. Parker and C. Bangham, pers. commun.]. It is likely that CTL response is contributing to control of the infection. However, the virus can eventually escape and the CTL response can decline, contributing to the development of AIDS. The role of the CTL response compared to other components should be reflected by an influence of HLA type on outcome. There are preliminary indications that the A1-B8-DR3 haplotype has a poor effect on outcome [37, 38] and there are data to suggest that B35 is also disadvantageous [39]. It will be harder to demonstrate protective effects, i.e. prolonged survival associated with some HLA types, because statistically these studies are much harder to do [41]. Furthermore, the interactions between a protective HLA type and a type that promotes escape mutants, perhaps B27 and B8 respectively, if present in the same patient, will have to enter into such calculations.

References

1 Cerottini J-C, Brunner KT: Cell-mediated cytotoxicity, allograft rejection and tumour immunity. Adv Immunol 1974;18:67–132.
2 Gardner I, Bowern NA, Blanden RV: Cell-mediated cytotoxicity against ectromelia virus-infected target cells. II. Identification of effector cells and analysis of mechanisms. Eur J Immunol 1974;4:68–72.
3 Zinkernagel RM, Doherty PC: H-2 compatibility requirement for T-cell mediated lysis of target cells infected with lymphocytic choriomeningitis virus. J Exp Med 1975;141:1427–1436.
4 Townsend A, Rothbard J, Gotch F, Bahadur B, Wraith D, McMichael A: The epitopes of influenza nucleoprotein recognized by cytotoxic T lymphocytes can be defined with short synthetic peptides. Cell 1986;44:959–968.
5 Townsend ARM, Gotch FM, Davey J: Cytotoxic T cells recognise fragments of influenza nucleoprotein. Cell 1985;42:457–467.
6 Bjorkman P, Saper M, Samraoui B, Bennett W, Strominger J, Wiley D: The foreign antigen binding site and T cell recognition regions of class I histocompatibility antigens. Nature 1987;329:512–519.
7 Bjorkman P, Saper M, Samraoui B, Bennett W, Strominger J, Wiley D: Structure of human class I histocompatibility antigen, HLA-A2. Nature 1987;329:506–511.
8 Falk K, Rotzschke O, Stevanovic S, Jung G, Rammensee HG: Allele-specific motifs revealed by sequencing of self peptides eluted from MHC molecules. Nature 1991; 351:290–296.
9 Lin Y, Asknoas BA: Biological properties of an influenza A virus-specific killer T cell clone. J Exp Med 1981;154:225–234.
10 McMichael AJ, Gotch FM, Noble GR, Beare PAS: Cytotoxic T-cell immunity to influenza. N Engl J Med 1983;309:13–17.
11 Yap KL, Ada GL: Transfer of specific cytotoxic T lymphocytes protects mice inoculated with influenza virus. Nature 1978;273:238–240.
12 Cannon MJ, Openshaw PJM, Ba A: Cytotoxic cells clear virus but augment lung pathology in mice infected with respiratory syncytial virus. J Exp Med 1988;168: 1163–1168.
13 Moss DJ, Rickinson AB, Pope JH: Long-term T cell-mediated immunity to Epstein-Barr virus in man. I. Complete regression of virus-induced transformation on cultures of seropositive donor leukocytes. Int J Cancer 1978;22:662–668.

14 Rowe M, Young LS, Crocker J, Stokes H, Henderson S, Rickinson AB: Epstein-Barr virus (EBV)-associated lymphoproliferative disease in the SCID mouse model: implications for the pathogenesis of EBV-positive lymphomas in man. J Exp Med 1991;173:147–158.

15 Buchmeier MJ, Welsh RM, Dutko FJ, Oldstone MBA: The virology and immunology of lymphocytic choriomeningitis virus infection. Adv Immunol 1980;30:275–331.

16 Plata F, Autran B, Martins LP, Wain-Hobson S, Raphael M, Mayaud C, Denis M, Guillon JM, Debré P: AIDS virus-specific cytotoxic T lymphocytes in lung disorders. Nature 1987;328:348–351.

17 Walker CM, Moody DJ, Stites DP, Levy JA: CD8+ lymphocytes can control HIV infection in vitro by suppressing virus replication. Science 1986;234:1563–1566.

18 Walker DB, Chakrabati S, Moss B, Paradis TJ, Flynn T, Durno AG, Blumberg RS, Kaplan JC, Hirsch MS, Schooley RT: HIV-specific cytotoxic T lymphocytes in seropositive individuals. Nature 1987;328:345–348.

19 Nixon DF, Townsend ARM, Elvin JG, Rizza CR, Gallwey J, McMichael AJ: HIV-1 gag-specific cytotoxic T lymphocytes defined with recombinant vaccinia virus and synthetic peptides. Nature 1988;336:484–487.

20 Walker BD, Flexner C, Paradis TJ, Fuller TC, Hirsch MS, Schooley RT, Moss B: HIV-1 reverse transcriptase is a target for cytotoxic T lymphocytes in infected individuals. Science 1988;240:64–66.

21 Culmann B, Gomard E, Kieny M-P, Guy B, Dreyfus F, Saimot A-G, Sereni D, Levy J-P: An antigenic peptide of HIV-1 NEF protein recognized by cytotoxic T lymphocytes of seropositive individuals in association with different HLA-B molecules. Eur J Immunol 1989;19:2383–2386.

22 Nixon DF, McMichael AJ: Cytotoxic T cell recognition of HIV proteins and peptides. AIDS 1991;5:1049–1059.

23 Gotch FM, Nixon DF, Alp N, McMichael AJ, Borysiewicz LK: High frequency of memory and effector gag-specific cytotoxic T lymphocytes in HIV seropositive individuals. Int Immunol 1990;2:707.

24 Hoffenbach A, Langlade-Demoyen P, Dadaglio G, Vilmer E, Michel F, Mayaud C, Autran B, Plata F: Unusually high frequencies of HIV-specific cytotoxic T lymphocytes in humans. J Immunol 1989;142:452–462.

25 Plata F, Dadaglio G, Chenciner N, Hoffenbach A, Wain-Hobson S, Michel F, Langlade-Demoyen P: Cytotoxic T lymphocytes in HIV-induced disease: implications for therapy and vaccination. Immunodefic Rev 1989;1:227–246.

26 Joly P, Guillon J-M, Mayaud C, Plata F, Theodorou I, Denis M, Debré P, Autran B: Cell-mediated suppression of HIV-specific cytotoxic T lymphocytes. J Immunol 1989;143:2193–2201.

27 Brinchmann JE, Gaudernack G, Vartdal F: CD8+ T cells inhibit HIV replication in naturally infected CD4+ T cells. Evidence for a soluble inhibitor. J Immunol 1990; 144:2961–2966.

28 Kannagi M, Masuda T, Hattori T, Kanoh T, Nasu K, Yamamoto N, Harada S: Interference with human immunodeficiency virus (HIV) replication by CD8+ T cells in peripheral blood leukocytes of asymptomatic HIV carriers in vitro. J Immunol 1990;64:3399–3406.

29 Nixon DF, Huet S, Rothbard JA, Kieny M-P, Delchambre M, Thiriart C, Rizza CR, Gotch FM, McMichael AJ: An HIV-1 and HIV-2 cross-reactive cytotoxic T cell epitope. AIDS 1990;4:841–845.
30 McMichael AJ, Askonas BA: Influenza virus-specific cytotoxic T cells in man; induction and properties of the cytotoxic T cell. Eur J Immunol 1978;8:705–710.
31 Zweerink HT, Askonas BA, Mllican D, Courtneidge SA, Skehel JJ: Cytotoxic T cells to type A influenza virus; viral haemagglutinin induces A-strain specificity while infected cells confer cross-reactive cytotoxicity. Eur J Immunol 1977;7:630–635.
32 Reitz MS, Wilson C, Naugle C, Gallo RC, Robert-Guroff M: Generation of a neutralization-resistant variant of HIV-1 is due to selection for a point mutation in the envelope gene. Cell 1988;54:57–63.
33 Phillips RE, Rowland-Jones S, Nixon DF, Gotch FM, Edwards JP, Ogunlesi AO, Elvin JG, Rothbard JA, Bangham CRM, Rizza CR, McMichael AJ: Human immunodeficiency virus genetic variation that can escape cytotoxic T cell recognition. Nature 1991;354:453–459.
34 Meyerhans A, Dadaglio G, Vartanian J-P, Langlade-Demoyen P, Frank R, Asjo B, Plata F, Wain-Hobson S: In vivo persistence of a HIV-1-encoded HLA B27 restricted cytotoxic T lymphocyte epitope despite specific in vitro reactivity. Eur J Immunol 1991;21:2637–2640.
35 Huet S, Nixon DF, Rothbard J, Townsend ARM, Ellis SA, McMichael, AJ: Structural homologies between two HLA B27 restricted peptides suggest residues important for interaction with HLA B27. Int Immunol 1990;2:311.
36 Jardetsky TS, Lane WS, Robinson RA, Madden DR, Wiley DC: Identification of self peptides bound to purified HLA-B27. Nature 1991;353:326–329.
37 Kaslow R, Duquensoy R, van Raden M, Kingsley L, Marrari MHF, Su S, Saah AJ, Detels R, Phair J, Rinaldo C: A1, Cw7, B8, DR3 HLA antigen combination associated with rapid decline of T-helper lymphocytes in HIV-1 infection. Lancet 1990; 335:927–930.
38 Steel CM, Ludlum C, Beatson D, Peutherer JF, Cuthbert RJG, Simmonds P, Morrison H, Jones M: HLA haplotype A1 B8 DR3 as a risk factor for HIV-related disease. Lancet 1988;i:1185–1188.
39 Pircher H, Moskphidis A, Rohrer U, Burki K, Hengartner H, Zinkernagel RM: Viral escape by selection of cytotoxic T cell-resistant variants in vivo. Nature 1990;346: 629–633.
40 Scorza-Smeraldi RS, Fabio G, Lazzarin A, Elsera NB, Moroni M, Zanussi C: HLA-associated susceptibility to acquired immunodeficiency syndrome in Italian patients with human immunodeficiency virus infection. Lancet 1986;ii:1187–1189.
41 Hill AVS, Allsopp CEM, Kwiatkowski D, Anstey NM, Twumasi P, Rowe PA, Bennett S, Brewster D, McMichael AJ, Greenwood BM: Common West African HLA antigens are associated with protection from severe malaria. Nature 1991;352:595–600.

Dr. Sarah Rowland-Jones, Molecular Immunology Group,
Institute of Molecular Medicine, John Radcliffe Hospital,
GB–Oxford OX3 9DW (UK)

Janossy G, Autran B, Miedema F (eds): Immunodeficiency in HIV Infection and AIDS.
EC/FERS/MRC Workshop on Immunodeficiency in HIV-1 Infections, Windsor, Surrey, 1991.
Basel, Karger, 1992, pp 228–234

A Soluble Inhibitor of Cytotoxic Functions

Behazine Sadat-Sowti, Patrice Debré, Brigitte Autran

Laboratoire d'Immunologie Cellulaire et Tissulaire, URA CNRS 0186,
CHU Pitié-Salpêtrière, Paris, France

The expansion of CD8+CD57+ lymphocytes has frequently been reported in persistent or recurrent viral infection or after kidney or bone marrow transplantation [1–3]. Several studies have similarly shown an increase of this minor T cell subset at early stages of HIV infection with a progressive expansion along with progression towards AIDS [4, 5]. The parallelism between the CD8+CD57+ PBL expansion and the expression of T cell activation markers on CD8 cells both in HIV infection and after transplantation suggest a link between these two phenomena, although the CD57 molecule itself is not considered to be an activation marker. When tested in vitro, a property of suppressing mitogen-induced B and T cell proliferation has been assigned to these cells [3, 6, 7]. Similarly, a transversal analysis of cytotoxic T cells (CTL) specific for HIV in the lungs of seropositive patients had allowed us to report an expansion of alveolar CD8+CD57+ T lymphocytes in AIDS patients and to demonstrate these cells' capability of inhibiting CTL function [8]. Blocking the effector phase of cytolytic processes can be a general feature of both antigen-specific and nonspecific killer cells and this process may be mediated by a soluble glycosylated molecule [9, 10]. Thus, CD8+CD57+ cells may contribute to T cell anergy observed in HIV infection and to the alteration of lymphocyte reactivity against allogeneic antigens. In this review the active suppressive role of these cells is documented together with the preliminary characterization of the suppressor factor involved.

Experimental Design

Patients. The cohorts of patients studied included both HIV-1 seropositive patients (100 individuals tested for HIV seropositivity by ELISA and Western blotting) and patients after having undergone bone marrow transplantation from HLA-identical siblings for hematologic malignancies (24 cases) or severe aplasic anemia (1 case). CMV infection was assessed on viral isolation in the blood or lungs.

Immunofluorescence Assay. PBL (5×10^5) were stained as previously described [7] with monoclonal antibodies (mAb) FITC-CD57 (anti-Leu7), CD8-PE (anti-Leu2a) (Becton-Dickinson, Grenoble, France). They were analyzed and then sorted using a Facstar (Becton-Dickinson) into two distinct CD8+CD57+ and CD57– cell populations. Their viability assessed by trypan blue exclusion test was >85%. The purity of sorted cells was >96%.

Immunomagnetic Separation. Avidin beads (Immunotech, Marseille, France) were coupled with biotinylated anti-CD57 (IOT10-NC1; Immunotech) by incubating overnight the mAb (50 μg) with 1-mg beads at 4 °C in PBS buffer. Cell suspensions were added to biotinylated avidin mAb bead complexes and mixed gently for 60 min at 4 °C. CD57+ cells were then isolated by a cobalt-samarium magnet (Dynal, Oslo, Norway). Positive and negative fractions contained >70% CD8+CD57+ cells and >5% CD8+CD57+ cells, respectively.

Inhibition of Cytolytic Activities. PBL from HIV seronegative donors were used for generating cytotoxic cells including alloreactive CTL, NK and LAK activities. As previously described, cytolytic effector cells were added to target cells at different killer-to-target (K/T) ratios to determine an optimal K/T ratio in preliminary standard chromium-release assay. Different dilutions of either PBL with high percentages of CD8+CD57+ cells or purified CD8+CD57+ cells or supernatants from CD8+CD57+ cells were added to the cytolytic assay and tested by standard 4-hour chromium release as previously described [7].

Supernatant Preparation. Supernatants (SN) were prepared from purified CD8+CD57+ and control CD57– cells resuspended at 5×10^6 cells/ml in serum-free culture medium and then incubated for 4 h at 37 °C without any stimulation. The cell suspensions were centrifuged at 500 *g*. The cells were discarded and the resulting SN were filtered through sterile 0.22-μ*M* filters.

CD8+CD57+ Lymphocytes Inhibit Specific and Nonspecific Cytolytic Activities

PBL samples which contained the highest percentages of CD8+CD57+ cells (>25%) were selected from 100 HIV-infected patients and 25 BMT patients. Among the HIV-infected patients particularly high percentages of

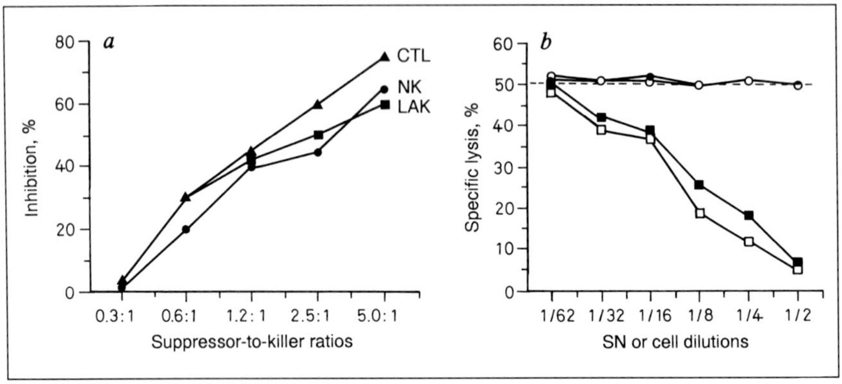

Fig. 1. Inhibition of CTL, NK and LAK activities by CD8+CD57+ PBL and SN isolated from AIDS patients. HLA-specific CTL, NK and LAK cells were tested at a 10/1 (K/T ratio) on EBV-infected B cell line, K562 and Daudi target cells, respectively. *a* Dilutions of PBL (25–45% CD8+CD57+ cells) were added at the onset of CRA to CTL (▲), NK (●) and LAK (■) cells. *b* Dilutions of sorted CD8+CD57+ (■) or CD57– (●) cells (2×10^6 cells) and CD8+CD57+ SN (□) or CD57– (○) were added to NK and LAK cytolysis.

CD8+CD57+ cells were observed in individuals with a low number of CD4+ cells. In BMT patients the increase in CD8+CD57+ cells was the highest among patients with active CMV infection without a direct association with graft-versus-host disease (GVH) [9, 10]. As shown in figure 1a, PBL samples containing >25% CD57+ cells inhibited allogeneic CTL activity as well as the NK or LAK cells mediated cytolysis, particularly when these cells were added at the onset of the cytolytic assays. The inhibition obtained was dose dependent and nonrestricted by HLA antigens, i.e. cells from histoincompatible donors could inhibit effector function. As demonstrated by cell-sorting experiments, only the CD8+CD57+ cells, but not the CD57– cells, were responsible for the inhibitory effect seen in the NK and LAK assays (fig. 1a). Similar inhibitory effects were also observed with the CD8+CD57+, but not with the CD57–, cells isolated from BMT patients (data not shown). In addition, a three-color fluorescence analysis performed on CD8+CD57+ PBL from both AIDS and BMT patients demonstrated that these cells displayed the CD3-TCR αβ complex, while both lacked the CD16 and CD56 NK cell-associated markers. This later finding is in agreement with the lack of cytolytic activity by purified CD8+CD57+

cells against NK-sensitive targets such as K562 and Daudi cells. Thus CD8+CD57+ cells are not NK cells.

Additional phenotypic analysis also reveals that the CD8+CD57+ subsets from both HIV seropositive and BMT patients display the CD29 molecule while 60–65% of these cells coexpress the CD45RA molecule. A cell surface expression of HLA-DR and CD38, but not the IL-2 receptor (CD25) or CD71 activation marker, can also be observed on 15–45% of CD8+CD57+ cells.

These observations delineate a common inhibitory mechanism which elicits an increase in CD8+CD57+ cells at the time when the immune system is activated by stimuli which are probably predominantly viral antigens. Moreover, these inhibitory properties suggest that CD8+CD57+ cells block a step of cytolytic processes which is common to allogeneic CTL, NK and LAK cells.

CD8+CD57+ Lymphocytes Release a Nonspecific Soluble Inhibitor of Cytolytic Activities

The inhibition of cell-mediated cytolysis by CD8+CD57+ cells does not require a cell-to-cell contact between killer cells and inhibitory cells [9]. SN were prepared from sorted CD8+CD57+ or CD57– PBL of HIV seropositive patients (fig. 1b) or BMT patients (data not shown) in 4-hour serum-free culture medium and were tested for their suppressor activity. Results showed that CD8+CD57+ SN could efficiently block NK- as well as LAK- and CTL-mediated cytolysis in a dose-dependent manner, whereas the control CD57– SN did not mediate any suppressor effect. Moreover, the inhibitory activity of SN was itself not lytic as assessed by the excellent viability after 4–72 h incubation (data not shown).

Different soluble factors such as PGE_2, TNFα, TNFβ and TGFβ were reported to inhibit immune responses in a variety of in vitro assays (11–15). However, the involvement of these mediators could be excluded in CD8+CD57+ SN since the inhibition of PGE_2 synthesis by indomethacin and ETYA had no effect on the suppressor activity of CD8+CD57+ cells. Similarly, the addition of neutralizing anti-TNFα, anti-IFNγ and anti-TGFβ mAb to SN did not affect the inhibition of cytolysis (data not shown), again indicating that the soluble inhibitor(s) released by CD8+CD57+ cells differ from these cytokines. Moreover, these SN do not contain detectable amounts of IFNγ, IL-2, IL-4 or IL-6 when assessed by ELISA.

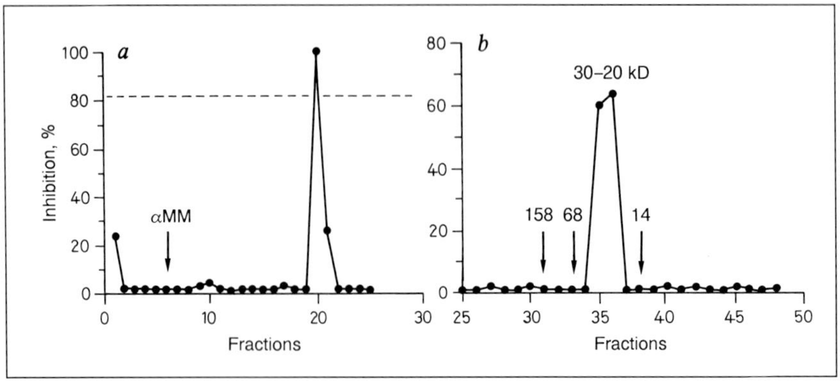

Fig. 2. a The CD8+CD57+ SN inhibitory activity is mediated by a Con A-binding molecule. CD8+CD57+ SN, concentrated fourfold, were applied to Con A-sepharose column presaturated with BSA. The Con A-binding molecules were desorbed by 1 *M* α-methyl-*D*-mannoside (αMM). Each fraction was tested as described. *b* Molecular weights of the inhibitory factors were estimated by size-exclusion chromatography (Superose 12). Protein known molecular weight standards were used for the calibration curve: aldolase (158 kD), BSA (67 kD) and ribonuclease (13.7 kD).

The inhibition of cytolysis was not abolished when the CD8+CD57+ SN were exposed to trypsin in concentrations ranging from 0.2 to 2.5 mg/ml but was abolished after exposure to pronase. Furthermore, the inhibitory activity was heat-resistant since it remained unaffected after 1 h at 80 °C treatment (data not shown). These data strongly argue in favor of a highly heat-resistant factor released by CD8+CD57+ cells. Thus, we investigated whether the inhibitory activity of CD8+CD57+ SN was glycosylated and could specifically bind to concanavalin A (Con A) lectin immunosorbent. A major decrease of CD8+CD57+ inhibitory activities was observed in the nonadsorbed fraction (fig. 2a). In contrast, 100% inhibition of LAK cytolysis was obtained with the fractions eluted by α-methyl-*D*-mannoside from the lectin immunosorbent indicating that the inhibitor present in CD8+CD57+ SN is able to bind to Con A and is glycosylated. The molecular weight of the CD8+CD57+ soluble inhibitor was estimated by size exclusion filtration chromatography (fig. 2b). CD8+CD57+ inhibitory activity was eluted as 20–30 kD molecule, although we cannot exclude that a molecule of smaller size is adsorbed onto a 20–30 kD carrier protein.

The Significance of These Findings

An increase of CD8+CD57+ T cells is observed in various immune deficiencies induced either in HIV disease or after allogeneic transplantation. In both of these conditions patients suffer from immunosuppression, and a number of questions remain unsolved. The exact relationship between the expansion of such cells and acquired immune deficiencies, viral infection and immune activation is still unknown. We do not know whether the emergence of cells with the CD8+CD57+ phenotype results from a specific stimulation of a minor T cell subset or reflects compensatory abnormalities of T cell differentiation at a time of the developing functional deficiency of other immune activities. Since CD8+CD57+ lymphocytes are capable of generating killer cells after in vitro stimulation but in the freshly separated populations these cells are not lytic, we favor the latter hypothesis. The fact that these cells spontaneously release a soluble glycosylated factor which is distinct from cytokines and prostaglandins and which is able to inhibit cell-mediated cytotoxic activities, suggests that CD8+CD57+ cells play an active role in immunoregulatory processes. The significance of such inhibitory CD8+ cells and their mediator in immunopathology remains to be determined. A further identification of the 20–30 kD molecule released by CD8+CD57+ T lymphocytes from long-term HIV-infected or BMT patients should allow to investigate the production of such a soluble compound in vivo and in order to understand its hypothetical role in the decline of cytolytic cell responses, and also to test the possibility of blocking this factor, if its contribution to immune reactivity can be proven to be deleterious.

References

1 Gratman JW, Loon AM, Oosteveer M, Naipal A, D'Amaro J, Witte T: Effect of cytomegalovirus infection on T lymphocyte-depleted bone-marrow transplantation. Exp Hematol 1987;15:803–808.

2 Legendre CN, Guttmann RD, Hou SK, Jean R: Two-color immunofluorescence and flow cytometric analysis of long-term renal allotransplant recipients: identification of a major Leu-7+:Leu3+ subpopulation. J Immunol 1985;135:1061–1066.

3 Leroy E, Calvo CF, Divine M, Gourdin MF, Beaujean F, Ben Aribia MH, Mishal Z, Vernant JP, Farcet JF, Senik A: Persistence of T8+/HNK1+ suppressor lymphocytes in the blood of long-term surviving patients after allogenic bone-marrow transplantation. J Immunol 1986;137:2180–2189.

4 Lewis D, Puck J, Babcock G, Rich R: Disproportionate expansion of a minor T cell subset in patients with lymphadenopathy syndrome and acquired immunodeficiency syndrome. J Infect Dis 1985;3:555–559.
5 Gupta S: Abnormalities in Leu2+7+ cells in AIDS, ARC and asymptomatic homosexuals. J Clin Immunol 1986;6:502–509.
6 Landray A, Gartland L, Clement T: Characterization of a phenotypically distinct subpopulation of Leu2+ cells which suppresses T cell poliferative responses. J Immunol 1983;131:2757–2761.
7 Clement LT, Grossi E, Gartland GL: Morphology and phenotypic features of the subpopulation of Leu2+ cells that suppress B cell differentiation. J Immunol 1984; 133:2460–2465.
8 Joly P, Guillon JM, Mayaud C, Plata F, Thedorou I, Denis M, Debre P, Autran B: Cell-mediated suppression of HIV-specific cytolytic T lymphocytes. J Immunol 1989;143:2193–2202.
9 Sadat-Sowti B, Debré P, Idzioreck T, Guillon JM, Hadida F, Okzenhendler E, Katlama C, Mayaud C, Autran B: A lectin-binding soluble factor released by CD8+CD57+ lymphocytes from AIDS patients inhibits T cell cytotoxicity. Eur J Immunol 1991;21:737–741.
10 Autran B, Leblond V, Sadat-Sowti B, Lefranc E, Got P, Sutton L, Binet JL, Debre P: A soluble factor released by CD8+CD57+ lymphocytes from bone marrow transplanted patients inhibits cell-mediated cytolysis. Blood 1991;77:2237–2241.
11 Garcia-Penarrubia P, Bankhurst AD, Koster FT: Prostaglandins from human T suppressor:cytotoxic cells modulate NK antibacterial activity. J Exp Med 1989;8: 601–605.
12 Rook AH, Kehrl JH, Wakefield L, Roberts AB, Sporn MB, Burington B, Lane H, Fauci A: Effects of transforming growth factor-beta on the functions of natural killer cells: depressed cytolytic activity and blunting of interferon responsiveness. J Immunol 1986;136:3916–3920.
13 Ruggerio V, Bablioni C: Synergistic anti-proliferative activity of interleukin-1 and tumor necrosis factor. J Biol Chem 1987;138:661–666.
14 Menu E, David V, Bensussan A, Chaouat G: Immunoactive products of human placenta II. Direct inhibition of non-MHC restricted cytolytic activity of human CD3 alpha-beta but not CD3 gamma-delta expressing T cell clones. J Reprod Immunol 1989;16:137–143.
15 Tsicopoulos A, Tonnel AB, Vorng H, Joseph M, Wallaert B, Kusnierz JP, Pestel J, Capron A: Lymphocyte-mediated inhibition of platelet cytotoxic function during Hymenoptera venon desensitization: characterization of a suppressive lymphokine. Eur J Immunol 1990;6:1201–1208.

Dr. Behazine Sadat-Sowti, Laboratoire d'Immunologie Cellulaire et Tissulaire,
URA CNRS 0186, CHU Pitié-Salpêtrière, 83, bd. de l'Hôpital,
F–75651 Paris Cedex 13 (France)

Janossy G, Autran B, Miedema F (eds): Immunodeficiency in HIV Infection and AIDS.
EC/FERS/MRC Workshop on Immunodeficiency in HIV-1 Infections, Windsor, Surrey, 1991.
Basel, Karger, 1992, pp 235–252

HIV-Specific Antibody-Dependent Cellular Cytotoxicity and Neutralization

Kristina Broliden, Per Anders Broliden, Britta Wahren

Department of Virology, Karolinska Institute and National Bacteriological Laboratory, Stockholm, Sweden

Antibodies can mediate lysis of virus-infected or tumor cells by binding to the target antigen on the cell surface. Effector cells will then bind to the Fc part of the antibody via its Fc receptor and kill the target cell. This mechanism is called antibody-dependent cellular cytotoxicity (ADCC) (fig. 1). An in vivo effect of ADCC in humans has been proposed against both tumors and viruses. For example, murine monoclonal antibodies have been used in antitumor therapy [1] and antibodies against HSV-mediating ADCC have protected neonatal mice from lethal infection [2]. The most beneficial role of ADCC is probably in early stages of infection by destroying the relatively small percentage of cells initially infected with HIV-1. However, the reaction may also contribute to the immunopathology of the disease by killing cells which are themselves not infected by HIV-1 but may bind free gp120 released from other cells.

Since HIV-1 can be transmitted either as free virus or associated cells, both the humoral and cell-mediated immunity must be active in an effective antiviral response. Both ADCC and antibodies which can directly neutralize the HIV-1, and therefore block its spread, may be critical in providing protective immunity against HIV-1. These antibodies can either be passively given or actively elicited by a vaccine prior to virus exposure. The aim of this review is therefore 3-fold. We wish to summarize (1) the recent results about both ADCC and neutralizing antibodies as well as (2) recent achievements to produce these potentially therapeutic reagents by modern biotechnological methods. In particular, we emphasize that

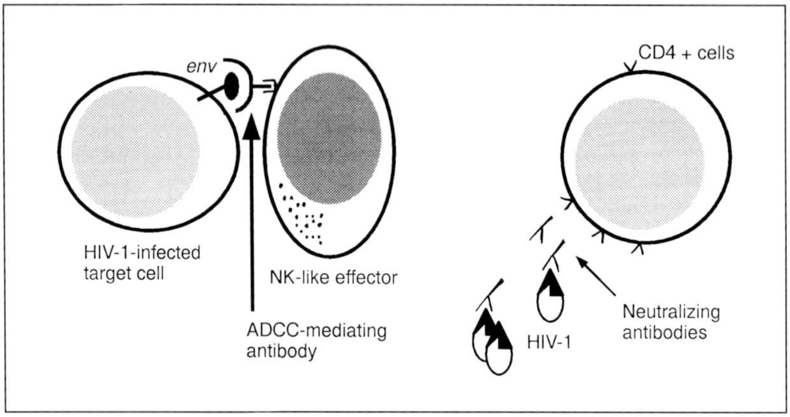

Fig. 1. Protective antibodies in HIV-1 infection mediate ADCC and virus neutralization. In this review the components of these defense systems, such as antibodies, effector cells and target cell antigens in ADCC as well as the neutralizing antibodies and the antigen epitopes recognized are summarized together with the therapeutic applications of these studies.

(3) by passive immunotherapy it seems to be possible to induce positive improvement in cellular immune responses such as responses to T-cell mitogens. It might therefore be possible to reverse the decline of T-cell function which is documented by Miedema et al., Janossy, Borthwick et al., and Christiansen et al., this volume.

ADCC-Mediating Antibodies and Effector Cells

ADCC activity has been found in about 90% of sera from HIV-1-infected persons when tested against target cells infected with several different HIV-1 strains [3]. Antibodies in some sera mediate ADCC against all strains whereas other sera lack ADCC activity completely or have ADCC against only a few HIV-1 strains. Although the frequency of ADCC positivity was similar in the different clinical stages of HIV-1 infection, the ADCC titers declined with disease progression [4]. This decrease was not related to a general decrease of HIV-1-specific IgG antibodies or to the CD4:CD8 lymphocyte ratio. The decrease in ADCC activity with disease progression has also been described in other reports [5, 6]. At variance with these results are studies that describe lack of correlation of ADCC and disease progression [7, 8] as well as a correlation of increasing ADCC with progressing disease [9]. In pediatric HIV-1 infection, presence of ADCC-mediating antibodies correlated significantly with a better clinical stage [10]. Although controversial, the findings of lower ADCC titers in more severe stages of HIV infection may indicate a protective role for the antibodies in delaying disease progression.

Since ADCC may have a role early in HIV infection by eliminating virus-infected cells, the time of appearance is important to investigate. The ADCC reaction has been described to be early and beneficial [11]. In a study with 10 sequentially followed HIV-1 antibody-positive men [12], ADCC appeared between 2 and 12 months after seroconversion as detected by HIV-$1_{HTLVIIIB}$-infected target cells. Strain-specific antibodies might develop even earlier which was clearly demonstrated in a laboratory worker who was accidentally infected with HIV-$1_{HTLVIIIB}$ and developed broadly reactive ADCC shortly after seroconversion.

HIV-1-specific ADCC-mediating antibodies have also been demonstrated in cerebrospinal fluid (CSF) [13]. It is yet unclear whether ADCC effector cells are present in the CSF in sufficient numbers and with normal function in order to mediate the ADCC reaction. Fewer subjects had ADCC-mediating antibodies in CSF than in sera, and the titers were about 100-fold lower in CSF than in sera. No correlation was found between the presence of ADCC-mediating antibodies in CSF and clinical status. The antibodies were in some cases proven to be intrathecally synthesized and not derived from passage through the blood-brain barrier. It remains to be established if ADCC limits the viral infection of the brain or, on the contrary, contributes to the histopathological changes.

HIV-2 antibody-positive sera demonstrated a higher frequency of ADCC reactivity and also a tendency for higher titers compared with HIV-1 antibody-positive sera [3, 14]. This is interesting since HIV-2 has been suggested to be less pathogenic or at least cause disease with longer latency than HIV-1. The higher frequency of ADCC may depend on several factors like stronger immunogenicity of HIV-2 or higher levels of expression of viral antigens on the target cell surface. The target region might also be more conserved. Alternatively, HIV-2-infected subjects develop more broadly reacting antibodies than those infected by HIV-1 as has been shown for virus neutralization [15].

The effector cells mediating ADCC are believed to be mainly NK/K cells which morphologically are defined as large granular lymphocytes.

Many HIV-1-infected subjects have normal effector cell functions, but as a group compared with healthy controls the function is clearly decreased [16]. The magnitude and frequency of dysfunction in antibody-dependent cytolysis with lymphocytes was also found to be greater among AIDS patients than among asymptomatic individuals [17]. In contrast, Ojo-Amaize et al. [7] suggest that effector cell activity remains high during disease progression. The HIV-1 specific ADCC might also be mediated by neutrophils [18]. As opposed to lymphocytes and mononuclear phagocytes, the neutrophils are not infected by HIV and therefore colony-stimulating factors could permit therapeutic regulation of neutrophils in AIDS. Also, natural antibodies may be useful in targeting HIV-infected cells for neutrophil cytotoxicity in vivo. Monocytes have been suggested to play a direct role in vivo in lysis or suppression of HIV-coated/infected cells and in the pathogenesis of AIDS [19]. A dysfunction of monocyte/macrophage activity in AIDS has however been proposed since a marked decrease in ADCC-CRBC (chicken red blood cells) activity was observed in AIDS patients [20].

Specificity of Target Cell Antigens in ADCC

The efficiency of ADCC activity against HIV-infected target cells thus depends on the presence and titer of ADCC-mediating antibodies and on effector cell function. In addition, ADCC depends on the HIV strain and the amount of viral antigen on the surface of target cells. Activated cells expressing HIV antigens on their surface are possible targets for ADCC whereas latently infected, resting cells may avoid recognition. The main target antigen for HIV-specific ADCC are probably the envelope glycoproteins due to several facts: (a) *gag* and *pol* gene products are not easily demonstrated on the target cell surface and are therefore unlikely to be involved in an efficient ADCC response; (b) sera with only *env*-specific antibodies mediate ADCC to the same extent as sera that also have *gag*- and *pol*-specific antibodies [14]; (c) the lack of ADCC cross-reactivity between HIV-1 and HIV-2 can be explained since their envelope glycoproteins are antigenically distinct; (d) HIV-1-specific monoclonal antibodies against *gag* and *pol* have not been demonstrated to mediate ADCC whereas *env*-specific monoclonal antibodies can mediate the activity.

For development of a future vaccine it is important to define which target epitopes are responsible for the ADCC activity and the broad cross-reactivity seen in some sera. Several studies have focused on specific target epitopes for ADCC. One of them is in the very C-terminal part of gp120 [21]. Two regions of gp41 have also been found to be possible targets for ADCC and are located in hydrophilic regions of the transmembrane protein [22]. Sera from mice that had been immunized with synthetic peptides from the HIV-1 *env* region also suggested that ADCC might be directed against several determinants of the transmembrane portion of the HIV-1 *env* glycoprotein [6]. Koup et al. [23] have used vaccinia virus vectors expressing envelope glycoproteins (gp160, gp120, gp41) or *gag* proteins (p55, p40, p24 and p17) in lymphoblastoid cell lines. Only the envelope glycoprotein complex gp120/gp41 was found to be target antigen for HIV-specific ADCC. Human monoclonal antibodies against gp41 mediated ADCC but not neutralization [24] and a murine monoclonal antibody directed to the V3 region of gp120 mediated both neutralization and ADCC [25]. Taken together, several ADCC epitopes have been suggested. It can be assumed that more epitopes can be targets for ADCC reactivity provided they are expressed on the cell surface in the appropriate conformation and concentration.

HIV-Specific Neutralization

HIV-1-specific neutralization by naturally occurring antibodies in the serum of individuals infected with HIV-1 was first described in 1985 [28, 29]. Assays for detection of neutralization include inhibition of viral infectivity, pseudotypes of VSV bearing the envelope antigens of HIV [29, 30], HIV plaque reduction on MT4 cells [31] and inhibition of multinucleated syncytium formation [29].

With regard to vaccine design it is interesting to investigate whether neutralizing antibodies can mediate ADCC. However, no direct correlation between these two activities has been demonstrated, suggesting that epitopes serving as targets for ADCC may be distinct from neutralizing epitopes [26, 27]. One common epitope though has been localized to the V3 region of gp120 since a monoclonal antibody with this specificity has been found to mediate both these functional activities.

The role of neutralizing antibodies in limiting the spread of HIV in vivo and the course of disease progression is a controversial issue. Initial reports described only very low neutralizing titers [28, 29, 32] in human sera but later studies using more sensitive HIV-1 strains describe titers higher than 30,000 [30]. Extensive studies have been performed where a correlation of neutralizing antibodies with earlier stages of HIV-1 infection was found [33] or not found [34, 35]. One of the studies consisted of prospectively monitored neutralizing activity in a cohort of homosexual men followed for 6 years [36]. High titers correlated with better clinical outcome and low or decreasing neutralizing titer signalled poor prognosis. These results differed from another longitudinal study where symptom-free subjects had significantly higher titer of anti-*gag* (p24) antibodies but only showed a trend towards an increasing neutralizing titer with time [37]. In children perinatally infected with HIV-1 a correlation between clinical status and neutralizing capacity was found [10, 33] which may indicate a protective function against vertical transmission of HIV and/or against disease development.

The protective role of neutralizing antibodies is also difficult to evaluate since the antibodies can form immune complexes with the virus. Thus, although the neutralizing antibody titer increases, this may not be rapid enough or the reactivity not sufficiently broad to control viremia. Furthermore, transient neutralization by antibodies will be of little value if the virus-antibody complex is simply delivered to vulnerable macrophages.

Strain and Type Specificity of Neutralizing Antibodies

Broad neutralization occurs against diverse strains of HIV-1 from patients in different parts of the world [30]. Some of the HIV-1 antibody-positive sera can cross-neutralize HIV-2 isolates [38] and some of the HIV-2 antisera cross-neutralize HIV-1 [15, 38]. During early infection in humans the response might be less broadly reactive. The kinetic of appearance of broadly reacting virus-neutralizing antibodies in human sera has been studied by using a heterologous isolate, HIV-$1_{HTLVIIIB}$. Virus-neutralizing antibodies against this isolate were detected 2–12 months after primary HIV-1 infection in 7 out of 10 individuals [12]. In contrast, strain-specific neutralizing antibodies to the autologous isolate have been shown to appear at the time of seroconversion [39]. In the same patients it was shown that sera collected at a later time during infection could still neutralize the first obtained autologous isolate but none of the following isolates. This may be explained by the fact that HIV-1 mutates its outer envelope during the course of infection and thus could escape the neutralizing effect of the antibodies [40–42]. Neutralization of viruses derived from molecular clones of HIV-1$_{HTLVIIIB}$ tested with a large panel of human and goat anti-HIV sera demonstrated strain-restricted neutralization [43]. The change of a single amino acid residue in gp120 was supposed to account for such 'clonal restriction' of neutralizing activity. It has also been proposed that HIV-1 isolates could be grouped into four classes according to their degree of sensitivity to serum neutralization and that representative viruses from each group should be combined for the development of an effective vaccine [44].

In contrast to the broadly reacting human sera, sera raised in animals to recombinant gp120 or highly purified, recombinant viral glycoproteins and synthetic peptides as well as purified gp120 from HIV-1 are usually strain-specific [30, 45–47]. Upon further boosts with recombinant gp120 cross-neutralizing antibodies could be generated [48]. It is not clear why immunization of animals with virus lysate or gp120 elicits a neutralizing response more or less specific to the strain used for immunization, while sera from infected persons are able to neutralize a wide range of isolates of HIV-1. A similar epitopic reactivity against the neutralizing V3 region of gp120 was shown in both cases, which may indicate a similar conformation of gp120 in native virus and as purified or recombinant protein. One explanation may be that the number of possible antibodies which may be produced to an antigen is high, and different antibodies to an antigen often bind to epitopes which overlap on the antigen surface. In this way, different antibodies can bind to a particular region without binding to exactly the same epitope. The neutralizing capacity of sera from infected individuals may thus be less sensitive to minor conformational differences of different strains.

Neutralizing Regions

Several determinants of HIV have been suggested as neutralization-inducing sites. The most common way of identifying such sites has been to immunize animals with native or recombinant proteins or with synthetic peptides. The antisera raised have then been tested for virus neutralization. Another way of determining regions important for neutralization is to correlate the presence of neutralizing capacity of sera from HIV-infected individuals with antibody responses to certain proteins or peptides.

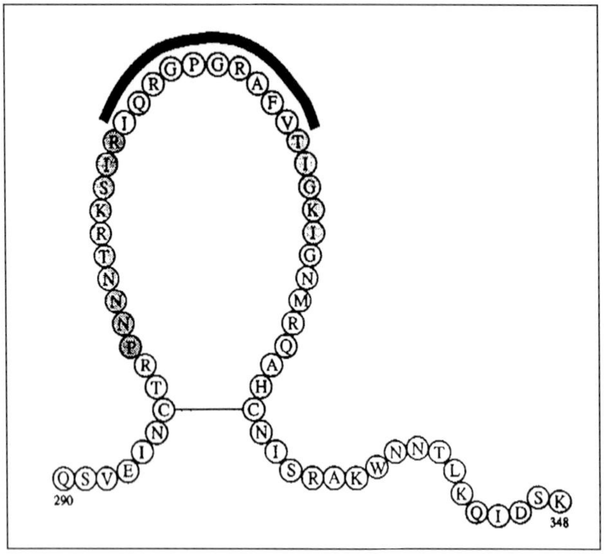

Fig. 2. The structure of the antigenic epitope on the gp120 envelop protein which is important in both neutralization and ADCC. From [57] with permission of publishers.

A number of regions on gp120 and gp41 have been defined for induction of HIV-1-neutralizing antibodies in animals by synthetic peptides. The majority are located on gp120 and focused on the V3 region [46, 49]. Peptides derived from conserved regions of gp120 and from the N-terminal end of gp41 were able to elicit neutralizing antibodies in mice [50, 51]. Additionally, two other regions were found to be involved in neutralization and were located on gp41. Monoclonal antibodies raised against one of them (a.a. 735-752) have been found to neutralize several different HIV-1 strains [52, 53]. Monoclonal antibodies reacting with the N-terminal end of p17 have also been reported to show an HIV-1 neutralizing property [54].

The neutralizing capacity generated by HIV-1 infection in humans also seems to be mainly associated with the V3 region (fig. 2). Böttiger et al. [38] found a correlation between reactivity to a.a. 296-331 and neutralization and Kenealy et al. [55] found that sera from HIV-infected individuals bound to a.a. 308-322, which also elicited neutralizing antibodies in animals. In a recent study [56] the conserved sequence GPGR was found to be the most important site for antibody binding to the V3 region in human sera. Thus, the monoclonal antibodies described above recognized the same binding site as the human antibodies which were produced against natural infection. In gp41 a new neutralizing region has recently been identified in a conserved part (a.a. 647-671) [57]. Peptides corresponding to this region were able to inhibit neutralization mediated by sera from HIV-1-infected individuals. Interestingly, neutralizing antibodies directed to this region were found to be cross-reactive within the HIV-1 group.

Synthetic peptides have been used to identify continuous antigenic sites on the external envelope glycoprotein gp125 of HIV-2. It was found that human seroreactivity occurred at a few distinct linear antigenic sites distributed at similar positions on the molecule as those in HIV-1 gp120 [58; Mannervik et al., pers. unpubl. data]. Whether these B-cell antigenic sites correspond to neutralizing regions remains to be seen. In another study [59] peptides representing regions of HIV-2 were evaluated for their immunogenic activity in guinea pigs. The peptides were selected to provide homologous representation of sites in the HIV-1 envelope proteins which were previously documented to have a particular immunogenic importance. Some of these antisera were found to be capable of mediating neutralization.

Antibodies in Passive Immunotherapy

Neutralizing antibodies confer immunity or less severe disease in most virus infections. The most common use is the immunoprophylaxis given for hepatitis A virus. Also with several children's diseases such as measles, mumps, rubella or varicella, specific immunoglobulins hae been proven effective [60]. With viruses that establish latent infections the use of immunoprophylaxis or immunotherapy is more debated. Cytomegalovirus-specific antibody infusions have been shown to decrease the rate of serious disease, while others have questioned the specific effect.

The antibody response to HIV consists of a range of antibodies with defined antigenic reactions and functional capacities. In vitro both ADCC- and NT-mediating antibodies have a virus-regulatory effect. These antibodies may disappear during later stages of the infection, and it is therefore motivated to substitute the host with highly functional antibodies.

Passive transfer of human plasma has been attempted [61]. In one or two infusions, 6 patients with advanced AIDS were given 55–500 ml of plasma with high titers of antibodies to p24 and neutralizing antibody. The p24 antigenemia cleared immediately and increased neutralizing antibody could be shown to appear in the serum of the recipient. The half-life of the donated antibodies was around 12 days. The absence of p24 antigen in serum of the recipient, however, lasted for 4–11 weeks. Symptoms were also significantly less during the observation time of 2 months postinfusion. CD4 cell numbers were low in all patients, but increased during 2 months and again decreased after reappearance of p24 levels in blood. The passive transfer of patient plasma was shown during this short-time follow-up to improve virological, immunological and clinical signs of disease, as well as having no identified side effects.

Monthly infusions of 500 ml of plasma from HIV-infected asymptomatic persons for a 3-month period were given to 10 HIV-infected and antigen-positive individuals [62]. Donors for the plasma were selected to be clinically healthy HIV-infected individuals with neutralizing titers of 320 or higher. The p24 antigen content was cleared and remained so during the 4 months of observation. The 10 recipients maintained or increased their antiviral titers. This could be shown to be partially due to increased antibody synthesis by the recipients. There was no change in CD4 cell counts. The treatment was well tolerated and the clinical status of 5 patients improved. The conclusion after continued monthly infusions of plasma was that ARC patients benefit in retaining or improving their clinical status and CD4 counts, while with patients who already have developed AIDS the infusions do not change the course of disease. A combined HIV-DNA and RNA PCR assay performed on plasma from the recipients showed a decreased reactivity comparing pretreatment with all posttreatment samples after the second infusion [63].

Six monoclonal antibodies that have been produced in mice to HIV-1 gp120 have been defined according to their neutralizing and ADCC-mediating capacity [25, 64]. All monoclonal antibodies were strongly dependent on the motif GPG of the V3 region of HIV-1 gp120. The two monoclonal antibodies with the highest neutralizing titers appeared to depend on QRGPGR on continued mapping, by deletion studies also the flanking a.a. I and A were shown to take part in the paratope reaction. Sequences from other HIV strains MN, RF and NY5 reacted strongly, while sequences of African isolates did react less well or not at all. A comparison with two other mouse monoclonal antibodies directed to this region was performed (table 1).

Two of these monoclonal antibodies have been used for passive immunization trials in 11 HIV-infected patients. The patients received

Table 1. 50% inhibition of syncytium-forming units (200) of HIV-1/IIIB by different monoclonal antibodies

Monoclonal antibody	Neutralization, μg/ml
110.4	90,000
0.5β	6
F58/H3	10

Neutralization of HIV-1/IIIB was measured by inhibition of syncytium formation on CEM-ss cells as described by Nara et al. [65]. Results obtained from Dr. J.F. Dedieu, Unité de Virologie Moléculaire, Institut Pasteur.

125 mg of purified monoclonal antibody once every 14 days and were monitored closely. Five of 7 patients' isolates appeared susceptible to neutralization by the monoclonal antibody used for these 7 patients [66]. A decrease of p24 values was seen in 9 of the 11 patients [Bratt et al., pers. unpubl. data]. Four patients developed anti-idiotypic responses. Serological responses to HIV-1 and other viruses were constant in the serum (but for antiviral responses see below). The immediate effects consisted of fever and chills of short duration, mainly in patients who had neutralizable virus.

The cellular assays of immune reactivity were also investigated with similar methods described by Clerici and Shearer, and Janossy, Borthwick et al., this volume. Three major observations were made which confirmed each other in terms of documenting substantial improvements. First, responses to recall antigens such as tetanus toxoid (TT), PPD and cytomegalovirus antigen (CMV) were enhanced in the majority of patients. Increased antigen-specific responses were documented in 6/11 subjects to TT, in 6/11 to PPD and 6/11 to CMV. Second, 3 patients responded to HIV-1 *gag* (p24) and 5 to HIV-1 envelop antigen pB1. Finally, in 10/11 patients a considerable increase was seen in lymphocyte responses to the mitogen phytohemagglutinin (PHA) (fig. 3).

Further investigations are necessary to investigate the cellular aspects of these improved responses. It is not known, for example, whether there are more cells activated in these cultures because of improved synthesis of cytokines or their receptors (see Sahraoui et al., this volume) or fewer cells are inactivated by mechanisms of cell destruction similar to apoptosis (see Gougeon et al., this volume). One of the most important aspects of these

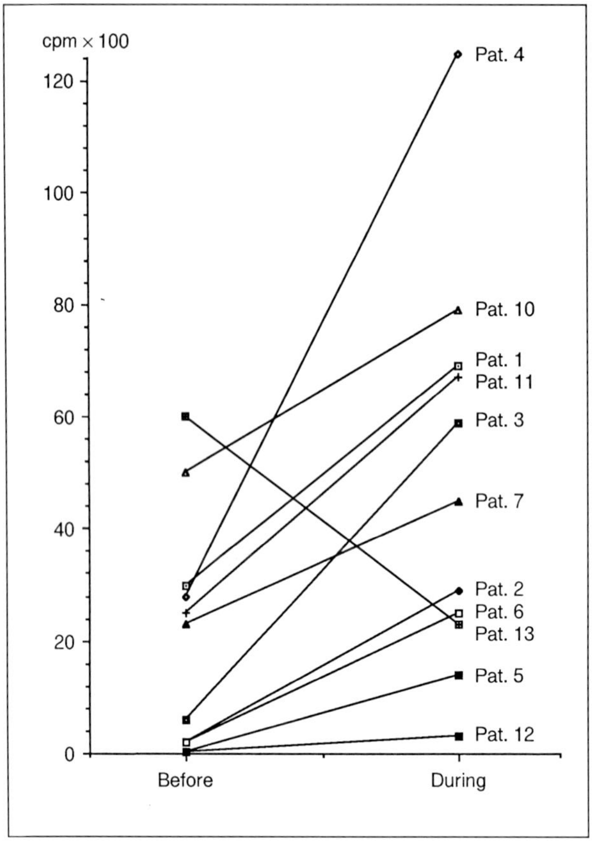

Fig. 3. Activation of peripheral lymphocytes from HIV-1-seropositive patients with the mitogen PHA before and during treatment with mouse monoclonal antibody against HIV-1 gp120. This antibody shows a high neutralizing and ADCC activity which depends upon the recognition of QRGPGR sequence of the gp120 V3 region (see also fig. 2). It is clear from these results that in 10/11 patients the PHA responses have improved while the patient received serotherapy.

observations above is that the monoclonal antibody utilized reacts with high affinity to the gp120 and neutralizes this viral protein. These clinical observations therefore implicate gp120 as a contributory factor to the development of severe immunodeficiency in vivo. The envelop gp120 is also immunosuppressive in vitro (see Cefai et al., this volume), and elimination of these molecules in their soluble form or as virus particles should improve the patients' immune responses. It is also desirable that human antibody of high affinity should be used for passive immunotherapy that is going to be continued for any extended period of time. The variable domains of the mouse monoclonal antibody F58/H3 have been cloned and sequenced and a mouse-human chimeric antigen-binding Fab fragment as well as a mouse-human chimeric IgG1 antibody product have been produced [67]. The chimeric Fab and IgG1 antibody bind to HIV-1 gp120, gp160 and peptides containing the GPGR sequence of HIV-1 V3. It is competent in both neutralization and ADCC. These products should be useful for continued evaluation of passive immunotherapy in infected persons.

Finally, it is important to point out that the same products might also be efficient in preventing transfer of virus from the infected pregnant mother to her child. In a model system using chimpanzees and HIV-1, it was demonstrated that a polyclonal virus-neutralizing IgG could prevent establishment of primary infection with the same virus strain as that used for production of antibody. A monoclonal antibody directed to the V3 region of that virus strain prolonged detectable infection from 4–6 to 14 weeks after challenge [68]. A chimeric mouse-human antibody directed to neighboring regions of the V3 region has been described [69]. The chimera as well as its parent mouse antibody neutralize HIV-1 infection with similar competence. Thus, antibodies that should have an effect on many HIV strains are available for human trials.

References

1 Seto M, Takahashi T, Nakamura S, Matsudaira Y, Nishizuka Y: In vivo antitumor effects of monoclonal antibodies with different immunoglobulin classes. Cancer Res 1983;43:4768–4773.

2 Kohl S, Loo LS: Protection of neonatal mice against herpes simplex virus infection: Probable in vivo antibody-dependent cellular cytotoxicity. J Immunol 1982;129: 370–376.

3 Ljunggren K, Biberfeld G, Jondal M, Fenyö EM: Antibody-dependent cellular cytotoxicity detects group and strain-specific antigens among HIV-1 and HIV-2 isolates. J Virol 1989;63:3376–3381.

4 Ljunggren K, Böttiger B, Biberfeld G, Karlsson A, Fenyö EM, Jondal M: Antibody-dependent cellular cytotoxicity-inducing antibodies against human immunodeficiency virus: Presence at different clinical stages. J Immunol 1987;139:2263–2267.

5 Rook AH, Lane HC, Folks T, McCoy S, Alter H, Fauci AS: Sera from HTLV-II/LAV

antibody-positive individuals mediate antibody-dependent cellular cytotoxicity against HTLV-III/LAV-infected T-cells. J Immunol 1987;138:1064–1067.

6 Blumberg RS, Paradis T, Hartshorn KL, Vogt M, Ho DD, Hirsch MS, Leban J, Sato VL, Schooley RT: Antibody-dependent cell-mediated cytotoxicity against cells infected with the human immunodeficiency virus. J Infect Dis 1987;156:878–884.

7 Ojo-Amaize E, Nishanian PG, Heitjan DF, Rezai A, Esmail I, Korns E, Detels R, Fahey J, Giorgi JV: Serum and effector-cell antibody-dependent cellular cytotoxicity activity remains high during human immunodeficiency virus disease progression. J Clin Immunol 1989;9:454–461.

8 Dalgleish A, Sinclair A, Steel M, Beatson D, Ludlam C, Habeshaw J: Failure of ADCC to predict HIV-associated disease progression or outcome in a haemophiliac cohort. Clin Exp Immunol 1990;81.

9 Shepp DH, Chakrabarti S, Moss B, Quinnan GV Jr: Antibody-dependent cellular cytotoxicity specific for the envelope antigens of human immunodeficiency virus. J Infect Dis 1988;157:1260–1264.

10 Ljunggren K, Moschese V, Broliden PA, Quinti I, Giaquinto C, Fenyö EM, Wahren B, Rossi P, Jondal M: Antibodies mediating cellular cytotoxicity and neutralisation correlate with a better clinical stage in children born to HIV-infected mothers. J Infect Dis 1990;161:198–202.

11 Sawyer LA, Katzenstein DA, Hendry RM, Boone EJ, Vujcic LK, Williams CC, Zeger SL, Saah AJ, Rinaldo CR Je, Phair JP, Giorgi JV, Quinnan GV Jr: Possible beneficial effects of neutralizing antibodies and antibody-dependent, cell-mediated cytotoxicity in human immunodeficiency virus infection. AIDS Res Hum Retroviruses 1990;6:341–356.

12 Goudsmit J, Ljunggren K, Smit L, Jondal M, Fenyö EM: Biological significance of the antibody response to HIV antigens expressed on the cell surface. Arch Virol 1988;103:189–206.

13 Ljunggren K, Chiodi F, Broliden PA, Albert J, Norkrans G, Hagberg L, Jondal M, Fenyö EM: Presence of HIV-1 antibodies mediating cellular cytotoxicity and neutralisation in cerebrospinal fluid. AIDS Res Hum Retroviruses 1989;5:629–638.

14 Ljunggren K, Chiodi F, Biberfeld G, Norrby E, Jondal M. Fenyö EM: Lack of cross-reaction between HIV and HIV-related West-African strains. J Immunol 1988;140: 602–605.

15 Weiss RA, Clapham PR, Weber JN, Whitby D, Tedder RS, O'Connor T, Chamaret S, Montagnier L: HIV-2 sera cross-neutralize HIV-1. AIDS 1988;2:95–100.

16 Ljunggren K, Karlsson A, Jondal M, Fenyö EM: Natural and antibody-dependent cytotoxicity in different clinical stages of HIV infection. Clin Exp Immunol 1989;75: 184–190.

17 Tyler DS, Stanley SD, Nastala CA, Austin AA, Bartlett JA, Stine KC, Lyerly HK, Bolognesi DP, Weinhold KJ: Alterations in antibody-dependent cellular cytotoxicity during the course of HIV-1 infection: Humoral and cellular defects. J Immunol 1990;144:3375–3384.

18 Baldwin GC, Fuller ND, Roberts RL, Ho DD, Golde DW: Granulocyte- and granulocyte-macrophage colony-stimulating factors enhance neutrophil cytotoxicity toward HIV-infected cells. Blood 1989;74:1673–1677.

19 Jewett A, Bonavida B: Peripheral blood monocytes derived from HIV+ individuals mediate antibody-dependent cellular cytotoxicity. Clin Immunol Immunopathol 1990;54:192–199.

20 Bender BS, Auger FA, Quinn TC, Redfield R, Gold J, Folks TM: Impaired antibody-dependent cell-mediated cytotoxicity activity in patients with the acquired immunodeficiency syndrome. Clin Exp Immunol 1986;64:166–172.
21 Bolognesi DP: Natural immunity to HIV and its possible relationship to vaccine strategies. Microbiol Sci 1988;8:236–241.
22 Kowalski M, Potz J, Basiripour L, et al: Functional regions of the envelope glycoprotein of human immunodeficiency virus type 1. Science 1987;237:1351–1355.
23 Koup RA, Sullivan JL, Levine PH, Brewster F, Mahr A, Mazzara G, McKenzie S, Panicali D: Antigenic specificity of antibody-dependent cell-mediated cytotoxicity directed against human immunodeficiency virus in antibody-positive sera. J Virol 1989;63:584–589.
24 Gorny MK, Gianakakos V, Sharpe S, Zolla-Pazner S: Generation of human monoclonal antibodies to human immunodeficiency virus. Proc Natl Acad Sci USA 1989; 86:1624–1628.
25 Broliden PA, Ljunggren K, Hinkula J, Norrby E, Åkerblom L, Wahren B: A monoclonal antibody to HIV-1-mediating cellular cytotoxicity and neutralisation. J Virol 1990;64:936–940.
26 Böttiger B, Ljunggren K, Karlsson A, Krohn K, Fenyö EM, Jondal M, Biberfeld G: Neutralizing antibodies in relation to antibody-dependent cellular cytotoxicity-inducing antibodies against human immunodeficiency virus type 1. Clin Exp Immunol 1988;73:339–342.
27 Lyerly HK, Pratz JE, Tyler DS, Matthews TJ, Langlois AJ, Bolognesi DP, Weinhold KJ: Discordant virus neutralization and anti-HIV ADCC activity in HIV-seropositive patient sera; in Ginsberg H, et al (eds): New Chemical and Genetic Approaches to Vaccination. Vaccines 88. Cold Spring Harbor, Cold Spring Harbor Laboratory 1988, pp 323–326.
28 Robert-Guroff M, Brown M, Gallo RC: HTLV-III neutralising antibodies in patients with AIDS and AIDS-related complex. Nature 1985;316:72–74.
29 Weiss RA, Clapham P, Weber J, Cheinsong-Popov R, Dalgleish A, Carne A, Weller I, Tedder RS: Neutralization of human T-lymphotropic virus type III sera of AIDS and AIDS-risk patients. Nature 1985;316:69–72.
30 Weiss RA, Clapham P, Weber J, Dalgleish A, Lasky J, Berman P: Variable and conserved neutralization antigens on human immunodeficiency virus. Nature 1986; 324:572–575.
31 Harada S, Purtilo D, Koyanagi Y, Sonnabend J, Yamamoto N: Sensitive assay for neutralizing antibodies against AIDS-related viruses. J Immunol Methods 1986;92: 177–181.
32 Groopman JE, Benz PM, Ferriani R, Mayer K, Davis Allan J, Weymouth JA: Characterization of serum neutralization response to the human immunodeficiency virus. AIDS Res Hum Retroviruses 1987;3:71–85.
33 Robert-Guroff M, Oleske JM, Connor EM, Epstein LG, Minefor AB, Gallo RC: Relationship between HTLV-III neutralising antibodies and clinical status of pediatric acquired immunodeficiency syndrome (AIDS) and AIDS-related complex cases. Pediatr Res 1987;21,6:547–550.
34 Wendler I, Bienzle U, Hunsmann G: Neutralizing antibodies and the course of HIV-induced disease. AIDS Res Hum Retroviruses 1987;3:157–163.
35 Faulkner-Valle GP, DeRossi A, Gassa OD, Chieco-Bianchi L: LAV/HTLV-III neu-

tralizing antibodies in the sera of patients with AIDS, lymphadenopathy syndrome and asymptomatic seropositive individuals. Tumori 1986;72:219.

36 Robert-Guroff M, Goedert J, Naugle C, Jennings A, Blattner W, Gallo R: Spectrum of HIV-1 neutralizing antibodies in a cohort of homosexual men: Results of a 6-year prospective study. AIDS Res Hum Retroviruses 1988;4:343–350.

37 Weber J, Weiss R, Roberts C, Weller I, Tedder R, Clapham P, Parker D, Duncan J, Carne L, Pinching A, Cheingsong-Popov R: Human immunodeficiency virus infection in two cohorts of homosexual men: Neutralising sera and association of anti-gag antibody with prognosis. Lancet 1987;i:119.

38 Böttiger B, Karlsson A, Andreasso P, Nauclér A, Mendes Costa C, Norrby E, Biberfeld G: Envelope cross-reactivity between human immunodeficiency virus types 1 and 2 detected by different serological methods: Correlation between cross-neutralization and reactivity against the main neutralizing site. J Virol 1990;64:3492.

39 Albert J, Abrahamsson B, Nagy K, Aurelius E, Gaines H, Krook A, Nyström G, Fenyö EM: Rapid development of isolate-specific neutralizing antibodies after primary HIV-1 infection and consequent emergence of virus variants which resist neutralization by autologous sera. AIDS 1990;4:107.

40 Reitz MS, Wilson C, Naugle C, Gallo RC, Robert-Guroff M: Generation of a neutralization-resistant variant of HIV-1 is due to selection for a point mutation in the envelope gene. Cell 1988;54:57–63.

41 Kinney-Thomas E, Weber J, McClure J, Clapham P, Singhal MC, Shriver MK, Weiss RA: Neutralizing monoclonal antibodies to the AIDS virus. AIDS 1988;2:25–29.

42 McKeating JA, Gow J, Goudsmit J, Pearl L, Mulder C, Weiss R: Characterization of HIV-1 neutralization escape mutants. AIDS 1989;3:777.

43 Looney DJ, Fischer A, Putney S, Rusche J, Redfield R, Burke D, Gallo R, Wong-Staal F: Type-restricted neutralization of molecular clones of human immunodeficiency virus. Science 1988;241:357–359.

44 Cheng-Mayer C, Homsy J, Evans LA, Levy JA: Identification of human immunodeficiency virus subtypes with distinct patterns of sensitivity to serum neutralization. Proc Natl Acad Sci USA 1988;85:2815–2819.

45 Matthews TJ, Langlois AJ, Robey WG, Chang NT, Gallo R, Fischinger P, Bolognesi D: Restricted neutralization of divergent human T-lymphotropic virus type III isolated by antibodies to the major envelope glycoprotein. Proc Natl Acad Sci USA 1986;83:9709–9713.

46 Palker TJ, Clark ME, Langlois A, Matthews TJ, Weinhold KJ, Randall RR, Bolognesi DP, Haynes BF: Type-specific neutralization of the human immunodeficiency virus with antibodies to env-encoded synthetic peptides. Proc Natl Acad Sci USA 1988;85:1932–1936.

47 Nara PL, Robey WG, Arthur L, Asher D, Wolff A, Gibbs C Jr, Gajdusek D, Fischinger F: Persistent infection of chimpanzees with HIV: Serological responses and properties of re-isolated viruses. J Virol 1987;61:3173–3180.

48 Berman PW, Groopman JE, Gregory T, Clapham PR, Weiss RA, Ferriani R, Riddle L, Shimasaki C, Lucas C, Lasky LA, Eichberg JW: Human immunodeficiency virus type 1 challenge of chimpanzees immunized with recombinant envelope glycoprotein gp120. Proc Natl Acad Sci USA 1988;85:5200–5204.

49 Rusche JA, Javaherian K, McDanal C, Petro J, Lynn DL, Grimaila R, Langlois A, Gallo RC, Arthur LO, Fischinger PJ, Bolognesi DP, Putney S, Matthews TJ: Antibodies that inhibit fusion of human immunodeficiency virus-infected cells bind a 24-amino acid sequence of the viral envelope, gp120. Proc Natl Acad Sci USA 1988; 85:3198–3202.
50 Ho DD, Sarngadharan M, Hirsch M, Schooley R, Rota T, Kennedy R, Chanh T, Sato V: Human immunodeficiency virus neutralizing antibodies recognize several conserved domains on the envelope glycoproteins. J Virol 1987;61:2024–2028.
51 Ho DD, Kaplan J, Rackauskas I, Gurney M: Second conserved domain of gp120 is important for HIV infectivity and antibody neutralization. Science 1988;239:1021–1023.
52 Chanh T, Dreesman G, Kanda P, Linette G, Sparrow J, Ho D, Kennedy R: Induction of anti-HIV neutralizing antibodies by synthetic peptides. EMBO J 1986;11: 3065–3071.
53 Dalgleish A, Chanh T, Kennedy R, Kanda P, Clapham P, Weiss R: Neutralization of diverse HIV-1 strains by monoclonal antibodies raised against a gp41 synthetic peptide. Virology 1988;165:209–215.
54 Papsidero LD, Sheu M, Ruscetti F: Human immunodeficiency virus type-1 neutralizing monoclonal antibodies which react with p17 core protein: characterization and epitope mapping. J Virol 1989;63:267–272.
55 Kenealy W, Matthews T, Ganfield M, Langlois A, Waselefsky D, Petteway S: Antibodies from human immunodeficiency virus-infected individuals bind to a short amino acid sequence that elicits neutralizing antibodies in animals. AIDS Res Hum Retroviruses 1989;5:173–182.
56 Broliden PA, Mäkitalo B, Åkerblom L, Ljunggren K, Rosen J, Utter G, Jondal M, Norrby E, Wahren B: Identification of amino acids in the V3 region of gp120 critical for virus neutralization by HIV-1-specific antibodies. Immunology 1991;73:371–376.
57 Broliden PA, von Gegerfelt A, Clapham P, Rosen J, Fenyö EM, Wahren B, Broliden K: Identification of human neutralizing regions of the human immunodeficiency virus type-1 envelope glycoproteins. Proc Natl Acad Sci USA 1991, in press.
58 Goudsmit J, Meloen RH, Brasseur R, Barin F: Human B-cell epitopes of HIV-2 transmembrane protein are similarly spaced as in HIV-1. J AIDS 1989;2:297–302.
59 Björling E, Broliden K, Bernardi D, Utter G, Thorstensson R, Chiodi F, Norrby E: Immune guinea pig sera to HIV-2 envelope synthetic peptides mediate neutralization and ADCC activity. Proc Natl Acad Sci USA 1991;88:6082–6086.
60 Pennington JE: Intravenous immunoglobulin therapy in serious infections. J Hosp Infect 1988;12(suppl D):1–2.
61 Jackson GG, Perkins JT, Rubenis M, Paul DA, Knigge M, Despotes JC, Spencer P: Passive immunoneutralisation of human immunodeficiency virus in patients with advanced AIDS. Lancet 1988;ii:647–652.
62 Karpas A, Hill F, Youle M, Cullen V, Gray J, Byton N, Hayhoe F, Tenant-Flowers M, Howard L, Gilgen D, Oates JK, Hawkins D, Gazzard B: Effects of passive immunization in patients with the acquired immunodeficiency syndrome-related complex and acquired immunodeficiency syndrome. Proc Natl Acad Sci USA 1988; 85:9234–9237.

63 Karpas A, Hewlett IH, Hill F, Gray J, Byron N, Gilgen D, Bally V, Oates JK, Gazzard B, Epstein JE: Polymerase chain reaction evidence for human immunodeficiency virus neutralization by passive immunization in patients with AIDS and AIDS-related complex. Proc Natl Acad Sci USA 1990;87:7613–7617.
64 Åkerblom L, Hinkula J, Broliden PA, Fridberger T, Rosen J, Ericsson-de la Crez M, Morein B, Wahren B: Neutralizing cross-reactive and nonneutralizing monoclonal antibodies to HIV-1 gp120. AIDS 1990;4:953.
65 Nara P, Hatch WC, Dunlop NM, Robey WG, Arthur LO, Gouda MA, Fischinger PJ: Simple, rapid, quantitative, syncytium-forming microassay for the detection of human immunodeficiency virus neutralizing antibody. AIDS Res Hum Retroviruses 1987;3:283–302.
66 Wahren B, Bratt G, Hinkula J, Gilljam G, Nordlund S, Broliden PA, Åkerblom L, Morein B, Sandström E: Monoclonal antibodies given as passive treatment to HIV-infected individuals; in Cinquième Colloque des Cent Gardes, Inst Pasteur, 1990, pp 263–267.
67 Marks JD, Wahren B, Gilljam G, Hinkula J, Winter G: Cloning of an HIV-1 neutralizing V3 specific monoclonal antibody and expression as a mouse-human chimaeric antigen-binding fragment and antibody. VIIth Int Conf AIDS, Florence 1991, abstr 1162, p 132.
68 Emini EA, Nara P, Schleif W, Lewis JA, Davide JP, Lee DR, Kessler J, Conley S, Matsushita S, Putney S, Gerety RJ, Eichberg JW: Antibody-mediated in vitro neutralization of human immunodeficiency virus type 1 abolishes infectivity for chimpanzees. J Virol 1990;64:3674–3678.
69 Liou RS, Rosen EM, Fung M, Sun W, Sun C, Gordon W, Chang N, Chang TW: A chimeric mouse-human antibody that retains specificity for HIV gp120 and mediates the lysis of HIV-infected cells. J Immunol 1989;143:3967–3975.

Kristina Broliden, MD, Department of Virology, Karolinska Institute and National Bacteriological Laboratory, S–105 21 Stockholm (Sweden)

Janossy G, Autran B, Miedema F (eds): Immunodeficiency in HIV Infection and AIDS. EC/FERS/MRC Workshop on Immunodeficiency in HIV-1 Infections, Windsor, Surrey, 1991. Basel, Karger, 1992, pp 253–264

Viral and Cellular Requirements for HIV-1 Infection in Monocytes[1]

Hanneke Schuitemaker, Neeltje A. Kootstra, Matthijs Tersmette, Frank Miedema

Department of Clinical Viro-Immunology, Central Laboratory of the Netherlands Red Cross Blood Transfusion Service and Laboratory for Experimental and Clinical Immunology of the University of Amsterdam, The Netherlands

As in other lentivirus infections, the cells of the monocyte/macrophages lineage are likely to be most important for the persistence and dissemination of HIV-1 throughout the body [1–5] and serve as reservoirs for the virus [6, 7]. In tissue compartments such as brain and lung, macrophages are the predominantly infected cells [1–4]. In the bone marrow and peripheral blood, however, only few cells among the monocyte/macrophage lineage harbor HIV-1 [8, 9]. It has been suggested that within this population a functionally and phenotypically distinct subset, referred to as dendritic cells, are preferentially involved (see Knight et al., this volume). The two major theories to explain the varied involvement of monocytes/macrophages are therefore (1) that a certain state of differentiation is required for the monocytes/macrophages to be susceptible to HIV-1 infection, and this stage of development occurs predominantly in tissues, and (2) that a separate lineage of dendritic cells is dominantly involved. As there are few immunological markers which selectively and positively recognize dendritic cells in tissues, these two possibilities are not readily distinguishable.

One approach to elucidate these questions is to induce cellular differentiation among monocytic cells with well-characterized cytokines. These highly reproducible systems can then be utilized to investigate susceptibility to virus exposure, and interpret the observations in the light of the

[1] This study was supported by a grant from the Netherlands Ministry of Public Health (No. 88005).

known actions by which these cytokines work. Enhancement of virus expression in monocytes/macrophages by the addition of cytokines such as IL-6 and TNF-α into the medium during the first 5 days of culture prior to infection has been described [10, 11] (see also Poli and Fauci, this volume). In the investigations below we further analyze whether blood monocytes can be made susceptible to HIV-1 by other cytokines such as IL-3 and GM-CSF. The rationale of these studies that these factors induce proliferation in a proportion of cells within the monocytic lineage, and the proliferative activity may lead to an increase in susceptibility to retroviral agents such as HIV-1.

The other partners in the HIV-1/monocytic interactions are the various HIV-1 isolates which are most frequently obtained from peripheral blood mononuclear cells (PBMC) of infected individuals by coculture with PHA-stimulated donor peripheral blood lymphocytes (PBL). In previous studies differences in monocytotropism among HIV-1 isolates have been documented [1, 2, 7, 12–14]. To further elucidate the pathogenic significance of monocytotropic HIV-1 variants, we tested a panel of isolates and biological clones with distinct biological phenotypes, originally obtained by coculture with PBL, for their ability to replicate in monocyte-derived macrophages (MPH) from different donors. These distinct biological properties have been defined as the syncytium-inducing (SI) capacity, replication rate and tropism for cell lines [15–19]. The reason for this interest is that in longitudinal studies an association was demonstrated between the clinical course of infected individuals and the biological phenotype of their sequential HIV-1 isolates [20]. The recovery of SI, H9 cell line tropic variants was correlated with subsequent rapid progression to AIDS [20, 21]. Furthermore, in primary isolation studies we compared additional biological properties of sequential isolates, obtained by coculture with either PBL or MPH from the same patient PBMC samples [22]. In these studies we demonstrate that monocytotropic HIV-1 isolates are present in the peripheral blood throughout the HIV-1 infection, and that the majority of these isolates has a non-SI (NSI), non-T-cell line tropic phenotype.

The Increased Susceptibility of Macrophages to HIV-1

To study the reasons between the susceptibility of blood monocytes and tissue macrophages we analyzed whether monocytes showed an innate resistance or could be made susceptible to HIV-1 by inducing their differ-

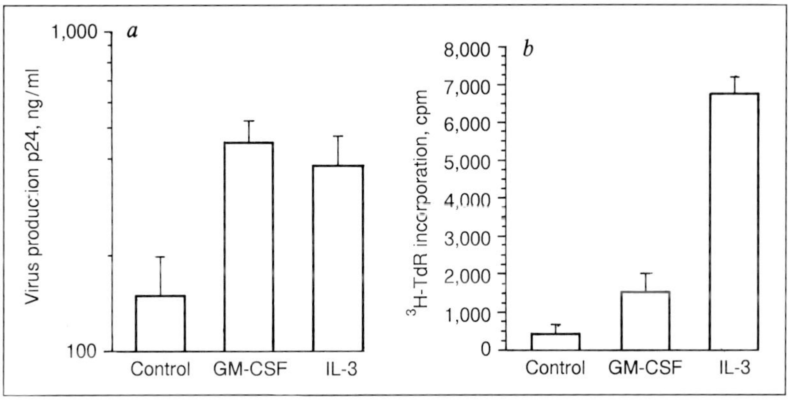

Fig. 1. a rIL-3 (10 ng/ml) and rGM-CSF (100 U/ml) enhance expression of monocytotropic HIV-1 strain HTLV-III Ba-L in purified peripheral blood monocytes. CSF were present during the first 5 days of culture. On day 5, monocytes were inoculated with a virus dose of 5×10^5 cpm/ml RT activity. Each bar represents the mean ± SD of 6 replicates (one representative example of 3 independent experiments). *b* rIL-3 (10 ng/ml) and rGM-CSF (100 U/ml) induce proliferation of 5-day cultured peripheral blood monocytes. CSF were administered directly at the start of culture and continuously present. Cultures were pulsed at day 4 with 7.3 kBq ^{3}H-thymidine, and ^{3}H-thymidine incorporation was determined at day 5.

entiation in culture. It was confirmed that fresh monocytes could not directly be infected with HIV-1 and only became susceptible to HIV-1 infection after a minimum of 2 days' culture. During this 2-day period cultured monocytes are most likely to be rendered susceptible to HIV-1 due to adherence to plastic and the consequent induction of differentiation [7]. The lack of HIV-1-infected monocytes in peripheral blood is thus most likely to be due to an immature differentiation state of monocytes.

We then studied the effect of colony-stimulating factors on HIV-1 infection in monocytes. MPH, infected after a 5-day culture in the presence of rIL-3 (10 ng/ml) or rGM-CSF (100 U/ml), demonstrated enhanced virus production (fig. 1a) [23, 24]. The mechanism of action by which this enhancement was established was further studied. First, we analyzed whether the observed increase in virus production after stimulation with rIL-3 and rGM-CSF was due to the activation of nuclear factors with binding sites in the HIV-1 long terminal repeat (LTR). To this end, monocytes cultured for 5 days in the presence of rIL-3 (10 ng/ml) or rGM-CSF

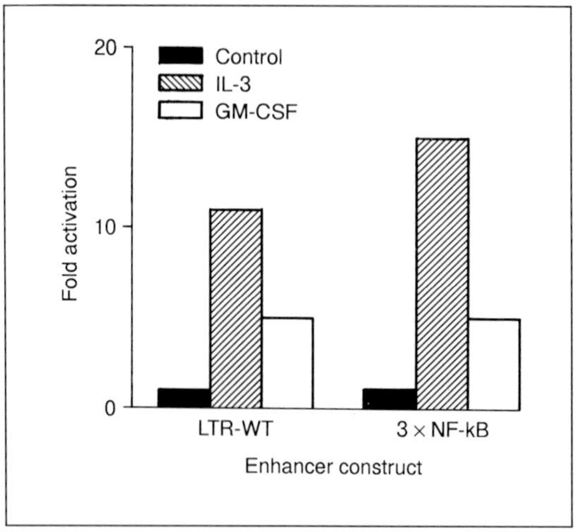

Fig. 2. rIL-3 (10 ng/ml) and rGM-CSF (100 U/ml) induce LTR-driven CAT activity associated with activation of NF-kB. Purified peripheral blood monocytes were cultured for 5 days in the presence or absence of CSF. At day 5, cells were transfected by electroporation [30] with 5 μg of the plasmid in which the complete HIV-1 LTR or only 3 NF-kB sites are linked to the reported gene CAT. 24 h after transfection, cell lysates were prepared and CAT activity was determined [30].

(100 U/ml) were transfected with a plasmid in which the complete HIV-1 LTR [25] or only two NK-kB sites [26] are linked to the reporter gene CAT. Enhanced LTR activity upon a 5-day CSF treatment could be demonstrated and the enhancement appeared to be associated with activation of NF-kB (fig. 2). Since the effect of these growth factors could be demonstrated only after a relatively long-term treatment, we suspect a secondary rather then a direct effect on the LTR activity.

IL-3 and GM-CSF are both growth factors that induce proliferation of hematopoietic stem cells. Their effects include the enhancement of the proliferations of peripheral blood monocytes [23, 24] (fig. 1b). Not only the ^{3}H-thymidine incorporation was increased but also 2–3 times higher numbers of monocytes/MPH could be recovered after culture.

Proliferation of monocytes is, however, still a debated issue. In contrast to the requirements for HIV-1 infection in T cells, HIV-1 infection of monocytes has been thought to be proliferation-independent [1]. Since we were clearly able to demonstrate low level of proliferation in monocytes,

Table 1. Association between biological HIV-1 variants with characteristic biological phenotype and clinical progression

HIV-1 variant	Syncytia in PBL	Replication rate	Host range	Prognosis	Lag time to AIDS, years	Survival with AIDS, years
SI	+	+++	PBL, sup-T1, H9	AIDS	1.2	1.0
Fast NSI	–	++	PBL (sup-T1)	AIDS	2.1	1.7
Slow NSI	–	+	PBL	AS	>4	>3

we studied to what extent this proliferation is a prerequisite for establishment of a productive HIV-1 infection in IL-3- and GM-CSF-treated cells. Treatment of monocytes with the proliferation-inhibiting agent mitomycin C at day 5 of culture 30 min prior to infection resulted in complete abolition of both proliferation and productive HIV-1 infection. This finding was also confirmed by irradiation of the monocytes (3,000 rad). This excludes the possibility that the lack of productive HIV-1 infection is attributable to the previously described DNA-polymerase inhibitory effect of mitomycin C. Mitomycin C treatment or irradiation of the monocytes 48 h after infection had no effect on HIV-1 expression indicating that proliferation is a prerequisite for productive infection rather than the expression of the already established virus.

Cell-Free Infection of MPH with Established HIV-1 Isolates

In longitudinal studies it was documented that the SI variants were not identifiable at the time of seroconversion but appeared only during the course of infection, preceding clinical deterioration [20]. In long-term asymptomatic individuals, NSI variants were detected from seroconversion onward [20]. These NSI HIV-1 variants were unable to replicate in the H9 cell line and showed variable replication rates. Furthermore, a correlation between the replication rate and the rate of CD4+ cell decline in vivo could be demonstrated [20]. Differences in biological phenotypes of HIV-1 variants and the correlation with disease progression are summarized in table 1.

Primary peripheral blood monocytes, >95% pure, were isolated from buffy coats obtained from seronegative plasmapheresis donors by density

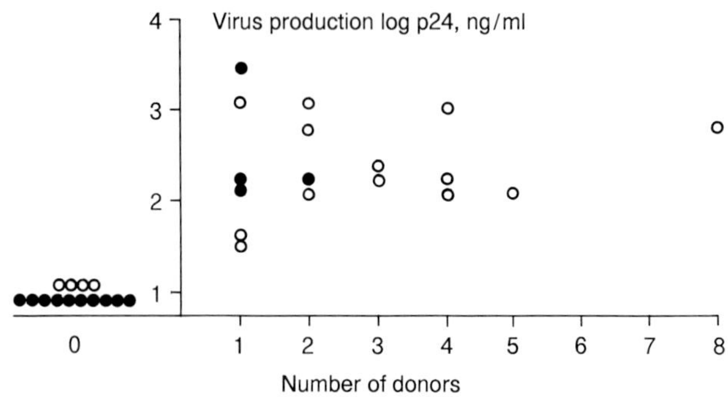

Fig. 3. Ability of NSI (○) and SI (●) variants to establish productive infection in MPH from 8 different donors. 5-day cultured MPH were infected with cell-free virus (inoculum 5 × 10^4 cpm of RT activity) originally obtained by coculture of PBMC from seropositive individuals with PHA-stimulated PBMC. Virus production in MPH cultures was measured repeatedly in a p24 antigen capture ELISA. Cultures were considered positive when at least two sequential samples gave values more than two times the background extinction value. For each virus, the number of donors whose MPH could be infected and the mean peak production are indicated.

Percoll gradient centrifugation followed by centrifugal elutriation. Monocytes were cultured for 5 days to obtain MPH. At day 5, cells were exposed to 17 NSI and 14 SI isolates. These isolates were previously obtained from blood mononuclear cells and extensively characterized by cocultivation on PHA blasts [27]. Unabsorbed virus was removed after 24 h and virus production in the cultures was monitored each week in a p24 antigen capture ELISA [27].

The two groups of virus isolates showed a significant difference (Fisher's exact test; p = 0.013) in their capacity to productively infect MPH (fig. 3). Both groups of virus isolates also demonstrated significant differences in the number of different donor MPH they could infect (Mann-Whitney U-Wilcoxon rank sum W test; p = 0.027). Most NSI isolates (13 of 17) were able to productively infect MPH, and the majority of these isolates could infect cells of more than 1 donor (median 3 of 8). Interestingly, all NSI isolates (n = 4) obtained from patients progressing to AIDS demen-

Table 2. Primary isolation of HIV-1 from sequential PBMC samples of seropositive individuals by cocultivation with either MDM or PBMC

Patient	Clinical status		Phenotype in PBL of isolates recovered in	
	start	end	PBL	MDM
349	AS	AS	NSI	NSI
186	SC	AS	NSI	NSI
232	SC	AS	NSI	NSI
331	SC	AS	NSI	NSI
105	AS	AIDS	SI	NSI
316	SC	AIDS	SI	NSI
571	SC	AIDS	SI	NSI
1082	SC	AIDS	SI	NSI
1145	SC	AIDS	NSI	NSI

tia [21] were able to infect MPH of at least 3 donors. The prototype monocytotropic variant HIV-1 Ba-L infected MPH from all 8 blood donors. From 14 SI isolates however, only 4 could infect MPH and only 1 of these isolates could infect monocytes of more than 1 donor (median 0 of 8).

Comparative Primary HIV-1 Isolation on PBMC and MPH

To analyze the prevalence and the biological phenotype of monocytotropic HIV-1 variants in the course of HIV-1 infection, virus was isolated from the patients' blood mononuclear cells, using either MPH or PBL as target cells. Sequential PBMC samples were used of 9 asymptomatic individuals, 5 of whom progressed to AIDS. For 7 individuals, the moment of seroconversion was known. From 6 of these 7 individuals the first isolate was obtained at the moment of seroconversion. The first isolate of patient 316 was from a sample obtained 1 year after seroconversion (table 2). Virus isolation was successful at 37/42 time points. The phenotype of these isolates was further investigated by cell-free passages to PBL. Monocytotropism was confirmed by cell-free transmission back to MPH. In all cases, except for the last isolate obtained from patient 105, isolates recovered by cocultivation with MPH were NSI in PBL (table 2).

The phenotype of these isolates was compared to the phenotype of HIV-1 isolates obtained from the same samples by cocultivation with PHA-stimulated PBL. In 5 of these patients (Nos. 186, 232, 331, 349 and 1145) in this way only NSI, non-H9-tropic isolates were obtained (data not shown). In 3 individuals a conversion of HIV-1 phenotype from NSI/non-T-cell line tropic to SI/T-cell line tropic [20] was observed in sequential isolates obtained by cocultivation with PBL. In addition, all isolates obtained by PBL cocultivation from subject 105 were of the SI phenotype.

All MPH-derived isolates were NSI in PBL, even when the PBL-derived isolate from the same sample was SI, and none of them could be transmitted to T-cell lines. These results demonstrate that virus isolation by coculture with MPH preferentially rescues HIV-1 variants with an NSI phenotype in PBL culture even when SI clones, which have a growth advantage in PBL culture [19], are present in the same patient sample. This again indicates that monocytotropic variants reside mainly in the population of NSI, non-T-cell line tropic HIV-1 clones.

The Significance of Macrophage Susceptibility to HIV-1

Here we have demonstrated that next to a certain state of differentiation, monocyte proliferation appears to be a prerequisite for HIV-1 infection. Macrophages in tissue are considered to be terminally differentiated cells and these cells probably have lost their replicative capacity [1, 3]. HIV-1 infection must therefore have occurred at a more early stage in ontogeny. Since monocytes in the peripheral blood in general do not harbor the virus [9], infection of cells of this lineage most likely has to occur upon diapedesis from the peripheral blood into the tissue compartment. The presence of HIV-1-infected T cells and important cytokines at the same site will create the optimal conditions for virus transmission from T cells to differentiating monocytes. Differentiated macrophages serve as a reservoir and from these cells new virus variants might be generated. The close cell-cell contact during the process of antigen presentation will create the optimal circumstances for virus transmission from macrophages to T cells. A suggestive indication for a frequent shuttling of virus between cells of the macrophage and the T-cell lineage is the isolation of monocytotropic HIV-1 variants from peripheral blood in the absence of HIV-1-infected monocytes. Separation of T cells and monocytes from peripheral blood demonstrate that the monocytotropic isolates originate from the peripheral blood T cells [unpubl. observations]. Rather than concluding that monocytotropism of these isolates is an in vitro artifact [29], we propose that these T cells only just have become infected by contact with infected tissue macrophages.

In a large proportion of seropositive individuals, the presence of SI variants has been demonstrated (see also Schellekens et al., this volume). Since a considerable group of HIV-infected individuals will have received the virus from such a SI variant carrier, it seems paradoxical that among asymptomatics only a few SI carriers can be identified. The appearance of SI variants during the course of infection and the loss of SI variants following seroconversion demonstrate that upon transmission to a new individual, SI variants disappear in the recipient. This phenomenon is most likely to be due to the selective elimination of the virus. As the host immune system is uncompromised and powerful HIV-1-specific cytotoxic responses are generated, the selective elimination of highly virulent variants may be the result of their higher expression in PBL [28]. The advantage of NSI variants, as demonstrated in this study, is their apparent monocytotropism, a mechanism by which they can 'hide' from immune surveillance. The presence of monocytotropic HIV-1 isolates during all stages of infection contributes to the hypothesis that monocytes, as in other lentivirus infections, are important for the persistence of infection.

The monocytotropism of NSI variants, and the general absence of monocytotropism in the SI variant population, indicates that NSI variants are not just functionally impaired HIV-1 mutants. Instead these variants may be endowed with specific properties that are lacking in the otherwise more efficiently replicating SI variants. More detailed molecular studies will be needed to precisely define the determinants for monocytotropism and the cause of the apparent linkage between these features and their lack of SI capacity in primary T cells.

Conclusions

Cells of the monocyte/macrophage lineage are considered to be important in HIV-1 infection and the reservoir and vehicle function of these cells contribute to persistence of infection and dissemination of the virus through the body. The influence of cytokines on HIV-1 infection and induction of susceptibility to HIV-1 in cellular interactions, perhaps associated with a burst of proliferative activity in MPH as they reach the tissue sites, should be one of the main efforts to study. This is because the controlling influence over this compartment in HIV-1 infection will subsequently lead towards a better check over virus production and development of disease.

References

1 Gartner S, Markovits P, Markovits DM, Kaplan MH, Gallo RC, Popovic M: The role of mononuclear phagocytes in HTLV-III/LAV infection. Science 1986;233:215–219.

2 Koyanagi Y, Miles S, Mitsuyasu RT, Merrill JE, Vinters HV, Chen ISY: Dual infection of the central nervous system by AIDS viruses with distinct cellular tropisms. Science 1987;236:819–822.

3 Gendelman HE, Orenstein JM, Baca LM, Weiser B, Burgers H, Kalter DC, Meltzer MS: The macrophage in the persistence and pathogenesis of HIV infection. Editorial review. AIDS 1989;3:475–495.

4 Koenig S, Orenstein HE, Gendelman JM, Dal Canto MC, Pezeshkpour GH, Yungbluth M, Janotta F, Aksamit A, Martin MA, Fauci AS: Detection of AIDS virus in macrophages in brain tissue from AIDS patients with encephalopathy. Science 1986;233:1089–1093.

5 Gendelman HE, Leonard JM, Dutko FJ, Koenig S, Khillan JS, Meltzer MS: Immunopathogenesis of human immunodeficiency virus infection in the central nervous system. Ann Neurol 1988;23(suppl):S78–S81.

6 Ziza JM, Brun-Vezinet F, Venet A, Rouzioux CH, Traversat J, Israel-Biet B, Barre-Sinoussi F, Godeau P: Lymphadenopathy-associated virus isolated from bronchoalveolar lavage fluid in AIDS-related complex with lymphoid interstitial pneumonitis. N Eng J Med 1985;313:183.

7 Gendelman HE, Orenstein JM, Martin MA, Ferrua C, Mitra R, Phipps T, Wahl LA, Lane HC, Fauci AS, Burke DS, Skillman D, Meltzer MS: Efficient isolation and propagation of human immunodeficiency virus on recombinant colony-stimulating factor 1-treated monocytes. J Exp Med 1988;167:1428–1441.

8 Folks TM, Kessler SW, Orenstein JM, Justement JS, Jaffe ES, Fauci AS: Infection and replication of HIV-1 in purified progenitor cells of normal human bone marrow. Science 1988;242:919–922.

9 Schnittman SM, Psallidopoulos MC, Lane HC, Thompson L, Baseler M, Massari F, Fox CH, Salzman NP, Fauci AS: The reservoir for HIV-1 in human peripheral blood is a T cell that maintains expression of CD4. Science 1989;245:305–308.

10 Poli G, Bressler P, Kinter A, Duh E, Timmer WC, Rabson A, Justement JS, Stanley S, Fauci AS: Interleukin-6 induces human immunodeficiency virus expression in infected monocytic cells alone and in synergy with tumor necrosis factor-α by transcriptional and post-transcriptional mechanisms. J Exp Med 1990;172:151–158.

11 Poli G, Kinter A, Justement JS, Kehrl JH, Bressler P, Stanley S, Fauci AS: Tumor necrosis factor-alpha functions in an autocrine manner in the induction of human immunodeficiency virus expression. Proc Natl Acad Sci USA 1990;87:782–785.

12 Popovic M, Gartner S: Isolation of HIV-1 from monocytes but not T lymphocytes. Lancet 1987;ii:916.
13 Cheng-Mayer C, Weiss C, Seto D, Levy JA: Isolates of human immunodeficiency virus type 1 from the brain may constitute a special group of the AIDS virus. Proc Natl Acad Sci USA 1989;86:8575–8579.
14 Gartner S, Markovits P, Markovits DM, Betts RF, Popovic M: Virus isolation from and identification of HTLV-III/LAV-producing cells in brain tissue from a patient with AIDS. JAMA 1986;256:2365.
15 Asjo B, Albert J, Karlsson A, Morfeld-Manson L, Biberfeld G, Lidman K, Fenyo EM: Replicative properties of human immunodeficiency virus from patients with varying severity of HIV infection. Lancet 1986;ii:660–662.
16 Von Briesen H, Becker WB, Henco K, Helm EB, Gelderblom HR, Brede HD, Rubsamen-Waigmann H: Isolation frequency and growth properties of HIV variants: multiple simultaneous variants in a patient demonstrated by molecular cloning. J Med Virol 1987;23:51–66.
17 Evans LA, McHugh TM, Stites DP, Levy JA: Differential ability of HIV isolates to productively infect human cells. J Immunol 1987;138:3415–3418.
18 Cheng-Mayer C, Seto D, Tateno M, Levy JA: Biologic features of HIV-1 that correlate with virulence in the host. Science 1988;240:80–82.
19 Tersmette M, De Goede REY, Al BJM, Winkel IN, Gruters RA, Cuypers HTM, Huisman HG, Miedema F: Differential syncytium-inducing capacity of human immunodeficiency virus isolates: frequent detection of syncytium-inducing isolates in patients with acquired immunodeficiency syndrome (AIDS) and AIDS-related complex. J Virol 1988;62:2026–2032.
20 Tersmette M, Gruters RA, De Wolf F, De Goede REY, Lange JMA, Schellekens PTA, Goudsmit J, Huisman JG, Miedema F: Evidence for a role of virulent HIV variants in the pathogenesis of AIDS obtained from studies on a panel of sequential HIV isolates. J Virol 1989;63:2118–2125.
21 Tersmette M, Lange JMA, De Goede REY, De Wolf F, Eeftink Schattenkerk JKM, Schellekens PTA, Coutinho RA, Huisman JG, Goudsmit J, Miedema F: Differences in risk for AIDS and AIDS mortality associated with biological properties of HIV variants. Lancet 1989;i:983–985.
22 Schuitemaker H, Kootstra NA, De Goede REY, De Wolf F, Miedema F, Tersmette M: Monocytotropic human immunodeficiency virus 1 variants detectable in all stages of HIV infection predominantly lack T-cell line tropism and syncytium-inducing ability in primary T-cell culture. J Virol 1991;65:356–363.
23 Schuitemaker H, Kootstra NA, Van Oers M, Van Lambalgen R, Tersmette M, Miedema F: Induction of monocyte proliferation and HIV expression by IL-3 does not interfere with anti-viral activity of zidovudine. Blood 1990;76:1490–1493.
24 Koyanagi Y, O'Brien WA, Zhao JQ, Golde DW, Gasson JC, Chen ISY: Cytokines alter production of HIV-1 from primary mononuclear phagocytes. Science 1988;241:1673–1675.
25 Arya SK, Guo C, Josephs SF, Wong-Staal F: Trans-activator gene of human T-lymphotropic virus type III. Science 1985;229:69–73.
26 Nabel G, Baltimore D: An inducible transcription factor activates expression of human immunodeficiency virus in T cells. Nature 1987;326:711–713.

27 Tersmette M, Winkel IN, Groenink M, Gruters RA, Spence P, Saman E, van der Groen G, Miedema F, Huisman JG: Detection and subtyping of HIV-1 isolates with a panel of characterized monoclonal antibodies to HIV-p24gag. Virology 1989;171: 149–155.
28 Asjo B, Barkhem T, Albert J, Chiodi F, Von Gegerfeldt H, Biberfeld P, Fenyo EM: HIV-1 strains with differences in replicative capacities are distinguished by in situ hybridization of infected cells (abstract). IVth International Congress on AIDS, 1988, abstr 1569.
29 Massari FE, Poli G, Schnittman SM, Psallidopoulos MC, Davey V, Fauci AS: In vivo T lymphocyte origin of macrophage-tropic strains of HIV. Role of monocytes during in vitro isolation and in vivo infection. J Immunol 1990;144:4628–4632.
30 Cann AJ, Koyanagi Y, Chen ISY: High efficiency transfection of primary human lymphocytes and studies of gene expression. Oncogene 1988;3:123–128.

Dr. Hanneke Schuitemaker, Department of Clinical Viro-Immunology,
Central Laboratory of the Netherlands Red Cross Blood Transfusion Service,
PO Box 9190, NL–1006 AD Amsterdam (The Netherlands)

Janossy G, Autran B, Miedema F (eds): Immunodeficiency in HIV Infection and AIDS.
EC/FERS/MRC Workshop on Immunodeficiency in HIV-1 Infections, Windsor, Surrey, 1991.
Basel, Karger, 1992, pp 265–277

Loss of Dendritic Cell Function: A Basis for Immunodeficiency?

Stella C. Knight, Steven Macatonia, Steven Patterson[1]

Clinical Research Centre, Harrow, Middlesex, UK

The variable latent period before HIV infection results in severe disease, and the precise cellular basis for development of extreme T cell deficiency, are two related aspects of HIV infection which are still difficult to explain. The loss in memory responses to recall antigens is seen in early HIV-1 infection. Nevertheless, this loss of function is a variable phenomenon which is frequently partial in a proportion of individuals and is not always apparent in all studies [1, 2]. The variability of this functional loss in terms of the prognosis of the patients has been described in the chapter by Miedema et al., this volume. This early loss in T cell secondary responses is observed at a time when T cells still respond well to stimulation with alloantigens in the mixed leukocyte reaction (MLR) [3, 4]. The responsiveness to the non-specific mitogen phytohaemagglutinin (PHA) is among the last responses to disappear as the disease progresses (see also Janossy, Borthwick et al., this volume). Although some evidence of loss of accessory cell function has been reported, studies on cells from monozygotic twins have suggested that the major defect in HIV infection is due to T cell defects rather than a failure of accessory cell function [1]. In this latter study, T cells from an HIV-1-infected patient failed to respond to recall antigens presented on accessory cells from a normal twin, while in

[1] S.M. is supported by the AIDS Directed Programme, UK. We are grateful to Norma Saunders for help in preparation of the manuscript.

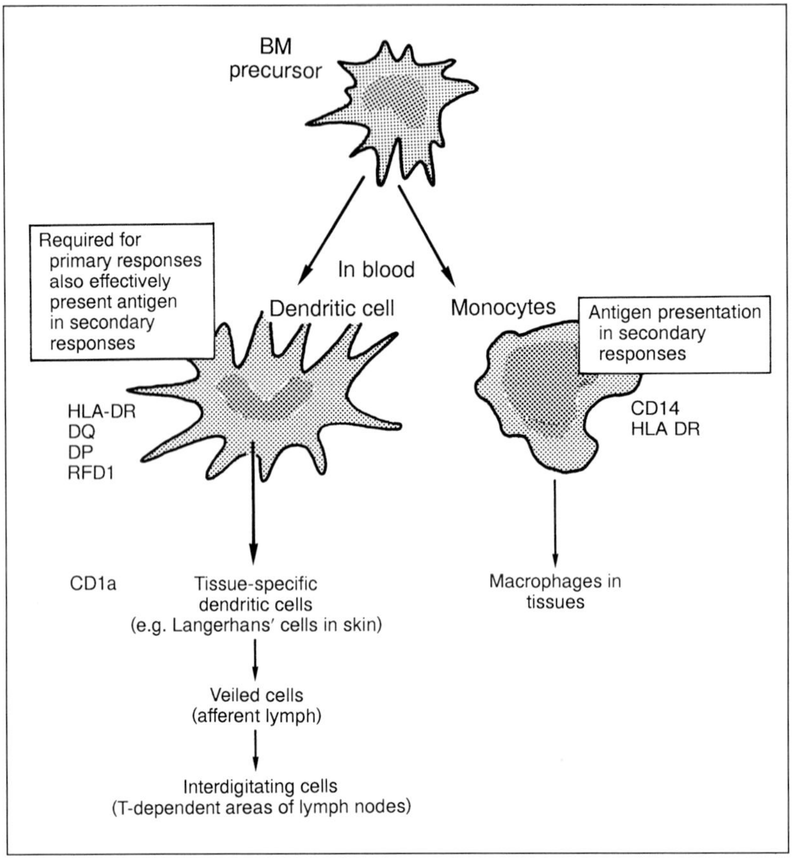

Fig. 1. Origins of dendritic cells. It has recently been shown that in the bone marrow dendritic cells and monocytes derive from a common precursor [32]. The main features of the dendritic and the monocytic lineages are shown.

contrast accessory cells from the same HIV-1-positive patient were able to present the recall antigens to the T lymphocytes of the normal twin. However, we will show here how this spectrum of immune defects could be explained in terms of loss in primary and secondary T cell stimulation by dendritic cells (DC) in the presence of normal antigen presentation by monocytic cells (MO). These concepts are based upon the heterogeneity of antigen presenting cells and their differential involvements in primary and secondary responses, as summarized in figure 1.

Antigen Presentation in Primary and Secondary Immune Responses

There is a distinction between the requirements for antigen presentation in primary and in secondary responses. DC are the most potent cells for stimulating primary responses. In studies in vitro the DC cells are the only cells to present antigen efficiently to stimulate primary T cell responses [5]. This is also seen in the primary MLR where DC are the effective antigen presenting population [6, 7]. The DC are also the major route of primary stimulation in vivo. The efficacy of these cells is documented by quantitative studies which reveal that DC are effective at stimulation using 1 per 4,000 lymphocytes in vitro [8] and as few as 200 cells sensitize a mouse to alloantigens in vivo [9]. By contrast, there is evidence that MO and other class II-bearing cells do not stimulate primary responses [5–7]. On the other hand, any class II-bearing cells with antigen, including MO, may be stimulatory for memory T cells, although it is not yet determined whether all or only recently activated memory T cells can be stimulated. Activated B cells are also good stimulators of secondary responses but it is not yet clear whether they are effective at stimulating primary responses to antigen [10, 11]. The differences between the primary and secondary responses are, however, not absolute. For example, DC are also able to stimulate secondary responses, and their superiority over MO on a cell for cell basis in presenting recall antigens can be clearly seen [5, 12]. In considering the effects of HIV on antigen-presenting cells, it is thus important to separate the effects on primary and secondary responses.

Dendritic Cells Are Infected by HIV-1

Following experiments showing the susceptibility of DC to infection with HIV in vitro [13–15] we showed that there is a high level of infection of DC in the peripheral blood of patients infected with HIV-1 as demonstrated using in situ hybridization techniques [16, 17] (fig. 2). These observations have recently been confirmed using PCR [Patterson et al., in preparation]. DNA from DC of peripheral blood of infected individuals was studied using a nested PCR technique with gag primers. An example comparing lymphocytes and enriched DC from an asymptomatic individual is shown in figure 3 where DNA equivalent to 5,000 cells is required to see lymphocyte positivity and the enriched DC are positive with 50 cells present. Infection of DC was thus demonstrated even in asymptomatic

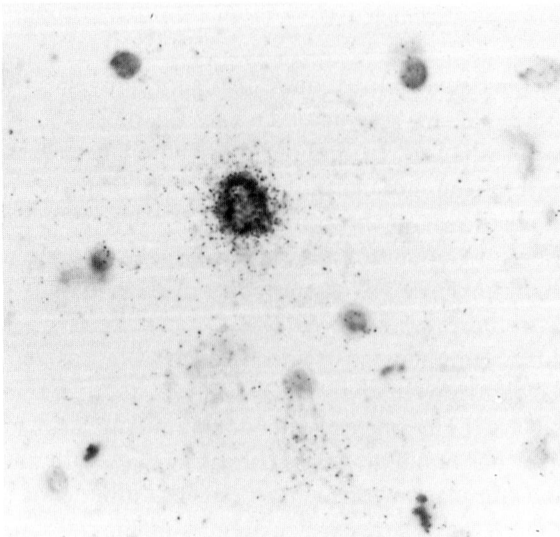

Fig. 2. Enriched dendritic cells from an HIV-infected individual were adsorbed onto 0.01% poly-*L*-lysine-coated slides and labelled with antibodies to CD19, CD14, CD16 and CD3 to label B cells, monocytes, NK cells and T cells using alkaline-phosphatase anti-alkaline phosphatase (APAAP) labelling. The DC were large 'fluffy' cells not labelling with the markers. In situ hybridisation identified HIV, DNA or RNA in the same samples [4]. Most HIV-positive cells were APAAP-negative DC. × 650.

individuals and despite the significant reduction in DC numbers in the peripheral blood of these patients [16]. The technique used for demonstrating the decrease in their number involved separation of mononuclear cells, incubation on plastic to remove macrophages and separation of cells over hypertonic metrizamide gradients. Enriched DC (on average 30% pure) could then be distinguished and counted from their typical morphology in light microscopy of wet preparations. Confirmation of this was made by the lack of staining using antibody labelling of contaminating cells (B, T, NK and MO) because there are no specific markers available for direct identification of DC. Further efforts are needed to establish direct counting of DC in blood since such techniques may give us prognostic tests in HIV-1 infected individuals. Our preliminary evidence in monitoring the effects of zidovudine therapy of individuals with stage IV disease also indicates that treatment may reverse the defects in numbers and function of DC [Macatonia et al., in preparation].

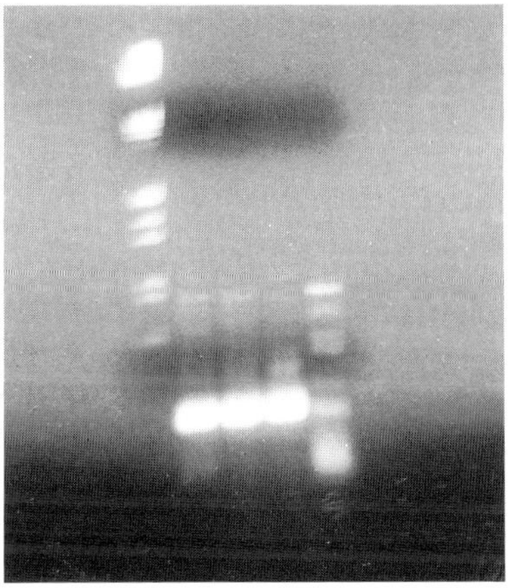

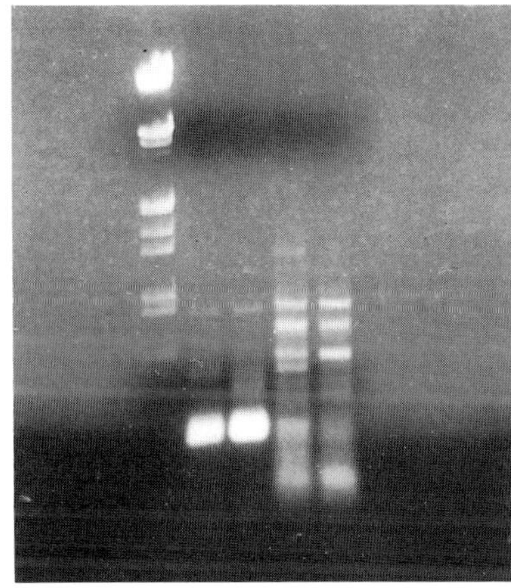

Fig. 3. Nested PCR on limiting dilutions of DNA extracted from dendritic and lymphoid cells of an HIV-infected patient in CDC stage II. Log dilutions of DNA equivalent to 5,000 to 5 cells were used with dendritic cells (*a*) and equivalent to 50,000 to 50 cells were used with lymphocytes (*b*). Dendritic cells and lymphocytes were purified as previously described [16] and the DNA extracted by incubating the cells for 1 h at 50 °C in 20 m*M* Tris pH 7.4 containing proteinase K (100 µg/ml) and SDS (0.5%). After a further 1 h incubation at 60 °C, the preparations were extracted with phenol/chloroform and the DNA precipitated with 0.3 *M* sodium acetate and ethanol. Nested PCR was performed on log dilutions of the DNA using oligonucleotide primers taken from the *gag* region of HIV-1 (forward primers: 1214–1234 and 1407–1425; reverse primers: 1628–1654 and 1668–1686, the coordinates are from HIV HXB2) as described by Simmonds et al. [33]. Samples were denatured for 1 min at 94 °C, annealed for 2 min at 55 °C and extended for 2 min at 72 °C. The first and nested PCRs were subjected to 30 and 40 rounds of amplification, respectively. The major product in positive samples after the second round of PCR was 247 bp in length and was visualized by staining with ethidium bromide after electrophoresis through a 1% agarose gel. Cells from seronegative control were processed in parallel.

In contrast to the evidence for infection and loss of cells of the DC lineage, there is little evidence of alterations in MO in circulating blood. The number of infected MO in peripheral blood detected by in situ hybridization was $<0.02\%$ [16] which is 2–3 orders of magnitude lower than the number of infected DC in our study. One reason for the lack of reproducibility of some of the results in the literature about MO involvement may be the variable DC contamination among cells in the MO category as investigated by the different groups. Langerhans' cells are the relatives of DC (fig. 1), and these cells in the skin are also infected in HIV-1-seropositive individuals [18–20] but, again, may be lost from the skin at advanced stages [21, 22].

The HIV-1 infection of MO in most tissues has been found to be at a very low level [23]. The exception to this is in the brain where microglial cells are reported to show much higher levels of infection [23]. However, there is now new evidence that microglial cells show features which distinguish them from most tissue macrophages. These bone-marrow-derived cells may therefore represent the DC lineage in the brain [24, 25]. Thus, resting microglia show constitutive expression of HLA-DR as well as CD1a, a characteristic feature which they share with the Langerhans' cells in skin (fig. 1). If the microglial cells are indeed primary antigen-presenting cells within the brain then this raises questions about their possible role in presenting HIV-1 antigens to induce inflammatory reactions within the brain. The high rate of infection in these cells would also be compatible with the observations of high infection levels in DC in peripheral blood of HIV-1-infected individuals [16, 17].

The higher levels of infection of DC than of MO usually observed is also reflected in the functional changes seen in these cells. When DC are separated from HIV-seropositive individuals they either fail completely to stimulate proliferation of allogeneic lymphocytes in the MLC or else they cause very low levels of stimulation (fig. 4) [16]. Their function in recruiting T cells into primary responses may thus be severely impaired. It must be emphasised that none of the HIV-infected individuals studied has received AZT therapy. Since we have shown that AZT reverses the defect in DC function, differences in the therapy could underlie differences between this study and others which suggests that MLC stimulating ability is not impaired. In one such study where cells from some patients showed good MLC stimulation [4], a read-out assay relying on lymphokine production was used. A dichotomy in the effect of HIV on lymphokine production and on proliferation, the former remaining intact but the latter being impaired could also account for the discrepancy between the MLC-stimulating activity in these two reports [4, 16]. In addition, in the description of patients with cells stimulating normal lymphokine production in allogeneic cells large numbers of stimulator cells (e.g. 1×10^6 per well) were used. This large number of cells may provide sufficient stimulating DC despite defects within the whole population. In our own studies very low

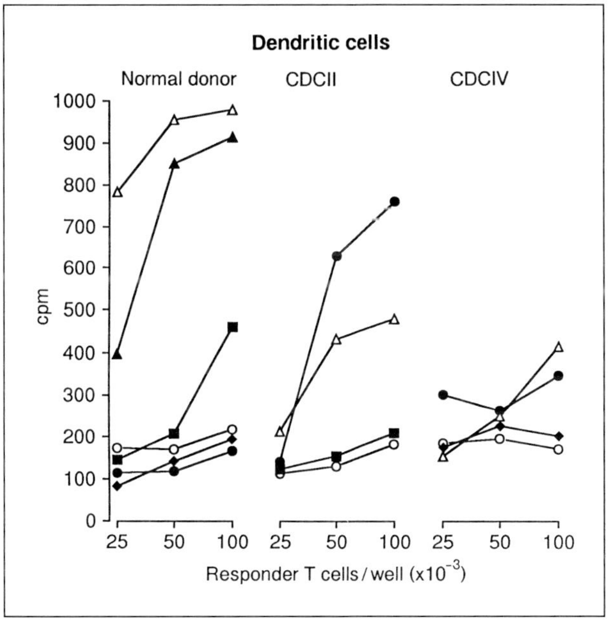

Fig. 4. Dendritic cell (DC)-induced T cell proliferation in HIV infection. Peripheral blood T cells (25 × 100 × 10^{-3}) from a normal uninfected donor, a CDC group II asymptomatic individual and a CDC group IV AIDS patient were cultured in 20 µl hanging drops with autologous or allogeneic DC (2000) as stimulator cells. Proliferation was assessed from the uptake of ^{3}H-thymidine on day 4 of culture (4, for cell separation, culture and harvesting details). ○ = Responder T cells without stimulators; ● = DC from normal donor; ▲ = allogeneic DC from a second normal donor; ■ = DC from CDC II individual; ♦ = DC from CDC IV individual; △ = autologous DC pulsed with influenza virus for 1 h at 37 °C before addition to cultures.

numbers of DC were used (e.g. 500–1,000) to stimulate 10^5 allogeneic cells. It is quite likely that if only small numbers of functional cells remained this would limit the responses. The phenomenon of low T cell response to recall antigens is demonstrated in figure 4 where stimulation by influenza virus and tetanus toxoid is severely blocked in the presence of DC taken from HIV-1-seropositive individuals. This contrasts with results using MO; those taken from asymptomatic individuals or from patients with persistent generalised lymphadenopathy stimulated normal secondary responses to these recall antigens [26]. These findings indicate that both MO and memory T cells were functional in these patients. Support for the normal function of resting T cells is also seen from the normal responsiveness of T cells from these stage II and III patients to stimulation by normal allogeneic DC in the MLR [16, 17]. Thus, in these patients, the major defect is in the DC function while MO and T cells appear to function normally. It is only in patients in stage IV disease (AIDS or AIDS-related complex) that the T cells fail to respond in any of the test systems used in our study.

Evolution of Immune Deficiency

The earliest defect in HIV infection is the partial loss of responsiveness to recall antigens which is lost to varying degrees in different studies. The defect in DC, which are more potent presenters of antigen in secondary responses than MO, would explain the losses in recall responses reported [1–3, 27, 28]. The distinction between the stimulation by DC or MO to the secondary stimulations observed was not made in these earlier reports. Despite some loss in recall responses, responses in MLR can remain intact. It is believed that, in the MLR, T cells can be activated by direct presentation of alloantigens by allogeneic DC. Such a scheme would account for the fact that responses by T cells to allogeneic stimulation remain unaffected in many patients. However, as the infection progresses, the responsiveness to allogeneic DC also decreases. We have evidence that the stimulation of lymphocytes by allogeneic DC is dependent on the presence of very small numbers of DC autologous with the responder T cells. Initial experiments demonstrating this point were performed in mice where cytotoxic antibodies specific for DC are available. As demonstrated in figure 5, addition of small numbers of responder-type DC to allogeneic MLR reactions enhanced the responses. This effect was seen when the responding T cells were partially depleted of DC by passage through a hypertonic metrizamide gradient and depleted of B cells and macrophages on nylon wool (fig. 5). A further treatment of responder cells with 33D1 (an anti-DC antibody) and complement caused a striking reduction in the MLR although the change in the total cell count was too slight to be detectable by routine counts using trypan blue dye exclusion and a counting chamber. The enhancement on adding back small numbers of DC was then very significant. In occasional experiments there was complete removal of the MLR by the antibody treatment to remove responder-type DC and reinstatement of stimulation by adding back 50–500 DC. This work suggests that there may be an absolute requirement for minute numbers of DC

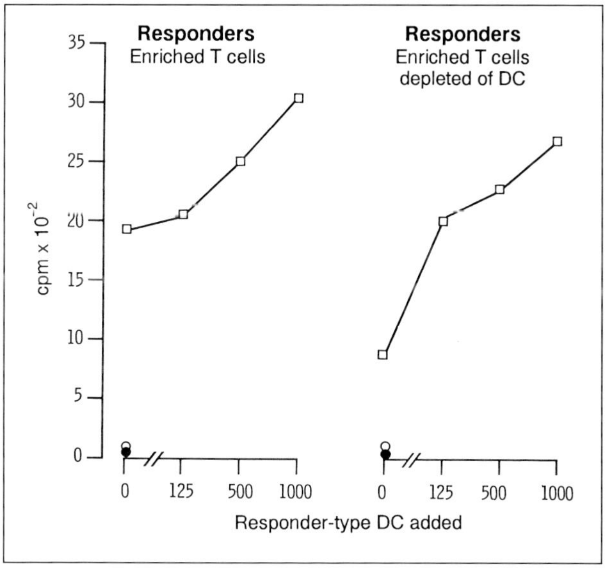

Fig. 5. Effect of responder-type dendritic cells on stimulation using allogeneic dendritic cells in the mixed leukocyte reaction. T cells were enriched from lymph nodes of CBA mice by removal of adherent cells on nylon wool and with and without depletion of dendritic cells using the anti-dendritic cell antibody 33DI plus complement. They were stimulated with allogeneic (B10) dendritic cells in the presence of different numbers of dendritic cells of responder type (CBA). Dendritic cells were isolated from lymph nodes using hypertonic metrizamide gradients [legend fig. 3, Macatonia et al., 5] and uptake of ^{3}H thymidine on day 3 of culture measured [legend fig. 6, Macatonia et al., 14]. Data from Macatonia and Knight [unpubl.]. ○ = 8 × 10^4 T cells (H-2^K); ● = 8 × 10^4 T cells + 1,000 syngeneic DC; □ = 8 × 10^4 T cells + 1,000 allogeneic DC (H-2b).

of responder type to allow responses in the allogeneic MLC. Preliminary studies of human cells indicate that similar effects can be seen. Such a requirement for very small numbers of DC may therefore mean that, as the DC defect in HIV infection becomes more severe, the responsiveness in MLR could also be lost. It is possible that a similar mechanism may apply to the loss of responses to mitogen which may also show some minimal requirement for APC.

In early HIV infection the fact that MO are still able to stimulate responses to recall antigens suggests that the major responses to antigens seen may be operating through memory T cells. However, it is now known that the function of memory T cells is severely handicapped in HIV-1 infection (see Clerici and Shearer, and Miedema et al., this volume). Additional mechanisms for the effective loss of functional T cells have also been described such as lytic infection with virus, cytotoxic killing or blocking of T cell receptor with gp120 [29].The loss of functioning memory T cells would be cumulative in the absence of an adequate primary recruitment mechanism by the DC. A logical explanation for the finding of an association between memory T cell defects and loss of immune function [30, 31] is thus also provided from the studies on infection of DC.

Results from studies of the twins where one had AIDS and the other was not infected can also be explained. When the memory T cell pool in an AIDS patient has been depleted then normal-presenting cells from the twin would be unable to stimulate recall responses. However, the stimulation of secondary responses in the memory T cell pool in the normal twin could be elicited by antigen presented by normal uninfected MO of the AIDS patients. The spectrum of loss of immune responses described in HIV-infected individuals can thus be described in terms of a major deficiency in the capacity of DC to recruit resting T lymphocytes into the memory pool.

Conclusions

We propose that infection of DC, critical antigen-presenting cells for primary recruitment of resting T cells may underlie the progressive loss of immune function (fig. 6). The proportion of DC positive for viral DNA by in situ hybridisation in HIV-seropositive patients can be 2–3 orders of magnitude higher than that of lymphocytes or MO [16]. It can also be demonstrated that these DC are partially depleted from the blood of HIV-1-positive asymptomatic individuals and that the remaining cells stimulate allogeneic lymphocytes to proliferate poorly or not at all. At the same time the ability of these cells to stimulate autologous responses to recall antigens is also severely depressed [26] (fig. 4). MO, which are only effective in stimulating memory T cells, function normally in presenting recall antigens to T cells [26] and T cells respond to normal allogeneic DC indicating that MO, memory T cells and naive T cells are still functionally intact in many asymptomatic individuals. However, in HIV-1 infection loss or functional impairment of T memory cells is also apparent (see Clerici and Shearer, and Miedema et al., this volume [30, 31]) and this handicap may be cumulative in the absence of an efficient primary T cell recruitment mechanism. The secondary activation via MO (from the patient or an

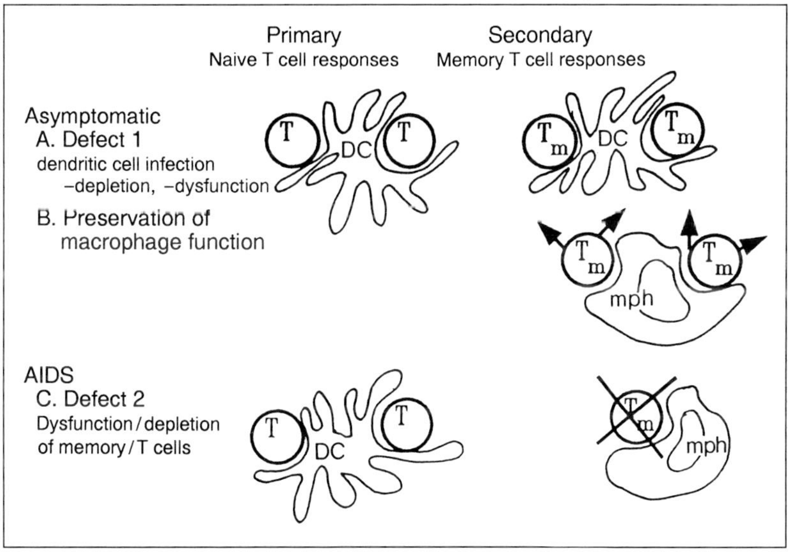

Fig. 6. Hypothesis: Defects of antigen presentation (defect 1) and memory cell function (defect 2) in progressing HIV-1 infection. For explanation see text. DC = Dendritic cell; mph = macrophage, T = naive T cell; T_m = memory T cell. Arrows indicate proliferating T_m cells.

identical twin [1]) will therefore be lost pari passu. MLR was previously believed to occur independently of responder-type DC. However, we show that very small numbers of responder-type DC may promote allogeneic MLR. The loss of MLR reactivity (and perhaps PHA responses) may, therefore, reflect increasing severity of DC loss of function. We speculate, therefore, that the sequential loss of immune functions in HIV infection may reflect a primary defect in recruitment of T cells into the memory pool by DC which becomes critical as pre-existing T memory cells become defective and removed.

References

1 Shearer GM, Clerici M: Early T-helper cell defects in HIV infection. AIDS 1991;5: 245–253.
2 Miedema F, Tersmette M, van Lier RAW: AIDS pathogenesis: A dynamical interaction between HIV and the immune system. Immunol Today 1990;11:293–297.

3 Shearer GM, Payne SM, Joseph LJ, Biddison WE: Functional T lymphocyte immune deficiency in a population of homosexual men who do not exhibit symptoms of acquired immune deficiency syndrome. J Clin Invest 1984;74:496–506.
4 Clerici M, Landay AL, Kessler HA, Zajac RA, Boswell RN, Muluk SC, Shearer GM: Multiple patterns of alloantigen presenting/stimulating cell dysfunction in patients with AIDS. Immunol 1991;146:2207–2213.
5 Macatonia SE, Taylor PM, Knight SC, Askonas BA: Primary stimulation by dendritic cells induces antiviral proliferative and cytotoxic T cell responses in vitro. J Exp Med 1989;169:1255–1264.
6 Inaba J, Steinman RM: Resting and sensitised T lymphocytes exhibit distinct stimulatory (antigen-presenting cell) requirements for growth and lymphokine release. J Exp Med 1984;160:1717–1735.
7 Young JW, Steinman RM: Dendritic cells stimulate primary human cytolytic lymphocyte responses in the absence of $CD4^+$ helper cells. J Exp Med 1990;171:1315–1332.
8 Knight SC, Farrant J, Bryant A, Edwards AJ, Burman S, Lever A, Clarke J, Webster ADB: Non-adherent, low density cells from human peripheral blood contain dendritic cells and monocytes both with veiled morphology. Immunology 1986;57:596–603.
9 Rico MJ, Streilein JW: Comparison of alloimmunogenicity of Langerhans' cells and keratinocytes from mouse epidermis. J Invest Dermatol 1987;89:607–610.
10 Metlay JR, Pure E, Steinman RM: Control of immune responses at the level of antigen-presenting cells: A comparison of the function of dendritic cells and B lymphocytes. Adv Immunol 1990;45:45–116.
11 Sprent J, Schaefer M: Antigen-presenting cells for unprimed T cells. Immunol Today 1989;10:17–22.
12 Macatonia SE, Patterson S, Knight SC: Primary proliferative and cytotoxic T cell responses to HIV induced in vitro by human dendritic cells. Immunology 1991;74:399–406.
13 Patterson S, Knight SC: Susceptibility of human peripheral blood dendritic cells to infection by human immunodeficiency virus. J Gen Virol 1987;68:1177–1181.
14 Macatonia SE, Patterson S, Knight SC: Suppression of immune responses by dendritic cells infected with HIV. Immunology 1989;67:285–289.
15 Patterson S, Gross J, Bedford P, Knight SC: Morphology and phenotype of dendritic cells from peripheral blood and their productive and non-productive infection with human immunodeficiency virus type 1. Immunology 1991;72:361–367.
16 Macatonia SE, Lau R, Patterson S, Pinching AJ, Knight SC: Dendritic cell infection, depletion and dysfunction in HIV-infected individuals. Immunology 1990;71:38–45.
17 Hughes R, Macatonia SE, Keat AS, Knight SC: The detection of human immunodeficiency virus in dendritic cells from the joints of patients with aseptic arthritis. Br J Rheumatol 1990;29:166–170.
18 Tsachler E, Groh V, Popovic M, Mann DL, Konrad K, Sofai B, Eron L, Veronese F, Wolff K, Stingl G: Epidermal Langerhans' cells. A target for HTLV-III/LAV infection. J Invest Dermatol 1987;88:233–237.
19 Rappersberger KS, Gartner S, Schenk P, Stingl G, Groh V, Tschachler E, Mann DP, Wolff K, Konrad K, Popovic M: Langerhans' cells are an actual site of HIV replication. Intervirology 1988;29:185–190.

20 Oxholm P, Helweg-Larsen S, Permin H: Immunohistological skin investigations in patients with the acquired immune deficiency syndrome. Acta Pathol Microbiol Immunol Scand 1986;94:113–116.
21 Belsito DV, Sanchez R, Baer MD, Valentine F, Thorbecke GJ: Reduced Langerhans' cell la antigen and ATPase activity in patients with the acquired immunodeficiency syndrome. N Engl J Med 1984;310:1279–1281.
22 Dreno B, Milpied B, Bignon J, Stadler JF, Litoux P: Prognostic value of Langerhans' cells in the epidermis of HIV patients. Br J Dermatol 1988;118:481–486.
23 Meltzer MS, Skillman DR, Gomatos PJ, Kalter DC, Gendelman HE: Role of mononuclear phagocytes in the pathogenesis of human immunodeficiency virus infection. Ann Rev Immunol 1990;8:169–194.
24 Lowe J, Maclennan KA, Powe DG, Pound JD, Palmers JB: Microglial cells in human brain have phenotypic characteristics related to possible function as dendritic antigen presenting cells. J Pathol 1989;159:143–149.
25 Hayes GM, Woodroofe MN, Cuzner ML: Microglia are the major cell type expressing MHC class II in human white matter. J Neurol Sci 1987;80:25–37.
26 Macatonia SE, Gompels M, Pinching AJ, Patterson S, Knight SC: Antigen presentation by macrophages but not by dendritic cells in HIV infection. Immunology, in press.
27 Miedema F, Petit AJC, Terpestra FG, Schattenkerk JKME, De Wolf F, Roos BJMA1, Lange JMA, Tanner SA, Goudsmit J, Schellekens PTA: Immunological abnormalities in human immunodeficiency virus (HIV)-infected asymptomatic homosexual men. HIV affects the immune system before $CD4^+$ T helper function depletion occurs. J Clin Invest 1988;82:1908–1914.
28 Lane HC, Depper JL, Greene WC, Whalen G, Waldmann TA, Fauci AS: Qualitative analysis of immune function in patients with acquired immunodeficiency syndrome. N Engl J Med 1985;313:79–84.
29 Rosenberg ZF, Fauci AS: The immunopathogenesis of HIV infection. Adv Immunol 1989;47:377–431.
30 Schnittman SM, Lane HC, Greenhouse J, Justement JS, Basel M, Fauci AS: Preferential infection of $CD4^+$ memory T cells by human immunodeficiency virus type 1: Evidence for a role in the selective T-cell functional defects observed in infected individuals. Proc Natl Acad Sci USA 1990;87:6058–6062.
31 Van Noesel CJM, Gruters RA, Terpstra FG, Shellekens PA, van Lier RAW, Miedema F: Functional and phenotypic evidence for a selective loss of memory T cells in asymptomatic HIV-infected men. J Clin Invest 1990;86:293–299.
32 Reid CDL, Fryer PR, Clifford C, Kirk A, Tikerpae J, Knight SC: Identification of hematopoietic progenitors of macrophages and dendritic Langerhans cells (DL-CFU) in human bone marrow and peripheral blood. Blood 1990;76:1139–1149.
33 Simmonds P, Balfe P, Peutherer JF, Ludlam CA, Bishop JO, Leigh Brown AJ: Human immunodeficiency virus-infected individuals contain provirus in small numbers of peripheral mononuclear cells and at low copy numbers. J Virol 1990;64: 864–872.

Dr. Stella C. Knight, Clinical Research Centre, Watford Road,
GB–Harrow, Middlesex HA1 3UJ (UK)

Janossy G, Autran B, Miedema F (eds): Immunodeficiency in HIV Infection and AIDS.
EC/FERS/MRC Workshop on Immunodeficiency in HIV-1 Infections, Windsor, Surrey, 1991.
Basel, Karger, 1992, pp 278–287

Cytokines and Immunoregulatory Factors in the Pathogenesis of HIV Infection: TNF Effects

Guido Poli, Anthony S. Fauci

Laboratory of Immunoregulation, National Institute of Allergy and Infectious Diseases, National Institutes of Health, Bethesda, Md., USA

Tumor necrosis factor-α/cachectin (here referred to as TNF-α) was originally described as both an antineoplastic peptide produced by activated macrophages during bacterial infection [1] and the mediator of cachexia associated with certain parasitic diseases [2]; soon thereafter, it was shown to be associated with important immunoregulatory functions. A cognate molecule produced by activated T lymphocytes was called lymphotoxin or TNF-β [1]. Although only approximately 30% homology occurs at the amino acid level between TNF-α and TNF-β, these molecules recognize the same surface receptors on target cells, and usually induce similar biological effects.

The potential role of TNF (either -α or -β) in the pathogenesis of infection with the human immunodeficiency virus (HIV), the causative agent of the acquired immunodeficiency syndrome (AIDS), was postulated on the basis of several pathological features. The depletion of CD4+ T cells, the common denominator of HIV infection [3], may be the consequence of an increased production of TNF and/or sensitization to its cytotoxic effect. Similarly, the profound cachectic state which characterizes the terminal stage of AIDS was certainly suggestive of increased levels of TNF [3]. However, TNF seems to play an additional and even more important role in the pathogenesis of HIV infection because of its ability of directly inducing virus expression in infected cells.

Does HIV Infection Lead to TNF Production?

Host infection by microbial agents may induce TNF production [1–3]. With regard to HIV infection, several studies have investigated this possibility both in vivo and in vitro. Increased levels of TNF usually associated with the most advanced stages of disease have been detected either in the plasma [4] or the cerebrospinal fluid (CSF) [5] of HIV-infected individuals. However, other studies [6] have not confirmed these findings. Because cytokines, like TNF, often exert their biological effects at the local sites of production in an autocrine/paracrine manner [2], body fluid levels of cytokine may not always accurately reflect the functional impact of these cytokines at the tissue level. Furthermore, the discrepancies between the two groups of studies could be potentially accounted for by the different specificities and sensitivities of the assays used. For example, bioassays can measure both TNF-α and TNF-β activities, whereas several studies using ELISA could exclusively detect TNF-α.

Perhaps of more relevance, several investigators have compared the production of TNF (usually TNF-α) in either peripheral blood mononuclear cells (PBMC) or PBMC subsets obtained from HIV-infected individuals or uninfected donors (table 1). In most cases [7–9], with only a few exceptions [10, 11], increased TNF production by the cells of infected individuals was seen. The monocyte fraction of PBMC is likely to be the principal source of TNF-α, but it must be pointed out that TNF-β production has not been investigated as intensively. In addition, B lymphocytes from infected individuals with hypergammaglobulinemia [12] and lung alveolar macrophages of AIDS patients with *Pneumocystis carinii* pneumonia [9] have also been found to secrete significant amounts of TNF-α.

HIV infection in vitro has been shown to result in the production of TNF from PBMC [13], CD4+ T cells [13], and monocyte-derived macrophages (MDM) [14]. Both inactivated HIV [14, 15] and purified gp120 envelope [15] have been shown to induce transient TNF-α production in MDM. Recombinant (r) soluble CD4 molecules (rsCD4) were shown to block TNF production in these experimental conditions [13, 14], whereas OKT4a monoclonal antibody (Mab) per se, but not OKT4 Mab, exerted an inductive effect [14]. These studies suggest that the interaction of the HIV envelope protein gp120 with the CD4 molecule may result in the induction of cytokine release. In contrast, Molina et al. [16] have not found evidence of a direct cytokine-inductive capability of HIV or gp120 although in ear-

Table 1. The role of TNF in the pathogenesis of HIV infection: in vivo studies

	Increased	Normal	Reduced
Plasma	+++	++	+*
CSF	+	++	n.d.
PBMC	++++	+	+
B cells	+	n.d.	n.d.
Lung alveolar macrophages	+	+	n.d.

Number of studies evaluating TNF presence in different body fluids or TNF production from cells isolated from HIV+ individuals as compared to HIV– controls (source: Medline; updated up to April 1991). See text for references.
* Reduced plasma levels of TNF were observed in HIV+ Zairian children as compared to HIV– children (C. Brown, G. Poli, A.S. Fauci, in preparation).
n.d. = Not described.

lier studies they reported an increased production of TNF-α and interleukin-1β (IL-1β) during acute infection of the monocytic cell line THP-1. The possibility that contaminant endotoxin (LPS) may have explained some of the observed effects was suggested [16]; however, differences among the viral strains or the envelope preparations used in the different studies may also account for these discrepancies [32].

Recently, it has been observed that B lymphocytes from HIV-infected individuals with hypergammaglobulinemia which constitutively secrete significant levels of TNF-α and IL-6 [12] can be further induced to secrete both immunoglobulin and cytokines by rgp120. In contrast, rgp120 did not affect cytokine production from B cells of either normal donors or hypergammaglobulinemic patients who had non-HIV-related immune disorders [33]. These studies suggest that in vivo priming of B cells during HIV infection may result in their ability to respond in vitro with increased cytokine secretion to stimulation by HIV or HIV envelope. Direct infection of circulating monocytes and B cells does not usually occur in most HIV-infected individuals [17] and, therefore, cannot be the explanation for the observed increased cytokine production. However, it is conceivable that autocrine and paracrine pathways of cytokine and HIV production may play an important role at the tissue level, where infected macrophages and T cells as well as activated B cells are present.

Does TNF Selectively Kill HIV-Infected Cells?

On the basis of previous observations in other viral infections, it has been proposed earlier that TNF may play an important role in clearing HIV-infected cells from the host, particularly in the presence of interferon-γ (IFN-γ) [18]. However, the results obtained by several investigators indicated that TNF actually activates HIV expression in both T cells and monocyte/macrophages. A potentially promising approach has been suggested recently by Kobayashi et al. [19] who showed that a murine Mab, named anti-FAS, exerts cytotoxic effects on TNF-sensitive target cells without inducing the immunoregulatory effects typical of this cytokine. When tested with HIV-infected cells, lysis was obtained in the absence of viral induction, whereas uninfected cells were not affected by the Mab [19].

We have also recently observed that the chronically HIV-infected promonocytic cell line U1 was much more sensitive to the lethal effect of the combination of TNF-α plus IFN-γ than the parental uninfected cell line U937. However, the analysis of several uninfected U937 clones revealed that the insensitivity of the bulk parental cell line was due to the presence and ultimate expansion of a small minority of clones resistant to the toxic effect of TNF-α plus IFN-γ treatment [P. Biswas et al., unpubl. observations]. Therefore, caution should be exerted in the extrapolation of certain in vitro results to the overall effect of TNF on HIV-infected cells.

TNF Induces HIV Expression in vitro: Cellular Studies

TNF has been shown to activate HIV expression or replication in a variety of cell systems, including acutely infected primary PBMC and CD4+ T cells [13], T cell lines [20], primary MDM [21], and chronically infected cell lines of both T [15, 22] and monocytic [23] lineages. TNF-α was found to be the predominant HIV-inductive activity secreted by both activated T cells and macrophages, although other cytokines such as IL-3, IL-4, IL-6, granulocyte-macrophage colony-stimulating factor (GM-CSF) and M-CSF have also shown up-regulatory effects on HIV expression in monocyte/macrophages [reviewed in 3, 24].

In acutely infected T cells and cell lines, TNF has been shown to dramatically enhance HIV-dependent syncytia formation [13, 19, 20], a phenomenon which has been proposed as a possible mechanism of CD4+ T cell depletion in vivo [3]. It remains to be established whether this effect is either a direct consequence of the increased levels of virus expression caused by TNF stimulation or involves additional effects of this cytokine on the capacity of infected cells to fuse with uninfected CD4+ target cells.

Analogous to its regulatory role in the homeostasis of the immune system [1, 2], TNF has been shown to act in an autocrine/paracrine manner in the up-regulation of HIV expression in both acutely infected primary PBMC [13], chronically infected promonocytic cell lines [23], and primary MDM [G. Poli et al., unpubl. observations]. Synergistic induction of virus expression has been described in infected cells costimulated with TNF-α and IL-6 or GM-CSF [21]. On the other hand, cytokines such as transforming growth factor-β (TGF-β) and interferons (IFNs) have been shown to exert potent suppressive effects on virus production [reviewed in 3, 24]. These observations suggest that virus expression in vivo is probably the result of cellular modulation by multiple exogenous signals, such as those provided by inductive and suppressive cytokines, in addition to the endogenous regulatory mechanisms exerted by accessory viral gene products such as *tat* and *rev* [reviewed in 25].

TNF Induces HIV Expression in vitro: Molecular Studies

At the molecular level, TNF has been shown to activate HIV expression by increasing its transcriptional rate. As prevoiusly described for the chemical agent phorbol myristate acetate (PMA), the interaction of TNF with its own surface receptor leads to the dissociation of the pleiotropic cellular transcription factor NF-κB from its inhibitor molecule I-κB, and to the subsequent translocation of activated NF-κB complexes from the cytoplasm to the cellular nucleus [reviewed in 26] (fig. 1). NF-κB complexes recognize specific consensus sequences present in the promoter/enhancer regions of several cellular genes as well as in the HIV long terminal repeats (LTR) [26, 27]. Binding of NF-κB to the HIV LTR leads to increased transcription of viral RNA [21, 25–27], which may account for the overall inductive effects of TNF on virus expression in the several models described above.

The importance of NF-κB-dependent transcription in HIV expression has been also investigated by using replication-competent viral clones mutagenized in their LTR sequences. Surprisingly enough, ablation of the

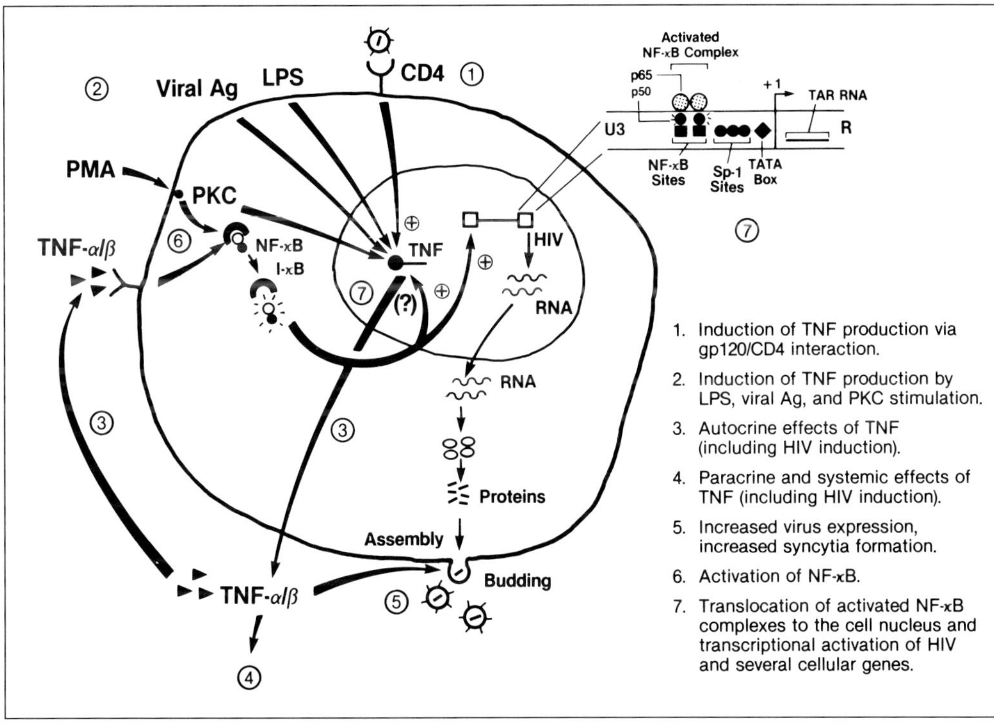

Fig. 1. The role of TNF in the pathogenesis of HIV infection.

two NF-κB consensus repeats from the virus LTR did not impair the ability of HIV clones to replicate in certain T cell lines and primary mitogen-stimulated PBMC [28]. Removal of the Sp-1 consensus domains (which remain between the NF-κB binding repeats and the TATA box, as shown in fig. 1) had variable effects on HIV replication [29], whereas deletion of both NF-κB and Sp-1 consensus sequences from the HIV LTR resulted in the total loss of replicative ability in all the target cells tested [28]. Of interest, TNF rescued some of the viral clones which were lacking the Sp-1 binding elements and were greatly impaired in their constitutive expression [29]. Therefore, a signficant level of redundancy characterizes the HIV LTR, whereas trans-complementation between different transcription factors (such as the NF-κB- and the Sp-1-dependent pathways) may occur. It is also conceivable that different transcription factors play a diversified role in contributing to HIV replication. For example, constitutive transcription factors, such as Sp-1, may be important in establishing baseline levels of virus expression during acute infection, whereas inducible transcription factors, such as NF-κB or NFAT-1 [30], may play a more important role in chronically infected cells carrying latent proviruses.

In addition to transcriptional mechanisms of virus up-regulation, the possibility that additional posttranscriptional steps of control are mediated by certain physiologic factors has been recently proposed [21]. IL-6 has been shown to induce expression of HIV proteins and virions in the absence of significant effects on the levels of newly transcribed viral RNA. In addition, IL-6 also synergized with TNF-α in increasing HIV transcription [21], whereas glucocorticoid hormones have been recently shown to act posttranscriptionally by increasing the stability of TNF-α-induced viral mRNAs [P. Bressler et al., submitted].

Therefore, in addition to the complex network of cytokines which can either up-regulate or suppress virus expression via interaction with cell surface receptors, intracellular networks (such as NK-κB, Sp-1, NFAT-1, and the glucocorticoid/glucocorticoid-receptor system) are likely to play an important role in the maintenance or disruption of the state of viral latency which characterizes HIV infection [3].

Conclusions and Perspectives

Only a few years ago the interaction of HIV with immune cells was thought to be restricted to the high affinity of one of its envelope proteins with the CD4 cell surface molecule. We have rapidly learned that human pathogenic retroviruses, such as HIVs and HTLVs, have evolved multiple levels of interdigitation with the human immune system. The ability of HIV to utilize transcriptional mechanisms of control of cellular gene expression, such as NF-κB, which are susceptible to the influence of cytokines (i.e. TNF) is a typical example. It is very likely that the 'TNF/NF-κB/HIV' cycle of virus expression is only one example of control of HIV expression by the immune system, and that additional levels of interaction exist and need to be more fully understood.

Focusing on the multiple roles of TNF in HIV infection, several questions remain to be answered. Although TNF represents an important factor capable of up-regulating HIV transcription in a variety of in vitro assays, its pathophysiological role in vivo has not yet been fully elucidated. The observation that plasma and/or CSF levels of this cytokine are elevated and hyperproduced by cells obtained from HIV+ individuals represent circumstantial evidence of the importance of TNF in HIV expression. However, additional studies are needed to further delineate the expression of this and other cytokines in infected tissues (for example, in lymphoid tissues [31, 34]) at different stages of disease. The availability of animal models of HIV infection (such as SIV-infected monkeys and SCID mice reconstituted with human hematopoietic cells) may represent potential tools to clarify this and other important questions of viral pathogenesis.

Whether TNF may also represent a mechanism of elimination of HIV-infected cells still needs to be elucidated. It is likely that the dissection of TNF cytotoxic function (hypothetically anti-HIV) from its immunoregulatory effects (very likely pro-HIV) will be required to accomplish this task. The possibility of using Mab directed against specific cell surface receptors which may selectively mediate one or the other TNF functions (i.e. the anti-FAS Mab) represents a potentially important approach. Alternatively, dissociation of the beneficial and detrimental effects of TNF should be investigated by pharmacological tools at the postreceptor levels of signal transduction.

References

1 Old LJ: Tumor necrosis factor. Sci Am 1985;258:59–75.

2 Sherry B, Cerami A: Cachectin/tumor necrosis factor exerts endocrine, paracrine, and autocrine control of inflammatory responses. J Cell Biol 1988;107:1269–1277.

3 Fauci AS: The human immunodeficiency virus: infectivity and mechanisms of pathogenesis. Science 1988;239:617–622.

4 Reddy MM, Sorrell SJ, Lange M, Grieco MH: Tumor necrosis factor and HIV p24 antigen levels in serum of HIV-infected populations. J AIDS 1988;1:436–440.

5 Grimaldi LME, Martino GV, Franciotta DM, Brustia R, Castagna A, Pristera R, Lazzarin A: Elevated alpha-tumor necrosis factor levels in spinal fluid from HIV-1 infected patients with central nervous system involvement. Ann Neurol 1991;29: 21–25.

6 Grunfeld C, Kotler DP, Shigenaga JK, Doerrler W, Tierney A, Wang J, Pierson RN, Feingold KR: Circulating interferon-α levels and hypertriglyceridemia in the acquired immunodeficiency syndrome. Am J Med 1991;90:154–162.

7 Lau AS, Livesey JF: Endotoxin induction of tumor necrosis factor is enhanced by acid-labile interferon-α in acquired immunodeficiency syndrome. J Clin Invest 1989;84:738–743.

8 Voth R, Rossol S, Klein K, Hess G, Schutt KH, Schroder HC, Meyer Zum Buschenfelde KH, Muller WEG: Differential gene expression of IFN-α and tumor necrosis

factor-α in peripheral blood mononuclear cells from patients with AIDS-related complex and AIDS. J Immunol 1990;144:970–975.

9 Krishnan VL, Meager A, Mitchell DM, Pinching AJ: Alveolar macrophages in AIDS patients: increased spontaneous tumour necrosis factor-alpha production in *Pneumocystis carinii* pneumonia. Clin Exp Immunol 1990;80:156–160.

10 Ammann AJ, Palladino MA, Volberding P, Abrams D, Martin NL, Conant M: Tumor necrosis factors alpha and beta in acquired immunodeficiency syndrome (AIDS) and aids-related complex. J Clin Immunol 1987;7:481–485.

11 Haas JG, Reithmuller G, Ziegler-Heitbrock HW: Monocyte phenotype and function in patients with the acquired immunodeficiency syndrome (AIDS) and AIDS-related disorders. Scand J Immunol 1987;26:371–379.

12 Rieckmann P, Poli G, Kehrl JH, Fauci AS: Activated B lymphocytes from human immunodeficiency virus-infected individuals induce virus expression in infected T cells and a promonocytic cell line, U1. J Exp Med 1991;173:1–5.

13 Vyakarnam A, McKeating J, Meager A, Beverley PC: Tumor necrosis factors (α, β) induced by HIV-1 in peripheral blood mononuclear cells potentiate virus replication. AIDS 1990;4:21–27.

14 Merril JE, Koyanagi Y, Chen ISY: Interleukin-1 and tumor necrosis factor α can be induced from mononuclear phagocytes by human immunodeficiency virus type 1 binding to the CD4 receptor. J Virol 1989;63:4404–4408.

15 Clouse KA, Robbins PB, Fernie B, Ostrove JM, Fauci AS: Viral antigen stimulation of the production of human monokines capable of regulating HIV-1 expression. J Immunol 1989;143:470–475.

16 Molina JM, Scadden DT, Amirault C, Woon A, Vannier E, Dinarello CA, Groopman JE: Human immunodeficiency virus does not induce interleukin-1, interleukin-6, or tumor necrosis factor in mononuclear cells. J Virol 1990;64:2901–2906.

17 Schnittman SM, Psallidopoulos MC, Lane HC, Thompson L, Baseler M, Massari F, Fox CH, Salzman NP, Fauci AS: The reservoir for HIV-1 in human peripheral blood is a T cell that maintains expression of CD4. Science 1989;245:305–308.

18 Wong GH, Krowka JF, Stites DP, Goeddel DV: In vitro and anti-human immunodeficiency virus activities of tumor necrosis factor-alpha and interferon-gamma. J Immunol 1988;140:120–124.

19 Kobayashi N, Hamamoto Y, Yamamoto N, Ishii A, Yonehara M, Yonehara S: Anti-Fas monoclonal antibody is cytocidal to human immunodeficiency virus-infected cells without augmenting viral replication. Proc Natl Acad Sci USA 1990;87:9620–9624.

20 Matsuyama T, Yoshiyama H, Hamamoto Y, Yamamoto N, Soma G-I, Mizuno D, Kobayashi N: Enhancement of HIV replication and giant cell formation by tumor necrosis factor. AIDS Res Hum Retrovir 1989;5:139–146.

21 Poli G, Bressler P, Kinter A, Duh E, Timmer WC, Rabson A, Justement S, Stanley S, Fauci AS: Interleukin-6 induces human immunodeficiency virus expression in infected monocytic cells alone and in synergy with tumor factor α by transcriptional and posttranscriptional mechanisms. J Exp Med 1990;172:151–158.

22 Folks TM, Clouse KA, Justement J, Rabson A, Duh E, Kehrl JH, Fauci AS: Tumor necrosis factor α induces expression of human immunodeficiency virus in a chronically infected T-cell clone. Proc Natl Acad Sci USA 1989;86:2365–2368.

23 Poli G, Kinter JS, Justement JS, Kehrl JH, Bressler P, Stanley S, Fauci AS: Tumor necrosis factor α functions in an autocrine manner in the induction of human immunodeficiency virus expression. Proc Natl Acad Sci USA 1990;87:782–785.

24 Poli G, Fauci AS: The role of monocyte/macrophages and cytokines in the pathogenesis of HIV infection. Pathobiol, in press.

25 Cullen BR, Green WC: Regulatory pathways governing HIV-1 replication. Cell 1989;58:423–426.

26 Lenardo MJ, Baltimore D: NF-κB: A pleiotropic mediator of inducible and tissue-specific gene control. Cell 1989;58:227–229.

27 Duh EJ, Maury WJ, Folks TM, Fauci AS, Rabson AB: Tumor necrosis factor α activates human immunodeficiency virus type 1 through induction of nuclear factor binding to the NF-κB sites in the long terminal repeat. Proc Natl Acad Sci USA 1989;86:5974–5978.

28 Leonard J, Parrott C, Buckler-White AJ, Turner W, Ross EK, Martin MA, Rabson AB: The NF-κB binding sites in the human immunodeficiency virus type 1 long terminal repeat are not required for virus infectivity. J Virol 1989;63:4919–4924.

29 Parrot C, Seidner T, Duh E, Leonard J, Theodore TS, Buckler-White A, Martin MA, Rabson AB: Variable role of the long terminal repeat Sp1-binding sites in human immunodeficiency virus replication in T lymphocytes. J Virol 1991;65:1414–1419.

30 Lu Y, Touzjian N, Stenzel M, Dorfman T, Sodroski JG, Haseltine WA: The NFκB independent cis-acting sequences in HIV-1 LTR responsive to T-cell activation. J AIDS 1991;4:173–177.

31 Emilie D, Peuchmaur M, Maillot MC, Crevon MC, Brousse N, Delfraissy JF, Dormont J, Galanaud P: Production of interleukins in human immunodeficiency virus-1-replicating lymph nodes. J Clin Invest 1990;86:148–159.

32 Clouse KA, Cosentino LM, Weih KA, Pyle SW, Robbins PB, Hochstein HD, Natarajan V, Farrar WL: The HIV-1 gp120 envelope protein has the intrinsic capacity to stimulate monokine secretion. J Immunol 1991;147:2892–2901.

33 Rieckmann P, Poli G, Fox CH, Kehrl JH, Fauci AS: Recombinant gp120 specifically enhances tumor necrosis factor-α production and Ig secretion in B lymphocytes from HIV-infected individuals but not from seronegative donors. J Immunol 1991;147:2922–2927.

34 Pantaleo G, Graziosi C, Butini L, Pizzo PhA, Schnittman SM, Kotler DP, Fauci AS: Lymphoid organs function as major reservoirs for human immunodeficiency virus. Proc Natl Acad Sci 1991;88:9838–9842.

Dr. Guido Poli, Laboratory of Immunoregulation, National Institute of Allergy and Infectious Diseases, National Institutes of Health, Bldg. 10, Room 11B-13, Bethesda, MD 20892 (USA)

Epilogue:
Immunological Monitoring of HIV-1 Infection

The observations above confirm the complexity of events leading to the deterioration of the immune system during HIV-1 infection – as has recently been pointed out by both retrovirologists and cellular immunologists [1–3]. The main conclusions of the meeting can be marshalled into five comments as follows:

(1) Although HIV-1 disease is caused by viral infection, changes in the immune systems contribute to and may even determine the speed of progression to AIDS. Thus the disease is viewed from a dual virological and immunological standpoint. The immunological changes first alter the CD4 lymphoid system and subsequently influence a much wider range of cells. It is assumed that an optimal antiviral treatment, if given early during the course of disease, would prevent and even reverse the immunological damage – and this is indeed a main aim. But in reality the *duality in AIDS pathogenesis* may lead to a duality in diagnosis, patient management and trial design: the virological methods alone, elegant and precise as they are, are probably inadequate to fully predict clinical events at the stage of disease progression when immunological diagnostic tests might also be required. It would also be surprising if antiviral therapy would not need to be supplemented by immunological interventions in order to correct some of the severe imbalances of immune parameters. If this pragmatic approach is taken then it also follows that the virological diagnostic methods, such as the analysis of p24 antigen, plasma viraemia and the biological phenotyping of isolate, will remain optimal for monitoring the effects of antiviral therapy – and immunodiagnosis will be most relevant to measure effects of putative immunotherapy.

Such comments would, however, remain speculative until guidelines for the immunological approach are worked out – and for this reason, in this symposium the first steps have been taken to document the severity of immune disorders as well as the relevance of immunological monitoring in patients with HIV-1 infection by utilizing a rich variety of methods. The observations summarized above complement other texts which cover the AIDS knowledge base [4]. These other projects frequently do not deal with the immunological aspects of AIDS pathogenesis in sufficient depth and sometimes even obscure the real issues.

(2) Viewed from this angle, the *clinical importance of CD4 counts* as a surrogate marker for AIDS diagnosis is clear. Details about the normal CD4 values and the technical achievements of flow cytometry are presented above. These observations support the suggestions made by CDC [5] and the World Health Organization [6] to introduce CD4 counts below $200/mm^3$ for the laboratory diagnosis of AIDS in order to back up the clinical definition of this condition. It has also been documented above that in countries concerned about adequate resource allocations the CD4 counts help predict and plan the costs of preventive treatment for those HIV-1-seropositive patients who have not developed AIDS. One potentially important source of inaccuracy, the calculation of absolute CD4 counts from two separate measurements made on two different instruments (i.e. an absolute counter and a flow cytometer) will also soon be assessed and avoided when flow cytometers capable of absolute CD4 counting in samples of whole blood arrive. In addition to the baseline CD4 numbers, the *rate of decline in CD4 levels* may also be a relevant prognostic parameter. Thus CD4 counting, coordinated with the measurements of serum load and the biological phenotype of the HIV-1, goes a long way for establishing appropriate monitoring of patients at risk for AIDS development.

(3) Nevertheless, one of the most important recurring themes of the meeting presented above has been the documentation of *severe immunological functional defects in asymptomatic patients* even before their CD4 counts drop to abnormally low levels. These changes of T cell reactivity, expected to be of direct clinical relevance since they are causally related to HIV-1-induced immunodeficiency, can be demonstrated with a wide variety of methods (fig. 1). These lymphocyte defects are readily detectable in unstimulated as well as in anti-CD3- and PHA-activated cultures, provided that the tests are standardized to reflect quantitative aspects, and are partly reflected in an increased apoptotic death rate among the lympho-

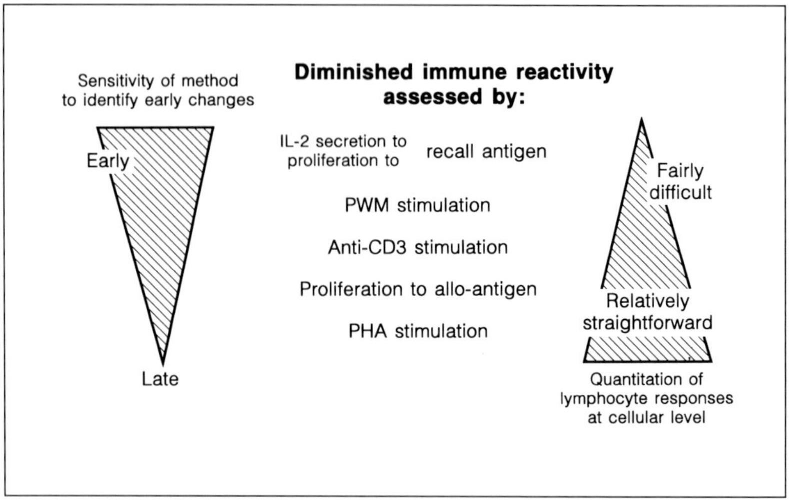

Fig. 1. The quantitative aspects and the onset of handicapped immunological responses detected by lymphocyte transformation assays utilized by the various authors in this book. All tests shown here reveal severe functional defects even in a large proportion of *asymptomatic* HIV-1-seropositive individuals but the assays differ in other respects. Some tests, such the responses to recall antigens, show abnormality particularly early as discussed by Clerici and Shearer above. Defects in responses to pokeweed mitogen (PWM) [discussed by Christiansen et al.] and to anti-CD3 stimulation develop at mid-course [see Miedema et al., and Gruters and Miedema]. Responses to phytohaemagglutinin (PHA) identify severe defects relatively late, unless the assay is set up in a quantitative manner, as shown by Janossy, Borthwick et al., when the investigation reveals the unresponsiveness of major subsets within both the CD4+ and CD8+ populations. All these tests have their roles in clinical diagnosis.

cytes taken from HIV-1-positive individuals. These observations strongly suggest that the functional handicap observed in both the CD4+ and CD8+ T cell populations is already determined in the patients in vivo, rather than being a mere result of cytokine deficiency or defects in cellular interactions in vitro. At this basic level, the relevant questions that remain unanswered include the mechanism(s) in vivo which trigger the alterations in lymphocyte physiology, whether or not these are the same for both CD4 and CD8 T cells, and whether the long-term administration of antiviral drugs slows down or accelerates the lymphoid cells' disorder.

(4) The conference also documents the *clinical applications of these lymphocyte performance assays.* Some of the techniques which have been

discussed are simple enough to reach the level of routine clinical applications and remain very inexpensive assays when compared to some of the molecular methods recently introduced for patient monitoring. These tests are arguably necessary for assessing immune performance in a clinical setting. In fact, some of the assays are already applied as 'whole blood tests' [7]. The optimum choice may, however, depend upon the precise aims of a given study. The anti-CD3 and PHA tests, for example, are likely to yield information about both CD4 and CD8 deficiency because these mitogens are unselective. At later stages of HIV-1 infection, however, they reflect mainly CD8+ lymphocyte behaviour as these cells dominate due to inverted CD4/CD8 ratios. On the other hand, pokeweed and other mitogens [8] preferentially activate the CD4 populations and may remain more selective for this subset. These latter tests and also antigen-stimulated cultures are also influenced by monocyte behaviour and reveal additional abnormalities of cellular collaborations. Consequently, such investigations may yield particularly sensitive tests of immunodeficiency, as has been emphasized before. Uniform standardization of these assays for widespread clinical use is perhaps a suitable venue for a practical workshop.

(5) Finally, the *increasing knowledge about lymphocyte physiology* in the fields of cytokines/receptors and communication molecules has designated new interests for investigation in AIDS development. These include cytokine abnormalities, reinterpreted in the light of the different biosynthetic activities of 'naive' and 'primed' T cell populations, and also the expression of well-characterized functional molecules on the membrane of various T lymphocyte subsets. Such a display, or the lack of it, can be analysed by multiparameter analysis on flow cytometers – and then interpreted in relation to the cells' functional performance. 'Markers' such as the IL-2 receptor β chain (p70) and CD28 antigen are just two examples which might be included in further studies on the lymphocyte subpopulations of HIV-1-positive individuals. These studies not only provide working hypotheses for explaining the defects seen but are likely to yield practical tests with diagnostic potential.

In the near future, exchange and standardization of adequate and suitable techniques among clinical immunologists are required to develop some of these assays for routine use.

George Janossy
The Royal Free Hospital
School of Medicine
London, UK

Frank Miedema
Red Cross Blood Transfusion
Service, Amsterdam
The Netherlands

Brigitte Autran
CH Pitié–Salpêtrière
Paris, France

References

1 Marx J: Clue found to T cell loss in AIDS. Science 1991;254:798–800.
2 Duesberg P: HIV is not the cause of AIDS. Science 1988;241:514–518.
3 Fauci AS: The human immunodeficiency virus: infectivity and mechanism of pathogenesis. Science 1988;239:617–622.
4 Cohen PT, Sande MA, Volberding PA (eds): AIDS Knowledge Base. Waltham/Mass, The Medical Publishing Group, 1990-2 (also available in CD-ROM format).
5 CDC: 1992 revised classification system for HIV infection and expanded AIDS surveillance case definition for adolescents and adults. In press.
6 WHO: Interim proposal for a WHO staging system for HIV infection and diseases. Weekly Epidemiol Rec 1990;65:221–224.
7 Schellekens PT, Roos MT, De Wolf F, Lange JM, Miedema F: Low T-cell responsiveness to activation via CD3/TCR is a prognostic marker for acquired immunodeficiency syndrome (AIDS) in human immunodeficiency virus-1 (HIV-1)-infected men. J Clin Immunol 1990;10:121–127.
8 Adriaansen HJ, Osman C, van Dongen JJ, Bresser JH, Kappetijn TCM, Hooijkaas H: Immunological marker analysis of mitogen-induced proliferating lymphocytes using BrdU incorporation or screening of metaphases. Staphylococcal protein A is a potent mitogen for CD4+ lymphocytes. Scand J Immunol 1991;32:687–694.

Subject Index